Developments in Bioenergetics and Biomembranes Volume 6

Other volumes in this series:

STRUCTURE AND FUNCTION OF MEMBRANE PROTEINS

STRUCTURE AND FUNCTION OF MEMBRANE PROTEINS

Proceedings of the International Symposium on Structure and Function of Membrane Proteins held in Selva di Fasano (Italy), May 23-26, 1983.

Editors

E. Quagliariello

F. Palmieri

1983

ELSEVIER SCIENCE PUBLISHERS
AMSTERDAM · NEW YORK · OXFORD

Published by:
Elsevier Science Publishers B.V.
P.O. Box 211
1000 AE   Amsterdam, The Netherlands

Sole distributors for the USA and Canada:
Elsevier Science Publishing Company Inc.
52 Vanderbilt Avenue
New York, N.Y. 10017

ISBN: for this volume: 0-444-80540-0
ISBN: for the series: 0-444-80015-8

Library of Congress Cataloging in Publication Data

International Symposium on Structure and Function of
    Membrane Proteins (1983 : Selva di Fasano, Italy)
    Structure and function of membrane proteins.

    (Developments in bioenergetics and biomembranes ;
v. 6)
    Bibliography: p.
    Includes index.
    1. Membrane proteins--Congresses. I. Quagliariello,
Ernesto. II. Palmieri, F. (Ferdinando) III. Title.
IV. Series. [DNLM: 1. Membrane proteins--Congresses.
2. Structure-activity relationship--Congresses. W1
DE997VK v.6 / QU 55 I674s 1983]
QP552.M44I57 1983   574.87'5   83-16565
ISBN 0-444-80540-0 (U.S.)

Printed in The Netherlands

PREFACE

The purpose of this International Sympsoium on "Structure and Function of Membrane Proteins" at Selva di Fasano was to review the present status of research into the structure-function relatioship of membrane proteins. Structure was what the organizers had primarily in mind, although they were aware that knowledge in this field is still embryonic. However, membrane proteins are no longer those oily creatures floating around in the lipid bilayer or in ill-defined detergent micelles. Membrane proteins can be put into well-ordered arrays of either two-or three-dimensional order, reflecting the well-defined monodisperse structure of these entities. Wereas, prokaryotic membrane proteins, particularly of the exotic varieties, are often highly expressed, making solubilization and purification less important, more difficulties have been encountered in the isolation of intact membrane proteins from eukaryotic membranes. Only in the last ten years have these problems been solved, in principle, through our understanding of the appropriate handling of detergents, in particular of the non-ionic variety of these. As a result, numerous solubilized membrane proteins are available in both monodisperse and native forms.

It should now be possible to obtain more information on the structure of these membrane proteins. Structural research comprises, first of all, the elucidation of the amino acid sequence, the folding and the overall shape, leading up to ultimate, atomic, resolution which requires crystals suitable for X-ray crystallography. Progress in elucidating the primary structure of membrane proteins has come from improved handling of the hydrophobic peptides, and also from the use of cloned c-DNA sequences.

Two-dimensional arrays constituted the first step in ordering membrane proteins and their successful application is now relatively widespread. Three-dimensional crystals for X-ray crystallography are available in only two instances, both of which have been discussed at this meeting. From primary sequences, hypothetical folding predictions are made, with the focus on the ubiquitous transmembrane α-helical segment. These predominantly theoretical models can be useful for explaining sidedness and assigning active sites, but will ultimately be replace by physical data. The other approach to structure determination is based on

monodisperse, soluble proteins or protein micelles using scattering
methods for X-ray, light and neutron beams. Structural information can
also be obtained by using fluorescence probes and spin labels as long as
specific localization and assignments at the protein can be made.

Structure and function are reciprocally linked. We wish to determine
the structure in order to understand the function, and the mechanism of
action will be understood only by our knowledge of the atomic structure.
The gathering of data on the function of membrane proteins prior to
knowledge of their structure is valuable for characterizing and defining
the proteins. Once the structure is known, another stage of research
will penetrate to the functional assignments of the structure.

We have witnessed in this symposium the various stages of this very
lively and effective research on biomembrane proteins. New enthusiasm
and scope for research has been created through obtaining the first
crystals of membrane proteins and it is to be hoped that some years from
now another meeting, possibly in the same room, might witness again the
great progress which we expect in this exciting field.

CONTENTS

PHYSICAL METHODS FOR THE STRUCTURE – FUNCTION RELATIONSHIP

© 1983 Elsevier Science Publishers B.V.
Structure and Function of Membrane Proteins,
E. Quagliariello and F. Palmieri editors.

GRAMICIDIN A TRANSMEMBRANE CHANNEL: KINETICS OF PACKAGING IN
LIPID MEMBRANES.

LANFRANCO MASOTTI[1], PAOLO CAVATORTA[2], ALBERTO SPISNI[1], EMANUELA
CASALI[1], GIORGIO SARTOR[1], IVONNE PASQUALI-RONCHETTI[3], ARTHUR SZABO[4].
[1]Institute of Biological Chemistry, University of Parma, Viale
Gramsci, 14, 43100 Parma, (Italy); [2]Institute of Physics-GNCB,
University of Parma, Via D'Azeglio, 85, 43100 Parma, (Italy);
[3]Institute of General Pathology, University of Modena, Via Campi,
165, 41100 Modena, (Italy); [4]Division of Biological Sciences,
National Research Council, 100 Sussex Dr., Ottawa, Ontario K1A OR6,
(Canada).

INTRODUCTION

   Biological membranes are recognized to act as anchoring point
for bound enzymes, to provide a medium where proteins, substrates
and products of enzyme reactions of limited solubility in water
become soluble, and where multienzyme complexes are organized. Furthermore they define cellular compartments where substrates, products and effectors of metabolic reactions are separated and control
their selective transport between compartments. In particular, selective ion movement across membranes is essential for the processes
of energy transduction and cell excitability.

   Several attempts have been made to characterize native channels
in membranes in terms of structure and mechanism (1,2), but because
of their complexity, attention has also been focused on model systems capable to function as selective channels for ions. Gramicidin
A'(GA'), a hydrophobic polypentadecapeptide, has been shown to be
able to form helical structures with an internal pore of about 4 Å
in diameter, that can accomodate monovalent cations such as Na^+ or
K^+ (3-10).

   It has been demonstrated (11) that GA' can be incorporated into
lysolecithin micelles and that it assumes a left-handed helical
configuration able to transport monovalent cations selectively (12).
Moreover we have shown that the formation of the channel is associated with a reorganization of the lipid phase in a bilayer structure, wherein channel aggregates are embedded (13,14).

   Having recognized the importance of a supramolecular organization
for the channel activity (15) our attention has been focused on the

studies concerning the kinetics of the incorporation and organization of GA' in the lipid phase.

MATERIALS AND METHODS

Lysolecithin was obtained from Sigma Chemical Company, St. Louis,
Mo., and checked for purity by nuclear resonance and thin layer
chromatography, Gramicidin was purchased from ICN Pharmaceuticals,
Cleveland, Ohio, as a mixture of 80% Gramicidin A, 6% Gramicidin
B, and 14% Gramicidin C, and was used without further purification.
The mixture is referred to as Gramicidin A' (GA').

Phospholipid was suspended in water and heated at 70°C up to
22 hours, as previously reported (15), either in the presence or in
the absence of GA'. Two samples with different GA'/lipid ratios
were prepared: the first with 6 mg GA', the second with 3 mg GA'
per 25 mg of lysolecithin. Aliquots were withdrawn at times :
0, 10, 30, 60, 120, 180, 270 minutes and 21 hours.

Static fluorescence measurements were carried out as previously
described (15). Lifetime measurements were performed using a
Spectra Physics apparatus with a pulsed laser source (1 psec).

Absorption measurements were performed using a Perkin Elmer
576 Spectrophotometer, equipped with a termostatically controlled
cuvette holder at temperature of 30±0.5°C. A molar extinction
coefficient of 22500 $mol^{-1}cm^{-1}$ at 281 nm in CH_3OH was used to
calculate the GA' concentration.

Circular Dichroism measurements were performed on a JASCO J-500
Spectropolarimeter equipped with a microprocessor unit for spectra
accumulation. The samples were diluted with water and run using
cuvettes with 0.2 mm pathlength. The ellipticities were calculated
using a mean molecular weight per residue of 124.5.

For electron-microscopic studies, the sample were left to equil-
ibrate at room temperature before being processed. The specimens
were examined in a Siemens Elmiskop 1A and Philips 410 electron
microscopes. The magnifications were calibrated by optical diffraction of catalase crystals. For negative staining see caption
to Fig.4.

RESULTS AND DISCUSSION

In order to verify if the initial amount of GA' used in the

experiments can influence the velocity of incorporation of the
polypeptide into the lipid system, the two concentrations reported in the Materials and Methods section were chosen. The
percent of GA' incorporated, taking as 100% the actual initial
amount of the polypeptide, shows that the rate of incorporation
is somewhat proportional to the initial amount of GA'.

The fluorescence emission maximum, as shown in Fig. 1, is shifted
from 342±1 nm at the time t=0 to 328±1 nm at the time t=22 hours
of incubation. Such a blue shift tipically indicates a change
in the polarity of the environment of tryptophane residues, from
a polar to a non polar one.

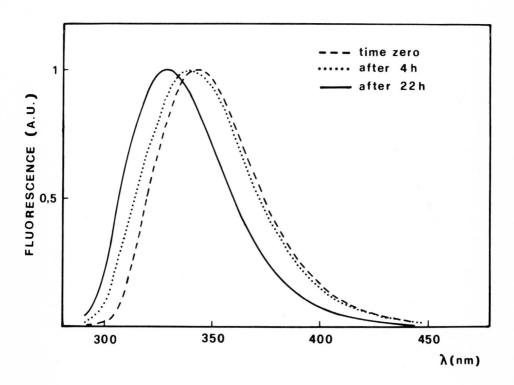

Fig. 1. Fluorescence spectra of GA' incorporated into the lipid system at
different times of incubation.

The time course of absorbance, corrected for light scattering, during incorporation, Fig. 2, clearly shows that the GA' incorporates into the lipid system. In fact at time t=0', the amount of GA' incorporated is just 6% of the total amount added and reaches the 80% at t=22 hours.

At the same time the quantum yield decreases from a value of 0.7 for both the samples at time t=0', to 0.06 at t=22 hours. However, the high values of the quantum yield at short times might be overestimated, being the absorbance not entirely corrected for light scattering and "obscured" absorbance (16).

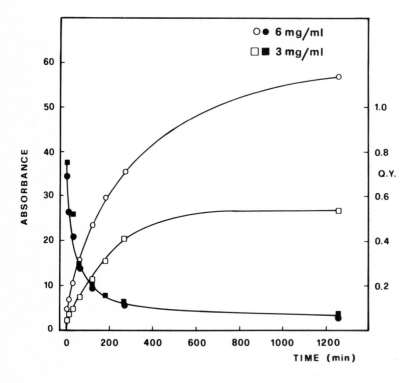

Fig. 2. Absorbance (open symbols)and quantum yield (Q.Y.)(full symbol) dependence on time of incubation.

The CD spectra, shown in Fig. 3, demonstrate that also at the time t=0', some of the polypeptide has already adopted an helical structure. At subsequent times the CD pattern is similar to that

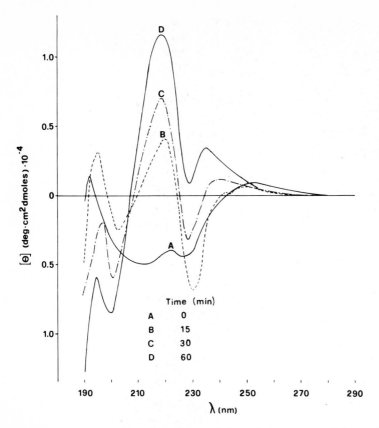

Fig. 3. Circular Dichroism spectra of G.A' incorporated into the lipid systems at different times of incubation.

reported previously (11), until after 2 hours is consistent with that of a left-handed helix, and it is stable with time.

The lifetime measurements (see Table 1) seem to indicate that at the time t=0' two types of tryptophane are detected, of which one predominates. The lifetime does not change with wavelength indicating that both types of tryptophane experience the same environment. Up to 90' the lifetimes change very little. At longer incubation times, shorter lifetimes are measured at 335 nm, while at longer wavelength again no change is observed.

The ultrastructural studies of the samples during incorporation

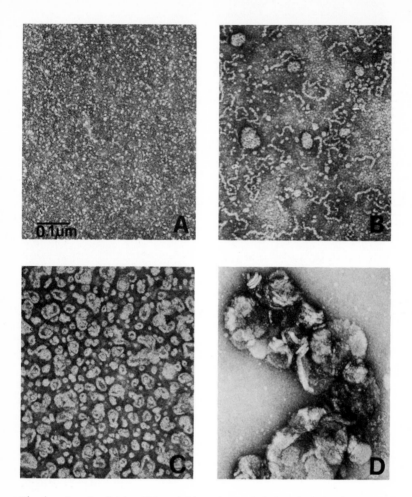

Fig.4a. Lysolecithin (25 mg/ml) and Gramicidin A' (6 mg/ml) in water after incubation for 15 minutes at 70°C. The suspension was diluted 1 : 400 and stained on a copper grid with 1% uranyl acetate. The sample consisted of 7-8 nm wide micelles.

Fig.4b. Same material as in Fig.4a after incubation for 60 minutes at 70°C. The sample, diluted and stained as in Fig.4a, appeared as a mixture of 7-8 nm wide micelles, of spherical particles with diameter ranging from 9 up to 12 nm, of a few large aggregates, and of rows of micelles; the rows exhibited various length and curvature.

Fig.4c. Same material and technical procedure as in Fig.4a after incubation for 90 minutes at 70°C. Small vesicle - like structures can be seen; the majority of them are not well organized as vesicles, and curved and almost sealed phospholipid "membranes" represent the main feature of the suspension. Similar images were obtained after 150 and 180 minutes of incubation.

Fig.4d. Same material and technical procedure as in Fig.4a. The specimen was incubated for 3 days at 70°C. Together with micelles of various size, a conspicuous amount of phospholipid was organized as characteristic vesicles; the diameter of these vesicles was very variable.

TABLE 1

FLUORESCENCE LIFETIMES OF GRAMICIDIN A' DURING THE INCORPORATION IN LYSOLECITHIN

Time (min)	λ 335 nm				λ 380 nm			
	τ_1	τ_2	F_1	F_2	τ_1	τ_2	F_1	F_2
0	7.50	1.89	.83	.17	8.00	2.50	.87	.13
90	7.43	1.82	.76	.25	7.90	2.30	.86	.14
240	6.45	1.50	.59	.41	7.00	1.86	.74	.26
1440	3.52	1.00	.50	.50	7.80	2.45	.70	.30

Lifetimes (τ) are given in ns. F is the fraction of each component.

are shown in Fig. 4. The lysolecithin micelles have the typical diameter of about 8 nm (13). Electron Microscopy detects an early incorporation of GA' and after 30' of incubation larger micelles, sometimes organized in rows, are observed. This phenomenon becomes very apparent after 60 and 90 minutes of incubation. The length of the rows and the curvature of the folds are variable, and there is a tendency to form circular structures. It is possible that the curvature is due to the incorporation of GA' and to its asymmetric distribution in the micelle which has become enlarged during incubation. After about 3 hours an increase of the folding of the "chains" is observed that can precede the fusion of the two ends. This could be the mechanism by which at longer times the membrane structure is finally achieved.

The early dramatic decrease of the quantum yield values combined with the near constance of the lifetimes seems to indicate that a static fluorescence quenching is being detected. This fact can be due to intermolecular interactions between GA' molecules as also shown by the aggregation process monitored by Electron Microscopy. The CD data though tell us that the polypeptide is also, in the same range of time, undergoing noticeable conformational changes. Possibly these two processes are responsible of the fluorescence behaviour.

After 90' of incubation the CD spectrum becomes that of a left-handed helix and the pattern remains unchanged with time. The correspondent fluorescence lifetimes begin to decrease indicating that as the channels become more dispersed in the lipid matrix, the quenching of the tryptophanes is also due to a dynamic process. In conclusion these studies contribute to shed some more light on the nature of Trp-Trp interactions during the process of channel incorporation. They also show that at the very beginning of membrane formation the polypeptide already has an ordered secondary structure. Relevant also to the general study of biomembranes is the observation that the aggregational process of polypeptides is the result of hydrophobic interactions as shown by the temperature-dependent incorporation.

AKNOWLEDGEMENTS

This work has been supported by grants of CNR and MPI, Rome, and CRPA, Reggio Emilia, Italy.

REFERENCES

1. Goodall, M., Lades, G. (1972), Nature 237,252.

2. Villegas, R., Villegas, G.M., Barnola, F.V., Racher, E. (1977) Biochem. Biophys. Res. Comm. 79,210.

3. Ulbricht, W. (1977) Ann. Rev. Biophys. Bioeng. 6,7.

4. Hille, B. (1979) Biophys. J. 22,283.

5. Ovchinnikov, Y.A. (1979) Eur. J. Biochem. 94,321.

6. Hladky, S.V., Urban, B.W., Haydon, D.A. (1979) in Membrane Transport Processes (Stevens, T.F., Tsien, R.W., Eds.), Vol.3, 89-103. Raven Press.

7. Sandblom, J., Heisenman, G., Neher, E. (1977) J. Membrane Biol. 31,383.

8. Läuger, P. (1979) Biochim. Biophys. Acta 552,143.

9. Urry, D.W., Spisni, A., Khaled, M.A., Long, M.M., Masotti, L. (1979) Int. J. Quantum Chem. Quantum Biol. Symp. 6,289.

10. Urry, D.W., Spisni, A., Khaled, M.A. (1979) Biochem.Biophys.Res.Comm.88,940.

11. Masotti, L., Spisni, A., Urry, D.W. (1980) Cell Biophys. 2,241.

12. Urry, D.W. (1973) in "Conformation of Biological Molecules and Polymers" (Jerusalem Symp. on Quantum Chem. and Biochem.) Vol.5, 723.

13. Pasquali-Ronchetti, I., Spisni, A., Casali, E., Masotti, L., Urry, D.W. (1983) Bioscience Reports 3,127.

14. Spisni, A., Pasquali-Ronchetti, I., Casali, E., Lindner, L., Cavatorta, P., Masotti, L., Urry, D.W. (1983) Biochim. Biophys. Acta, in press.

15. Cavatorta, P., Spisni, A., Casali, E., Lindner, L., Masotti, L., Urry, D.W. (1982) Biochim. Biophys. Acta 689,113.

16. Urry, D.W. (1972) Biochim. Biophys. Acta 265,115.

ELUCIDATION, BY LOW ANGLE X-RAY AND NEUTRON DIFFRACTION AND BY HIGH RESOLUTION NUCLEAR MAGNETIC RESONANCE, OF SOME STRUCTURAL ASPECTS OF THE INTERACTION OF BEE VENOM MELITTIN WITH PHOSPHO-LIPIDS.

ROBERTO STROM[1,2], FRANCA PODO[3], CARLO CRIFO'[1,2], COSTANTINO SALERNO[2], CARMEN BERTHET[4] AND GIUSEPPE ZACCAI[5]

[1]Institute of Biological Chemistry of the University of Rome, [2]C.N.R. Center for Molecular Biology and [3]Istituto Superiore di Sanità, Rome, Italy; [4]European Molecular Biology Laboratory and [5]Institut Laue Langevin, Grenoble, France.

INTRODUCTION

Melittin, the major bee venom peptide, possesses the peculiar property of combining[1] a high water solubility ($>$ 60 mM) to a strong affinity ($<$1 μM) for lipid membrane systems, upon which it exerts "lytic" effects[2-4]. Its aminoacid sequence has vast homologies with sequences of leader peptides of eukaryotic proteins[5] and of neuropeptides of the endorphin family[6,7].

Gly-Ile-Gly-Ala-Val-Leu-Lys-Val-Leu-Thr-Thr-Gly-Leu-Pro-Ala-Leu-
 1   5   10   15
-Ile-Ser-Trp-Ile-Lys-Arg-Lys-Arg-Gln-Gln-NH$_2$
       20   25

Primary structure of melittin[8]

MELITTIN CONFORMATION AND ASSOCIATION STATE IN SOLUTION.

Melittin conformation, as evaluated from its circular dichroism spectrum in the 200-240 nm region, has been shown to undergo dramatic changes according to the interactions of the peptide with the solvent components[9-11], being an unordered random coil in water at neutral pH and at low salt concentration, a right-handed helix if placed at high ionic strength or in the presence of divalent (or multivalent) anions, or at alkaline pH, or when interacting with detergents or with phospholipids. In aqueous solution, a monomer-to-tetramer transition usually parallels these conformational changes[12,13], this correlation being however possibly lost when the peptide interacts with partially hydrophobic structures[1,14].

The possibility of a more complex situation -- such as the existence of more than two extreme conformations, monomeric random coil or tetrameric helix -- could however be evidenced by the use of high resolution NMR and by photon correlation spectroscopy[15]. By analysing e.g. how the ^{1}H-NMR spectrum of melittin in D$_2$O, shown in the lower part of fig.1, is progressively modified by addition of phosphate ions into the upper spectrum of the same

fig.1 (the binding of phosphate ions to melittin being itself
monitored by ^{31}P-NMR), the variations of single features of these
spectra can be divided[15,16] into "early" and "late" ones (Table 1),
the former being possibly associated to the conversion of the
flexible random coil into a more compact structure at the level
of the single polypeptide chain, the latter (such as, in particular, the marked upfield shift of the signal from the CH_3 protons
of Ile2, caused by the spatial proximity to this aminoacid of the
Trp19 ring) indicating the occurrence of inter-chain interactions
in a tetrameric structure.

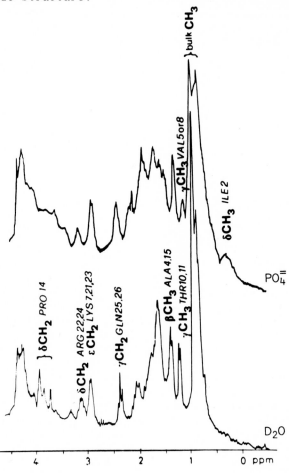

Fig.1. 200 MHz H-NMR spectra at 24°C of 4 mM melittin in D O
(lower profile) and in the presence of 13.75 molar excess of
phosphate buffer, pH7.0 (lower profile).Some resonances of particular interest are indicated (from ref.17)

TABLE 1
MODIFICATIONS OF MELITTIN ^{1}H-NMR SPECTRUM UPON PHOSPHATE BINDING
(from ref.16)

Frequency range (ppm)	"Early modifications (appearing already upon binding of 1 ion)" Chemical group(s) and signal modification	"Late" modifications (starting after more than 1 ion is bound) Chemical group(s) and signal modification
0.3		Ile2δCH$_3$;upfield shift
0.6-1.0	terminal CH$_3$'s;broaden	
1.1		Val5,8γCH$_3$;downfield shift
1.4-1.6	Ala4,15CH$_3$;downf.shift	
1.6-1.9	CH$_2$'s; downfield shift	
2.3-2.5		Gln25,26γCH$_2$;broadening
3.1-3.2		Arg22,24δCH$_2$;reduced int.
3.6-3.9	Pro14δCH$_2$;shape change	
7.0-7.5	Trp19C4-H;upfield shift Trp19C7-H;downf.shift	Trp19C2-H;downfield shift Trp19C5-H;upfield shift

Photon correlation spectroscopy measurements of the translational diffusion characteristics of melittin show, on the other hand,
that the highly flexible monomer, which exists in quasi-neutral
aqueous solutions of the peptide in the absence of added salts,
is converted, as the ionic strength is slightly raised by addit-
tion of NaCl, to a more compact structure, which can be approximated to a prolate ellipsoid of 13x13x44 A^3, such as can be
expected from a single polypeptide chain; under these conditions,
the ellipticity values at 220 nm indicate that little or no helix
is present. Only at high ionic strength or upon addition of divalent anions, where a right-handed helical conformation can be
detected by circular dichroism, does melittin form larger structures, amenable to oblate ellipsoids of 26x42x42 A^3,similar if
not identical to the tetrameric forms crystallized by Eisenberg
et al.[18] from ammonium sulfate.
These conclusions are substantiated by experiments on the kinetics of the conformational transition observable as varaiation
in circular dichroism: as shown in fig.2, the increase of negative
ellipticity upon mixing of melittin with divalent anions can be
divided into a very fast initial step ($t_{1/2} \leq$20msec) followed by
a relatively slower, quasi-first-order process, the rate constant
of which is non-linearly dependent on melittin concentration (fig.

TABLE 2
REDUCED DIFFUSION COEFFICIENTS AND EQUIVALENT HYDRODYNAMIC RADII
OF MELITTIN AS DETERMINED BY PHOTON CORRELATION SPECTROSCOPY
(from ref.15, slightly modified). Ellipticity values at 220 nm
are reported in the last column.

Melittin conc. (mg/ml)	Salt	pH	$D_{20,w}$ (cm^2/s)	r_h (Å)	θ_{220} (deg.cm^2/dmole)
11	pure H_2O	6.2	42 (x10^{-7})	5.4	-3.1 (x10^{-3})
11	0.03 M NaCl	6.3	21.7	9.2	-4.8
10	0.15 M NaCl	6.4	21.6	9.3	-6.1
5	0.15 M NaCl	6.3	19.2	10.5	-5.8
5	1.50 M NaCl	6.3	12.6	15.8	-15.6
10	0.05 M Na_2SO_4	6.3	11.2	17.8	-15.8

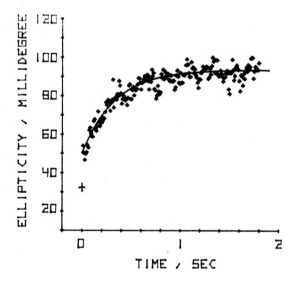

Fig.2. Dicrograph tracing upon rapid mixing of melittin (0.1mg/ml)
with 0.075 M Na_2SO_4.

3).Such a kinetics is fairly well accounted for by an inter-pep-
tide association step following the binding of the divalent
anions to each protomer. If relatively dilute monovalent anions
are substituted to the divalent ones, the fast initial event
largely predominates, the slower process becoming almost negligible (Table 3).

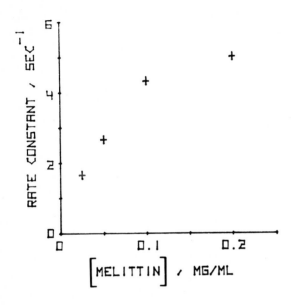

Fig.3. Dependence, on
melittin conc., of the
rate constant for the
"slow" process of helix
formation upon mixing
with 0.15 M Na_2SO_4.

TABLE 3.
DEPENDENCE OF KINETIC PARAMETERS OF CONFORMATIONAL CHANGE ON
SALT CONCENTRATION.
Experiments as in fig.2; melittin conc.:0.1 mg/ml; $\Delta\theta_{fast}$ and $\Delta\theta_{slow}$
indicate the variations in ellipticity of the fast and slow
processes; k_{slow} is the rate constant (in sec) of the latter.

Nature of salt	Salt conc.		Ionic strength	$\Delta\theta_{fast}$	$\Delta\theta_{slow}$	k_{slow}
NaCl	250	mM	0.250	15	5	10
Na_2SO_4	18.7	mM	0.056	10.3	14.5	1.4
Na_2SO_4	37.5	mM	0.112	8.6	20.3	1.5
Na_2SO_4	75	mM	0.225	18.1	43.5	2.9
Na_2SO_4	100	mM	0.300	20.9	50.6	3.4
Na_2SO_4	150	mM	0.450	32.4	47.6	4.3
Na_2SO_4	200	mM	0.600	32.8	62.3	6.3
Na_2SO_4	300	mM	0.900	33.2	80.6	6.4

MELITTIN CONFORMATION UPON INTERACTION WITH PHOSPHOLIPIDS.

From the kinetic point of view, the interaction of melittin with either natural or model membranes is a complex process. Fig.4 shows e.g. that, even in the relatively simple case of an aqueous solution of melittin mixed with sodium dodecylsulfate micelles, the peptide conformation, as revealed by its circular dichroism signal at 220 nm, is modified through at least three different steps.

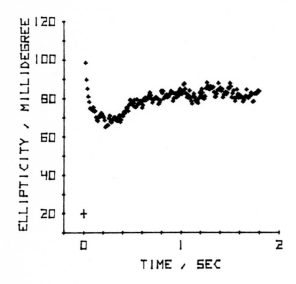

Fig.4. Dicrograph tracing upon rapid mixing of melittin (0.1 mg/ml) with sodium dodecylsulfate (0.125 mg/ml in H_2O)

At equilibrium, the circular dichroism spectrum of melittin interacting with phospholipids or with detergents is typical of a right-handed helix. ^{1}H-NMR reveals however the presence of distinct features, different from those found in aqueous solutions at high ionic strength or in the presence of divalent anions. By using perdeuterated dipalmitoylphosphatidylcholine (interference by phospholipid signals being thus negligible), it was found that the glutamine and lysine residues of the C-terminal portion of the peptide were strongly immobilized.The upfield shift of Ile2 CH , indicative of spatial proximity of this residue to the Trp19 ring of a different polypeptide chain, was moreover absent, without therefore any NMR evidence for a self-aggregation of the peptide (Fig.5).

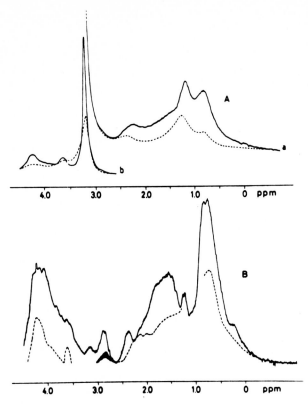

Fig.5. A)100 MHz ^{1}H-NMR spectra at 32°C of per deuterated dipalmitoyl phosphatidylcholine, in the absence (---) or presence (——) of melittin (phospholipid-to-peptide ratio=5:1). Spectra a) are 5x as compared to spectra b).

B) Comparison of the 100 MHz ^{1}H-NMR spectra at 32°C of 4mM melittin in 0.15M phosphate (——) and interacting with perdeuterated diparlmitoylphosphatidylcholine (---), the latter after substraction of the spectrum of phospholipids alone.

STRUCTURAL ASPECTS OF MELITTIN INTERACTION WITH PHOSPHOLIPID VESICLES.

Lanthanide shift reagents, which, in vesicular structures, can discriminate between inner and outer choline N-methyl groups, allow further investigation, by ^{1}H-NMR, of the effects of melittin on the permeability of phospholipid vesicles. Addition of the peptide to such vesicles suspended in a Pr^{3+}-containing medium caused a progressive merging of the $N^+(CH_3)_3$ signals from the inner and outer compartments, indicating that Pr^{3+} ions progressively penetrated the vesicles. This phenomenon could be reversed by addition of EDTA, and restored with more Pr^{3+}: the vesicles appear therefore to have, despite the increased permeation rate, an inner aqueous compartment which was still distinct from the outer one. The rate constant of permeation was found to depend rougly on the fifth power of the melittin-to-phospholipid ratio, suggesting the possibility that several melittin protomers be required to induce the increase in permeability.

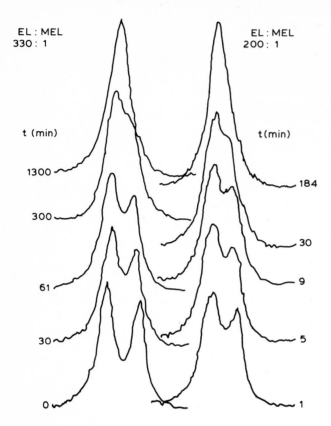

EL : MEL
330 : 1

EL : MEL
200 : 1

t (min)

1300

300

61

30

0

t (min)

184

30

9

5

1

Fig.6. Permeation
of 10 mM PrCl3 in
egg phosphatidyl-
choline vesicles
at 25°C, after ad-
dition of melittin.
The phosphatidyl-
choline:melittin
molar ratios (EL:MEL)
were 330:1 and 200:1;
^{1}H-NMR spectra were
recorded at various
times after addition
of melittin, as in-
dicated (modified
from ref.17)

An independent proof that melittin does not act by total disruption of membrane structure, but rather by generating aqueous
pathways in the context of otherwise quasi-intact phospholipid
bilayers, was obtained through neutron and x-ray low angle diffraction experiments. In oriented hydrated multilayers of C_6-mono-
deuterated dimiristoylphosphatidylcholine (d_6-DML), neutron diffraction analysis did not show, apart from a certain loss in resolution in the melittin-containing sample, any major perturbation
of the bilayer profile (Fig.7). Differential profiles revealed
however that, in melittin-containing samples, H-D exchangeable
protons were present down to almost the very center of the bilayer (fig.8A), and that melittin itself was present well within
the hydrophobic core of the alkyl chains region (fig.8B)

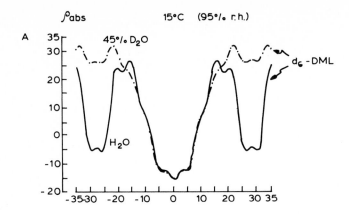

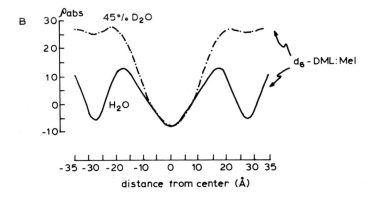

Fig.7. Neutron scattering density profiles from hydrated oriented multilayers of dimiristoylphosphatidylcholine (A) or of dimiristoylphosphatidylcholine + melittin mixtures, molar ratio=10:1 (B). Dimiristoylphosphatidylcholine was deuterated in position 6 of one of the acyl chains (d_6-DML). Diffraction patterns were obtained from both samples equilibrated in either H_2O (——) or 45% D_2O + 55% H_2O (—·—), at the temperature of 15°C and at 95% relative humidity. Scattering density was expressed in absolute units (10^{15}cm/A^3),by taking advantage of isomorphous H_2O-D_2O replacement and by assuming invariance of scattering density at the very center of the bilayer.

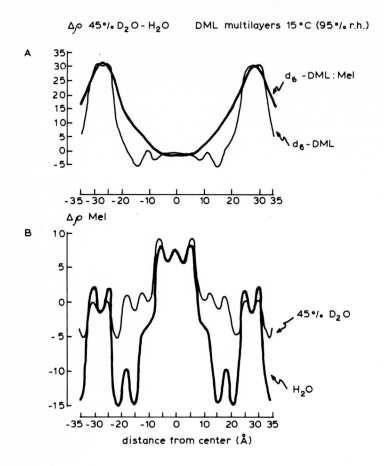

Fig.8A. 45% D_2O <u>minus</u> H_2O differential neutron scattering profiles of d_6-DML and d_6-DML+Mel hydrated oriented multilayers at 15°C and 95% relative humidity. The profile indicates the position of H-D exchangeable protons.

Fig.8B. d_6-DML+Mel minus d_6-DML differential neutron scattering density profiles in H_2O and in 45% D_2O + 55% H_2O, at 15°C and 95% relative humidity. The profile indicates the position of melittin in the bilayers.

Small angle x-ray diffraction, though essentially confirming
the neutron scattering profiles for samples above the transition
temperature of the phospholipids, added a novel problem concerning the structure of samples below this transition temperature.
As shown in Table 4, under these conditions there was, superim-
posed to the uni-directional lamellar pattern, a two-dimensional
hexagonal array of reflections, with a repeat distance smaller
than the lamellar one. Preliminary experiments with oriented
samples, aimed at ascertaining the direction of this unit cell
relative to the lamellar repeat, do not sofar support the hypothesis of an in-plane ordering of melittin residues, but seem
to indicate the coexistence of two well distinct phases, the
lamellar and the hexagonal one, both oriented normally to the
phospholipid-water interface.

TABLE 4.
SMALL ANGLE X-RAY DIFFRACTION PATTERN FROM NON-ORIENTED DML-MEL
HYDRATED SAMPLES.
Beam wavelength:1.542 A; sample-to-film distance:9.9 cm. Compos-
ition of the sample:DML/Mel/H_2O (25/10/9,w/w/w, or DML/Mel 10/1,
mole/mole,+10% H_2O).

TEMPERATURE:60°C diameter of reflection (intens.)	TEMPERATURE:15°C diameter of reflection (intens.)

6.4 (very strong)   d=48.4 A
12.8=6.4x2 (medium)
19.2=6.4x3 (weak)
25.5=6.4x4 (very weak)
	5.2 (very strong) d=59.2 A
	    7.4(very strong d=41.6 A
	10.4=5.2x2 (medium)
	    12.9=7.4x 3 (medium)
	15.5=5.2x3 and/or 7.4x2 (medium)
	20.8=5.2x4 (medium)
	    26.2=7.4x2 3 or 7.4x 13 (very
	                                     weak)

CONCLUSIONS.
The simultaneous use of several spectroscopic techniques,
aimed to the clarification of the structural basis for the interaction of melittin with phospholipid model membranes, has brought
to light the complex behaviour of this relatively small molecule,
which can change of both conformation and aggregation state, and
thus exist as well in aqueous solution and firmly bound to membranes. In the latter case, the overall evidence points toward
an insertion of melittin deep into the alkyl regions of the phospholipid bilayers, without however disregarding the importance
of the ionic interaction between the phosphate headgroups and the
cationic aminoacid residues, as an essential factor in anchoring
the C-terminal portion of melittin to the phospholipid-water inter
face. Due to the existence, on one face of the helical conformation of the peptide, of several hydrophilic residues (the
rest of the surface of the helix being essentially hydrophobic),

22

it may be envisaged that formation of aqueous channels, such as
are detected by the Pr   experiments, occurs by cooperation
of several peptide protomers. No spectroscopic evidence has
however sofar substantiated this hypothesis.

REFERENCES.
1. Knüppel,E., Eisenberg,D. and Wickner,W.(1979) Biochemistry,
   18,4177-4181.

2. Sessa,G.,Freer,J.H.,Colacicco,G. and Weissmann,G. (1969) J.
   Biol.Chem.,244,3575-3582.

3. Habermann,E. (1972) Science, 177,314-322.

4. De Grado,W.F., Musso,G.F.,Lieber,M.,Kaiser,E.T. and Kézoly,
   F.J. (1982) Biophys.J.,37,329-338.

5. Garnier,J.,Gaye,P.,Mercier,J.-C. and Robson,B. (1980) Biochimie,
   62,231-239.

6. Li,C.H. and Chung,D.(1976) Proc.Natl.Acad.Sci.U.S.A.,73,1145-
   1148.

7. Lazarus,L.H.,Ling,N. and Guillemin,R. (1976) Prof.Natl.Acad.
   Sci.U.S.A.,73,2156-2159.

8. Habermann.E. and Jentsch,J. (1967) Hoppe Seyler's Z.Physiol.
   Chem.,348,37-50.

9. Gauldie,J.,Hanson,J.M.,Rumjanek,F.D., Shipolini,R.A. and Ver-
   non,C.A. (1976) Europ.J.Biochem.,61,369-376.

10. Dawson,R.C., Drake,A.F., Helliwell,J. and Hider,R.C. (1978)
    Biochim.Biophys.Acta,510,75-86.

11. Strom,R., Crifò,C.,Viti,V., Guidoni,L. and Podo,F. (1978) FEBS
    Letters,96,45-50.

12. Talbot,J.-C. , Dufourcq,J., de Bony,S., Faucon,J.F. and Lussan,C.
    (1979) FEBS Letters,102,191-193.

13. Quay,S.C. and Condie,C.C. (1983). Biochemistry,22,695-700.

14. Lauterwein,J., Büsch,C., Brown,L.R. and Wüthrich,K. (1979)
    Biochim.Biophys.Acta,566,244-264.

15. Podo,F.,Strom,R.,Crifò,C. and Zulauf,M. (1982) Int.J.Peptide
    Protein Res.,19,514-527.

16. Strom,R.,Podo,F.,Crifò,C.,Berthet,C.,Zulauf,M. and Zaccai,G.
    (1983) Biopolymers,22,391-397.

17. Strom,R.,Podo,F.,Crifò,C. and Zaccai,G. (1982) in:Bossa,F.,
    Chiancone,E., Finazzi Agrò,A. and Strom,R.(Eds.),Structure
    and Function Relationships in Biochemical Systems, Adv.Exp.
    Medicine and Biology,Plenum Press, New York, pp.195-207.

18. Eisenberg,D., Terwilliger,T.C. and Tsui,F. (1980) Biophys.J.,
    32,252-254.

TIME-DEPENDENT FLUORESCENCE STUDIES ON THE LOCATION OF RETINAL IN BACTERIORHODOPSIN AND THE EFFECT OF BACTERIORHODOPSIN ON THE ORDER AND DYNAMICS OF THE LIPID PHASE

M.P. HEYN[1], N.A. DENCHER[1] AND M. REHOREK[2]
[1]Abt. Biophysik, FB Physik, Freie Universität, Arnimallee 14, D-1000 Berlin 33, FRG
[2]Dept. of Biophys. Chemistry, Biozentrum der Universität Basel, CH-4056 Basel, Switzerland

INTRODUCTION

   The light-driven proton pump bacteriorhodopsin (BR) contains as the chromophore one retinal bound to the ε-aminogroup of lysine 216 via a Schiff base. The chromophore is not only responsible for the primary step of light-absorption, but is most likely also involved in the proton translocation process. In structural studies using electron- and X-ray diffraction this essential component of BR could not be resolved. The lateral position of retinal in the plane of the membrane was determined using neutron diffraction and fluorescence energy transfer (1,2). Both studies concluded that retinal is located in the interior of BR. The experimental difficulties and uncertainties with these methods were considerable. In this study we also use fluorescence energy transfer to localize retinal within BR, but employ an entirely different approach as in (2). In reference 2 energy transfer was detected between reduced retinals as donors and normal retinals as acceptors both of which were fixed in position and orientation in the hexagonal lattice of the purple membrane. In the present study the purple membrane is solubilized and monomeric BR is reconstituted in dimyristoylphosphatidylcholine (DMPC) vesicles of known lipid to protein ratio. Well above the phase transition temperature of the lipids BR remains monomeric (3,4). When the fluorescent donor 1,6-diphenyl-1,3,5-hexatriene (DPH) is introduced into the lipid phase energy transfer occurs to the acceptor retinal of BR. The donors and acceptors are randomly distributed and diffusing in the plane of the bilayer (3) and energy transfer occurs when donors and acceptors get close enough together. The probability of energy transfer will thus increase with the surface density of acceptors. The rate of energy transfer is inversely proportional to the sixth power of the distance between donor and acceptor. If retinal is buried in the interior of BR energy transfer will therefore be less than in the case when retinal is located on the surface of BR and donor and acceptor can get very close together. This effect was used to determine the distance of closest approach between DPH and retinal within the framework of a model in which BR is approximated by a cylinder with retinal at the center. The reconstituted vesicles provide an excellent model system for testing current theories of energy transfer in two-dimensional random distributions of donors and acceptors (8). In order to do an energy transfer experiment properly, the mobility and orientation of the donor and acceptor transition dipole moment directions have to be known. For this purpose fluorescence depolarization experiments were performed which provide both dynamical information on the rate of rotation of the donor and structural information on the extent to which this rotational motion is restricted by the anisotropic environment. From such measurements the membrane viscosity and the order parameter of the lipid phase can be obtained as a function of the BR to lipid ratio. The effect of the intrinsic membrane protein BR on these two parameters of the lipid phase can thus be investigated. Such measurements are of considerable interest for a better understanding of lipid-protein interactions.

MATERIALS AND METHODS

DMPC/BR vesicles of variable lipid to protein ratio were prepared as described before (9). Energy transfer and fluorescence depolarization experiments were performed as described elsewhere (10) at a DPH to lipid ratio of 1:500. The chromophore retinal was destroyed by illumination with a 300 W Xenon lamp. No spectral contribution of retinal or its products was observable after 2 hours of illumination.

RESULTS

Location of retinal using energy transfer from DPH to bacteriorhodopsin

Fig. 1 shows the fluorescence decay of DPH in DMPC vesicles of

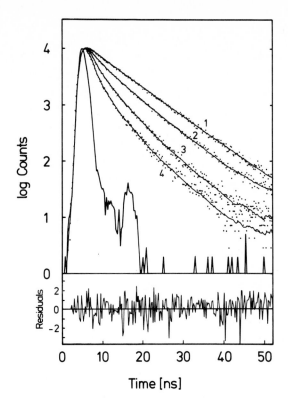

Fig. 1. Lamp pulse and fluorescence decays (....) of DPH in DMPC
vesicles as a function of the molar BR to DMPC ratio at 35°C. The
molar ratios are (top to bottom) 0 (sample 1), 1:331 (sample 2),
1:171 (sample 3), and 1:102 (sample 4). The data points were fitted with a sum of exponentials convoluted with the excitation
profile (_____). Bottom, the weighted residuals for sample 4.

varying BR content at 35°C. For sample 1, vesicles without BR,

the decay is single exponential with a lifetime of 8.2 ns. As the

amount of BR increases the decay becomes non-exponential and the

mean lifetime gradually decreases to 2.0 ns for the vesicles of

sample 4 (molar BR to DMPC ratio of 1:102). When the acceptor

retinal of BR was destroyed by bleaching the lifetime for each of

the samples returned to the value for the BR-free vesicles. These

measurements clearly indicate that energy transfer is the reason

for the decrease in lifetime with increasing surface density of

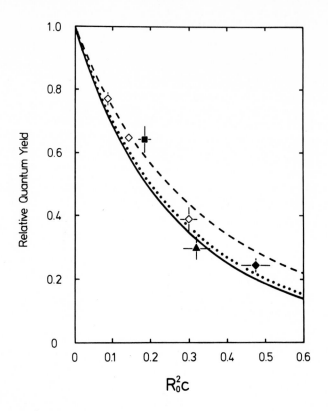

Fig. 2. Relative quantum yield of DPH (calculated from the ratio of the mean lifetimes in the presence and absence of acceptor) vs. the number of acceptors per R_O^2, (R_O^2c): sample 2 (■), 3 (▲), 4 (◆) (see legend of Fig. 1), and of sample 4 with partially bleached retinal (◇) at 35°C. The solid (____), the dotted (....), and the dashed curve (----) are calculated from Eq. 17 of reference 8 for $R_e = 0$, $R_e = 0.25\,R_O$, and $R_e = 0.5\,R_O$, respectively.

acceptors. The mean lifetime is directly proportional to the quantum yield. The relative quantum yields for samples 2,3 and 4, and for sample 4 with the acceptor retinal bleached to 63, 30 and 18% of the original amount are plotted in Figure 2 as a function of the surface density of acceptors c. The natural dimensionless variable characterizing the energy transfer is the product R_O^2c where R_O is the Förster radius (10). R_O^2c is the number of acceptor molecules within a R_O-neighborhood of a donor. Acceptors

within this range of the donors have a high probability of de-
exciting the donors by radiationless energy transfer. The relative quantum yield of the donor is thus expected to decrease with
increasing R_O^2c. In Figure 2 the theoretical predictions are shown
for the dependence of the quantum yield on R_O^2c for the model of
a membrane with donors and acceptors randomly distributed (8). BR
is randomly distributed in these DMPC vesicles at 35°C (3,4). The
surface density of acceptors can be calculated from a knowledge
of the lipid to protein ratio and from the known surface areas of
BR and DMPC. For the calculation of the Förster radius, R_O, the
overlap integral of the donor emission spectrum with the acceptor
absorption spectrum is required. Numerical integration yielded
the value of 7.3×10^{-14} cm^3M^{-1}. It is well known that the major
uncertainty in R_O is due to the angular factor K^2. In the present
case enough is known, however, about the orientation and mobility
of both donor and acceptor to calculate K^2. The orientation of
the transition dipole moment of the acceptor with respect to the
membrane normal has been measured (11). A value of 75° was used.
The acceptor does not wobble on the fluorescence time scale (12)
and BR as a whole rotates with a correlation time in the μs-range
(3). The donor is known to wobble rapidly on the ns time scale
around its mean position. The extent of wobbling can be determined from time dependent fluorescence depolarization measurements (see below). The end value of the decay of the fluorescence
anisotropy r_∞ is a direct measure of the order parameter S of the
donor. As we will see below the order parameter of the donor depends on the lipid to protein ratio and can be accurately determined for each sample. Using this information, the angular dependence of R_O can be calculated for each lipid to protein ratio
(10). For the four samples of Figure 2, the K^2 factor varies between 0.56 and 0.44 and the corresponding Förster radius R_O between 45.1 and 43.2 Å. Since energy transfer depends on the in-
verse sixth power of the distance between donor and acceptor, the
extent of energy transfer is very sensitive to the distance of
closest approach, R_e, between donor and acceptor. When this distance R_e becomes comparable to R_O the energy transfer will be
much reduced. The three curves of Figure 2 are theoretical predictions (8) for the decrease in the relative quantum yields with

R_0^2c for values of the distance of closest approach of $R_e = 0$, $R_e = 0.25\ R_O$, and $R_e = 0.50\ R_O$. The curves of Figure 2 were gene-rated on the basis of the theory of Wolber and Hudson, which as-sumes a random distribution of donors and acceptors and a cylin-drical symmetry of the excluded volume around donor and acceptor (8). In the present case this means that retinal is assumed to be at the center of a cylinder of radius R_e whose value can be expe-rimentally determined. On the basis of a careful comparison be-tween data and theory we find $R_e = 18 \pm 5$ Å. When BR is approxi-mated by an effective cylinder (13), one obtains a radius of about 17 Å. We may thus conclude that retinal is deeply buried within BR. A further critical test of the model used to analyze the data is provided by the decays shown in Figure 1. Using the excluded volume radius of $R_e = 18$ Å, it is possible to predict the highly non-exponential decays of curves 2,3 and 4 of Figure 1. The predicted curves are in satisfactory agreement with the observed decays (10).

Effect of BR on the dynamics and order of the lipid phase

The rotational motion of the rodlike molecule DPH can be used to determine the order parameter and viscosity of the lipid phase (14,15,16,17). For this purpose the decay of the fluorescence anisotropy of DPH is followed on the ns-time scale. The decay constants provide information on the mobility or viscosity of the lipid phase. The end value of the anisotropy decay, r_∞, which is reached at long times, is a measure of the order parameter of DPH in the lipid phase (14,15). One of the interesting questions in the area of lipid-protein interactions is how incorporation of intrinsic membrane proteins affects the order and viscosity of the lipid phase. The BR/DMPC vesicles described in the first part of this paper are ideally suited to answer this question. At each lipid to BR ratio the end value r_∞ and the decay constants can be determined to see if systematic effects in these parameters occur as the amount of BR is increased. Since energy transfer occurs, the time dependent decay of the anisotropy r(t) was measured both for vesicles with BR and vesicles with bleached BR in which the acceptor was destroyed by light. Figure 3 shows the decay of the anisotropy of DPH for pure DMPC vesicles at 35°C (curve 3) and

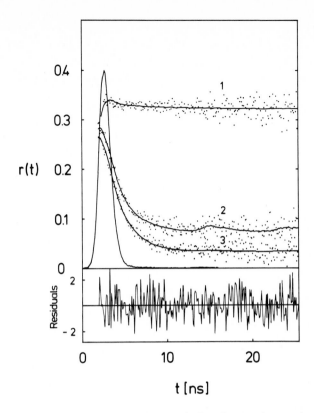

Fig. 3. Decay of the anisotropy r(t) of DPH in BR/DMPC vesicles. Curves 1 and 2, BR/DMPC vesicles of protein to lipid ratio 1:115, at 0.5 and 35°C respectively. Curve 3, protein-free DMPC vesicles at 35°C. Bottom, the weighted residuals for curve 3.

for BR/DMPC vesicles at 1:115 at 35°C (curve 2) and 0.5°C (curve 1). It is clear that at 0.5°C far below the phase transition of DMPC, when the lipids are in the gel phase, the label is practically immobilized. At 35°C the end value r_∞ for the vesicle with protein (curve 2) is higher than without protein (curve 3), indicating that the order parameter of the lipid phase is increased by the protein. The bump in the experimental points for curve 2 (shown before deconvolution of the data) is due to the lamp afterpulse (see lamp pulse Figure 1). The decay of r(t) was fitted with a sum of two exponentials plus a constant term (r_∞) and the best fits are shown in Figure 3 (solid lines). It is clear from this figure that the mean decay constant is about the same

in the sample with and without protein, indicating that the viscosity of the lipid phase is not much affected by the incorporation of the protein. The order parameter S can be calculated from
the end value r_∞ and the initial value r_0 by $S^2 = r_\infty/r_0$ (14,15).
The results for S for many different lipid to protein ratios are
collected in Figure 4. At 35°C, S increases from 0.29 for the
pure lipids to 0.61 at the highest protein to lipid ratio (1:52).
The results demonstrate convincingly that the order parameter of
the lipid phase increases significantly by the incorporation of
BR. For comparison the order parameter is also plotted in Figure
4 at 0.5°C, when the lipids are in the gel phase. At this temperature the lipids are already highly ordered and incorporation of
the protein has within experimental error no effect. The decay
constants of r(t) were used to calculate the viscosity (16).

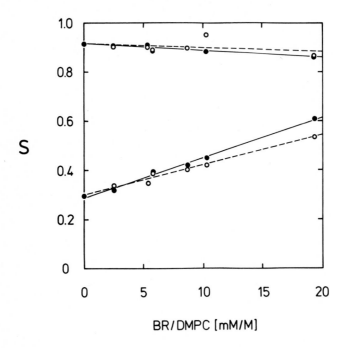

Fig. 4. Order parameter S of DPH in BR/DMPC vesicles at 35°C as a
function of the molar protein to lipid ratio (●). Open symbols:
bleached vesicles. Lower data set 35°C, upper data set 0.5°C.

Within experimental error the viscosity is not affected by the protein. The absolute values of the viscosity of the samples at 35°C range between 0.1 and 0.3 Poise depending on the model used.

DISCUSSION

In accordance with neutron diffraction (1) and energy transfer (2) experiments on the purple membrane, the present experiments with monomeric BR indicate that retinal is located in the interior of the protein. The theory for energy transfer in two-dimensional random distributions of donors and acceptors (8) seems to provide a good description of the data. In agreement with previous steady-state fluorescence depolarization experiments (3, 4,14) the effect of incorporation of the intrinsic membrane protein BR on the properties of the lipid phase is mainly to increase the order parameter of the lipids. Deuterium NMR experiments with other systems reach the opposite conclusion. This discrepancy is most likely due to the different time scales of the two methods. The lipid viscosity determined from the DPH wobbling is an order of magnitude smaller than that obtained from the rotational diffusion of BR in the same DMPC vesicles (3). Different descriptions may be required for the rotational diffusion of small molecules such as DPH and large membrane proteins.

REFERENCES

1. King, G.I., Mowery, P.C., Stoeckenius, W., Crespi, H.L. and Schoenborn, B.P. (1980) Proc. Natl. Acad. Sci. 77, 4726.

2. Kouyama, T., Kimura, Y., Kinosita, K.Jr. and Ikegami, A. (1981) J. Mol. Biol. 153, 337.

3. Heyn, M.P., Cherry, R.J. and Dencher, N.A. (1981) Biochemistry 20, 840.

4. Heyn, M.P., Blume, A., Rehorek, M. and Dencher, N.A. (1981) Biochemistry 20, 7109.

5. Dencher, N.A. and Heyn, M.P. (1979) FEBS Lett. 108, 307.

6. Heyn, M.P., Kohl, K.-D. and Dencher, N.A. (1981) in: F. Palmieri et al. (Eds.), Vectorial Reactions in Electron and Ion Transport in Mitochondria and Bacteria, Elsevier/North-Holland Biomedical Press, Amsterdam, pp. 237-244.

7. Dencher, N.A., Kohl, K.-D. and Heyn, M.P. (1983) Biochemistry 22, 1323.

8. Wolber, P.K. and Hudson, B.S. (1979) Biophys. J. 28, 197.

9. Heyn, M.P. and Dencher, N.A. (1982) Methods Enzymol. 88, 31.

10. Rehorek, M., Dencher, N.A. and Heyn, M.P. (1983) Biophys. J. 43, in press.

11. Heyn, M.P., Cherry, R.J. and Müller, U. (1977) J. Mol. Biol. 117, 607.

12. Kouyama, T., Kimura,Y., Kinosita, K.Jr. and Ikegami, A. (1981) FEBS Lett. 124, 100.

13. Henderson, R. and Unwin, P.N.T. (1975) Nature 257, 28.

14. Heyn, M.P. (1979) FEBS Lett. 108, 359.

15. Jähnig, F. (1979) Proc. Natl. Acad. Sci. 76, 6361.

16. Kinosita, K.Jr., Kawato,S. and Ikegami, A. (1977) Biophys. J. 20, 289.

17. Lipari, G. and Szabo, A. (1980) Biophys. J. 30, 489.

INTERCONVERSIONS BETWEEN SUB-CONFORMATIONAL STATES OF THE NA/K PUMP, SHOWING
SIGMOID OR HYPERBOLIC ACTIVATION OF RB MOVEMENTS

S. J. D. KARLISH[1] AND W. D. STEIN[2]

[1]Biochemistry Department, The Weizmann Institute of Science, Rehovot, Israel
and [2]Institute of Life Science, Hebrew University, Jerusalem, Israel.

INTRODUCTION

Two major conformational forms of the non-phosphorylated Na/K pump have been
identified by a number of physical techniques (1,2,3,4). State E_1 is the conformation having high affinity Na and low affinity K binding sites facing the
cell interior. $E_2(K)_{occ}$ is a state containing occluded K ions (5). Classical
K-K exchange through the Na/K pump, as studied in red cells, requires both ATP
and inorganic phosphate, and has been thought to reflect interconversions between the E_1 and $E_2(K)_{occ}$ states and the phosphoenzyme E_2P in which K sites
face the cell exterior (3):

$$K_{in} + E_1 \text{ ATP} \rightleftharpoons E_1K \text{ ATP} \underset{\text{ATP}}{\rightleftharpoons} E_2(K)_{occ} \xrightarrow{\text{Pi}} E_2\text{-P}\cdot K \rightleftharpoons E_2P + K_{out}$$

ATP accelerates the rate of the conformation transition $E_2(K)_{occ} \xrightarrow{} E_1$ (4,6).

Recently, using phospholipid vesicles reconstituted with pig kidney (Na,K)
ATPase (7), we described new modes of passive K(Rb) movements through the pump
protein in the complete absence of ATP and inorganic phosphate (8). These
fluxes are very slow and include a Rb-Rb exchange, exchange of Rb with its con-
geners such as K, Cs, Na, Li, and in addition a net (as opposed to exchange)
movement of Rb ions. We accounted for these new modes of Rb movements by the
schemes depicted in fig. 1. Net Rb flux in the cytoplasmic to extracellular
direction involves the sequence: $E_1 \longrightarrow E_1Rb \longrightarrow E_2(Rb)_{occ} \longrightarrow E_2Rb \longrightarrow E_2 \longrightarrow E_1$.
Rb-Rb exchange involves oscillation between the forms $E_1 \rightleftharpoons E_1Rb \rightleftharpoons E_2(Rb)_{occ} \rightleftharpoons E_2Rb \rightleftharpoons E_2$,
while exchange of Rb from cytoplasmic sites for congener X at extracellular
sites involves the paths in the lower scheme. A detailed kinetic analysis of
Rb movements, together with a consideration of the rates of conformational
transitions measured directly on isolated (Na,K)ATPase, allowed us to infer
that the steps labelled a and d are fast, while those labelled b, c and f are
slow and e is very slow. Each of b, c, e and f is dominantly or partially
rate-limiting in the different new passive modes of Rb movements. We concluded
from the analysis that the form $E_2(Rb)_{occ}$ is a stable species, and therefore
the role of Rb occlusion is to minimize slippage and ensure tight coupling of
active cation fluxes to ATP hydrolysis. ATP and Pi are regulators in the sense
that they control the rate at which Rb leaves the occluded state in the

direction of forward pumping or reversal of the pump, respectively(9).

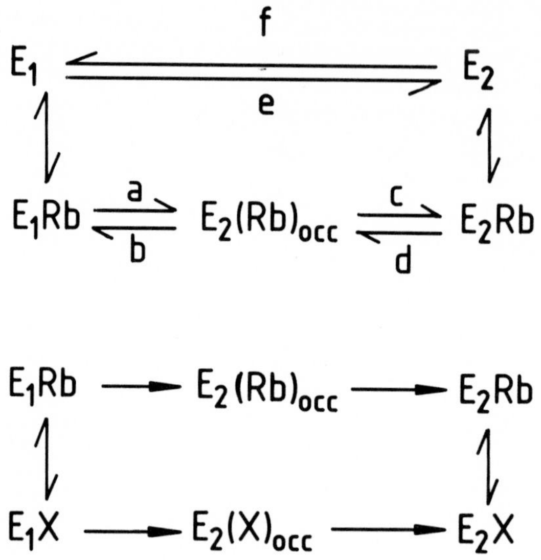

$$E_1 \underset{e}{\overset{f}{\rightleftharpoons}} E_2$$

$$E_1Rb \underset{b}{\overset{a}{\rightleftharpoons}} E_2(Rb)_{occ} \underset{d}{\overset{c}{\rightleftharpoons}} E_2Rb$$

$$E_1Rb \longrightarrow E_2(Rb)_{occ} \longrightarrow E_2Rb$$

$$E_1X \longrightarrow E_2(X)_{occ} \longrightarrow E_2X$$

Fig. 1. Schemes for passive Rb movements through the Na/K pump

Hyperbolic or sigmoid activation kinetics by Rb ions at cytoplasmic sites

A striking finding in our previous studies of Rb-Rb exchange was the fact that in the absence of ATP and Pi, there was a strictly hyperbolic dependence on concentration of Rb at cytoplasmic sites. Net Rb movement into Tris-loaded vesicles, was also dependent on the Rb concentration in a somewhat sigmoid fashion (8).

Since sigmoidicity of cation activation is such a characteristic feature of Na/K pump functions (10), usually related to the 3Na : 2K : 1ATP stoichiometry, we have looked at this feature in greater detail. The passive Rb movements we have described in the vesicles occur in the simplest of transport conditions, and thus provide a convenient system for investigating the requirements for the manifestation of a hyperbolic or sigmoid dependence on Rb concentration.

We first looked at the effects of different cations at the trans face, and measured the vanadate$_o$-inhibitable component of the total Rb uptake. Since vanadate is known to bind to the cytoplasmic surface of the Na/K pump (11),

we are dealing with a Rb flux from the cytoplasmic to the extracellular face of
pumps, which in the vesicles are oriented inside-out with respect to the normal
cellular orientation.

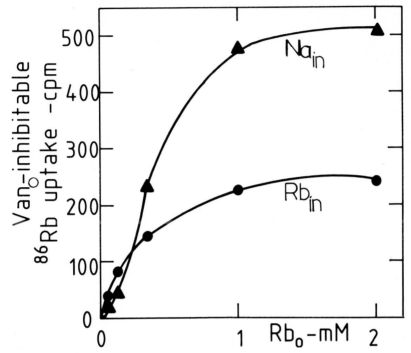

Fig. 2. Dependence on Rb_o concentration of Rb uptake into vesicles loaded with
Rb_i or Na_i. Vesicles containing 150 mM RbCl or NaCl were prepared from soya
bean phospholipid and pig kidney (Na,K)ATPase as described in ref. (7). The
vanadate$_o$-sensitive ^{86}Rb uptake was estimated over 4 minutes at different Rb_o
concentrations, essentially as described in fig. 3 of ref. (8).

(fig. 2) shows that when Na is the 'trans' ion, Rb at the cytoplasmic surface
shows a strongly sigmoidal concentration dependence, while for Rb at the 'trans'
face the strictly hyperbolic dependence is observed. The upper half of Table I
records the kinetic parameters derived from these curves and similar experiments
using vesicles loaded with either Li or Tris ions. The second, third and fourth
columns record the V_{max}, $K_{0.5}$ and Hill coefficients. The numerical values of
the Hill coefficients confirm that Rb-Rb exchange displays strictly hyperbolic
behaviour, Rb-Na exchange is strikingly sigmoidal, and shows that intermediate
behaviour occurs with Tris or Li ions at the extracellular face. Notice that
there is no correlation between the numerical value of the Hill coefficient and
the $K_{0.5}$ or V_{max} of the Rb fluxes into the different 'trans' ions.

The lower half of Table I records kinetic parameters for Rb-Rb exchange in the presence of ATP, and Mg ions. Mg alone and ATP alone affect the $K_{0.5}$ and V_{max} for Rb-Rb exchange (cf. also 7,12,13) but neither ligand alone affects appreciably the shape of the Rb concentration dependence curve. Again there is no correlation between the derived Hill coefficient and the V_{max} and $K_{0.5}$ values.

TABLE I

KINETIC PARAMETERS FOR VANADATE$_o$-INHIBITABLE [86]RB UPTAKE INTO VESICLES, IN DIFFERENT CONDITIONS

Data in the third, fourth and fifth columns were obtained from a number of experiments (brackets) like those in fig. 2. The recorded parameters represent the average values obtained by fitting the data by linear regression to the Hill equation $\log(v/V_{max} - v) = n \log s - \log K$. In the few cases where the V_{max} value could not be obtained reliably by 1/v and 1/S plots of the raw data, that V_{max} was chosen which gave the highest correlation coefficient in the linear regression analysis. The Rb-Rb exchange experiments in the lower half of the table were performed as in fig. 2, but with a suitable range of Rb$_o$ concentrations and the indicated concentrations of ATP, Pi and Mg ions. The fifth column of the top half of the table was performed essentially as in fig. 3.

Condition	V_{max} rel.	$K_{0.5}$-Rb$_o$ mM	Hill Coefficient	$K_{0.5}$-Na$_o$ mM at 1 mM Rb$_o$
Rb$_{in}$	1.0	0.67±0.08	1.02±.02 (8)	20-50
Tris$_{in}$	0.6	0.55±0.04	1.26±.03 (3)	∿ 3
Na$_{in}$	2.5	0.39±0.01	1.43±.05 (2)	∿1.5
Li$_{in}$	3.0	0.98±0.18	1.15±.04 (3)	∿ 3
Rb$_{in}$	1.0	0.67	1.01	
+ATP-3 mM	2.5	2.9	1.04	
+Mg$_o$-3 mM	0.6	1.4	1.04	

High or low affinity inhibition by Na ions at cytoplasmic sites

In conventional transport experiments, using red cells or the reconstituted vesicles in the presence of ATP and Mg it is invariably found that Na ions combine with cytoplasmic sites of the pump with a high affinity. We have found, however, that for the ATP - and Pi - independent passive Rb fluxes in the vesicles, the identity of the trans ion determines whether Na ions will inhibit Rb uptake from the cytoplasmic surface with a high or low apparent affinity. Fig. 3 shows the result of an experiment using again vesicles loaded with either Rb or Na ions, and the vanadate$_o$-inhibitable uptake of Rb, studied in the presence of increasing concentrations of the Na ions at the cytoplasmic face of the pump. The striking finding is that with Rb inside the vesicles, Na is a poor inhibitor of Rb movements from the cytoplasmic face of the pump. Only with Na inside the vesicles does Na at the cytoplasmic surface inhibit with the classical high affinity. The fifth column of Table I records apparent affinities derived from this experiment and similar experiments using vesicles loaded with Tris or Li ions. Again it is Na or Rb ions at the extracellular surface of the pumps which display the extremes of behaviour while Li and Tris occupy an intermediate position.

Two sub-conformations of the unphosphorylated Na/K pumps?

In this section we propose tentatively that there are two sub-conformations of the pump, one of which displays hyperbolic behaviour and the other sigmoid behaviour. We assume that sigmoid behaviour implies that more than one Rb ion must be bound to the pump protein and transported in each cycle. In fact Glynn and Richards have shown directly (5) that about 3 Rb ions are occluded per pump molecule in the form $E_2(Rb)_{occ}$, which is thought to participate in the passive Rb flux modes.

The simplest explanation of sigmoid behaviour is that only when say three Rb ions have bound to the form E_1 is there a concerted transition to the form $E_2(3Rb)_{occ}$. Thus, $E_1 + Rb \rightleftharpoons E_1Rb \rightleftharpoons E_12Rb \rightleftharpoons E_13Rb \rightleftharpoons E_2(3Rb)_{occ}$. In such a situation the affinities for Rb at all three sites may or may not be the same. If the affinities at the three sites happen to be very different, the Rb concentration at which half the maximal flux is observed, will reflect largely the binding constant of the lowest affinity site. However the sigmoid behaviour will be most pronounced in the region where the highest affinity site is being titrated. In this region, however, the measured signal (i.e. the magnitude of flux in our experiments) is extremely small and therefore largely inaccessible. In such a case the curve will appear hyperbolic. By contrast, where we find sigmoid behaviour, the implication would be that the affinities for the Rb binding sites are fairly close together.

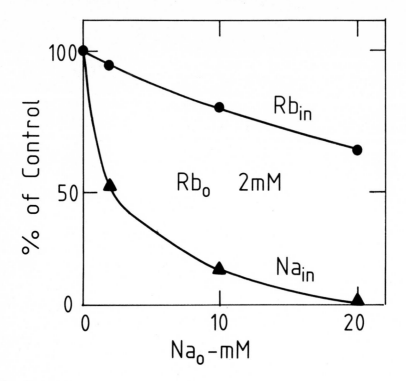

Fig. 3. Inhibition by Na_o of Rb uptake into vesicles loaded with Rb_{in} or Na_{in}. Vanadate-sensitive Rb uptake was measured over 2 mins at Rb_oCl, 2 mM and the indicated concentrations of $NaCl_o$, essentially as described in ref. (8).

We have been concerned that the different behaviour, sigmoid or hyperbolic might not reflect true changes in affinity for Rb bindingsites, that is a real change in the shape of the curves. The possibility might have been that in the different conditions, such as in fig. 2, we are not looking at ranges of Rb concentrations equally comparable to the relevant half-saturation concentration. An inspection of Table I shows that our concern here is unjustified. Even in conditions with very similar $K_{0.5}$ values, the shapes of the curves can be quite different as judged by the different Hill numbers. Conversely, similar Hill numbers can be obtained in conditions with markedly different $K_{0.5}$ values, compare say the control Rb-Rb exchange and that with ATP only. Thus we are not being misled by the position of the uptake curve on the Rb concentration axis. The difference between sigmoid and apparent hyperbolic behaviour is real.

There is an alternative explanation for the hyperbolic-sigmoid dichotomy. It is conceivable that in those conditions in which hyperbolic behaviour is observed, that the stoichiometry of transport has changed and only one Rb ion is bound at the cytoplasmic surface on form E_1. Although somewhat unlikely we cannot exclude this possibility at present.

The first possibility, that is both high and low affinity Rb binding sites on E_1, is consistent with the following data which however is not conclusive. There is good evidence both from fluorescence and transport studies for a low affinity Rb binding to E_1 with an intrinsic affinity in the range 50-100 mM (3,14,9, 13). [The apparent affinities in a transport experiment such as recorded in Table I, are given approximately by this intrinsic affinity divided by the equilibrium constant of the protein conformation change $E_1 K \leftrightarrow E_2 (K)_{occ}$, here some 100 - 1000 (8,14,15)]. Evidence has been reported for high affinity Rb binding with constants 1 μM and 30 μM (16) or 6 μM (17), but it is not known whether this binding occurs in the E_1 or E_2 conformations. There is some evidence for relatively high affinity Rb binding ($K_A \sim 60$ μM) in the presence of 2 mM ATP which should stabilize the E_1 form (18).

Irrespective of the exact explanation of the hyperbolic behaviour, the implication is that there are sub-conformations of at least the E_1 state of the pump protein with different cation binding characteristics. We will term these E_1 and E_1^* for sigmoid and hyperbolic behaviour, respectively. The results in this paper have shown that there is a close parallel between the sigmoid-hyperbolic dichotomy and the high or low efficacy of Na as an inhibitor at cytoplasmic sites in vesicles containing the different cations. It is therefore economic to assume that the change from E_1 to E_1^* state is accompanied by the decrease in Na binding affinity seen in fig. 3. By our hypothesis the sub-conformation E_1^* predominates with a high Rb at the extracellular surface, while E_1 predominates with a high Na at the extracellular surface. There is some precedent for allosteric effects of Na at extracellular sites. Work with red cells has shown that extracellular Na (i) inhibits an uncoupled Na efflux and accompanying ATP hydrolysis (20,21) (ii) reduces the apparent affinity for K at the extracellular surface and induces a strongly sigmoid dependence of K uptake on the extracellular K concentration (22,23) and (iii) appears to reduce the efficacy of vanadate as an inhibitor at cytoplasmic sites (24,25). Some or all of these phenomena might be manifestations of the same allosteric transitions that we have observed. Phenomena (ii) and (iii) would reflect effects of Na on E_2 forms of the pump and may therefore imply that one must consider also E_2 and E_2^* forms in addition to the E_1 and E_1^* forms.

Finally it is worth pointing out that the conditions favouring the postulated E_1^* to E_1 sub-conformational transition, namely a high extracellular are of course the normal physiological conditions. Since Na competes with K(Rb) at cytoplasmic sites (26) it could be this same factor that allosterically confers sigmoidal character on the Na activation of normal active Na/K exchange. The sigmoidal nature of the activation by cellular Na serves the purpose of allowing a steep increase in the rate of active Na pumping when cellular Na increases over a narrow concentration range.

ACKNOWLEDGEMENT

We thank Mrs. R. Goldshlegger for invaluable technical assistance.

REFERENCES

1. Jorgensen, P. L. (1982) Biochim. Biophys. Acta 694, 27-68.
2. Jorgensen, P. L. (1975) Biochim. Biophys. Acta 401, 399-415.
3. Karlish, S. J. D., Yates, D. W. and Glynn, I. M. (1978) Biochim. Biophys. Acta 525, 252-264.
4. Karlish, S. J. D. and Yates, D. W. (1978) Biochim. Biophys. Acta 527,115-130.
5. Glynn, I. M. and Richards, D. E. (1982) J. Physiol. 330, 17-45.
6. Post, R. L., Hegevary, C. and Kume, S. (1972) J. Biol. Chem. 247, 6530-6540.
7. Karlish, S. J. D. and Pick, U. (1981) J. Physiol. 312, 505-529.
8. Karlish, S. J. D. and Stein, W. D. (1982) J. Physiol. 328, 295-316.
9. Karlish, S. J. D., Lieb, W. R. and Stein, W. D. (1982) J. Physiol. 328, 333-350.
10. Glynn, I. M. and Karlish, S. J. D. (1975) Ann. Revs. Physiol. 37, 13-55.
11. Cantley, L. C., Resh, M. and Guidotti, G. (1978) Nature 272, 552-554.
12. Karlish, S. J. D. and Stein, W. D. (1982) J. Physiol. 328, 317-331.
13. Eisner, D. E. and Richards, D. E. (1983) J. Physiol. In press.
14. Karlish, S. J. D. (1980) J. Bioenergetics and Biomembranes 12, 111-136.
15. Beauge, L. and Glynn, I. M. (1980) J. Physiol. 299, 367-383.
16. Matsui, H. and Homareda, H. (1982) J. Biochem. (Tokyo) 92, 193-217.

17. Jorgensen, P. L. and Petersen, J. (1982) Biochim. Biophys. Acta 705, 38-47.

18. Jensen, J. and Ottolenghi, P. (1983) in Current Topics in Membranes and Transport. Eds. Hoffman, J. F. and Farbush, B. In press.

19. Jorgensen, P. L. and Petersen, J. (1979) in: (Na,K)ATPase Structure and Kinetics. Eds. Skou, J. C. and Norby, J. G. Academic Press, pp. 143-157.

20. Garrahan, P. J. and Glynn, I. M. (1967) J. Physiol. 192, 159-174.

21. Glynn, I. M. and Karlish, S. J. D. (1976) J. Physiol. 256, 465-496.

22. Garrahan, P. J. and Glynn, I. M. (1967) J. Physiol. 192, 175-188.

23. Cavieres, J. D. and Ellory, J. C. (1975) Nature 255, 338-340.

24. Beauge, L. A., Cavieres, J. D., Glynn, I. M. and Grantham, J. J. (1980) J. Physiol. 301, 7-23.

25. Beauge, L. and Berberian, S. (1983) Biochim. Biophys. Acta 727, 336-350.

26. Garay, R. P. and Garrahan, P. J. (1973) J. Physiol. 231, 297-325.

© 1983 Elsevier Science Publishers B.V.
Structure and Function of Membrane Proteins,
E. Quagliariello and F. Palmieri editors.

ESR STUDIES OF THE SULPHYDRYL GROUPS OF THE MITOCHONDRIAL PHOSPHATE
CARRIER BY MALEIMIDE SPIN LABELS.

E. BERTOLI[1*], I. STIPANI[1], F. PALMIERI[1], J. HOUSTEK[2], P. SVOBODA[2],
F.M. MEGLI[1] and E. QUAGLIARIELLO[1].
[1]Institutes of Biochemistry, Faculties of Pharmacy and Sciences,
University of Bari, 70126 Bari, Italy and [2]Institute of Physiology,
Czechoslovak Academy of Sciences, Prague 142 20, Czechoslovakia.

INTRODUCTION

The electroneutral phosphate translocator of the inner mitochondrial
membrane represents the major system of phosphate supply for oxidative
phosphorylation. The carrier contains free sulphydryl groups
which are essential for its function (1-2).

Information on the physico-chemical properties of the sulphydryl
groups of the native membrane-bound phosphate carrier can be
reached by using paramagnetic analogues of N-ethylmaleimide under
conditions in which they react with sulphydryl groups of the phosphate
carrier specifically. For this purpose maleimide spin labels of
different chain length are used in order to characterize the
microenvironment of the phosphate carrier sulphydryl groups in
situ as well as to sense the average depth at which they are
located with respect to the aqueous phase of the membrane.

MATERIALS AND METHODS

The following maleimide spin labels (MSL) with different length
(3) from Syva, Palo Alto (USA) were used: MSL_a (no spacer between
the nitroxide group and maleimide) 6.8 Å long, MSL_b (spacer
$-CO-NHCH_2CH_2-$) 11.6 Å long and MSL_c (spacer $-CO-NHCH_2CH_2OCH_2CH_2-$)
15.3 Å long.
Preparation of mitochondria. Rat liver mitochondria were isolated
in 0.25 M sucrose, 10 mM Tris-HCl, 1 mM EDTA, pH 7.4 (STE medium).
Maleimide spin labeling of the native phosphate translocator.

Procedure 1 (non-protected mitochondria). MSL were added to
mitochondria (5 mg protein/ml) at a concentration inhibiting

*Visiting Professor from the University of Ancona. This work was supported
by the Consiglio Nazionale delle Ricerche (Italy) and the Czechoslovak Academy
of Sciences.

approximately 50% of phosphate transport activity. After 2 min, the reaction was stopped by 2 mM cysteine followed (for ESR measurements) by centrifugation in STE medium supplemented with 0.2% BSA (to remove the non-specifically bound MSL) and 5 mM ferricyanide (to protect the nitroxide group against irreversible oxidation). The mitochondria were washed repeatedly until the supernatant was free of the spin label (as controlled by ESR spectra of concentrated supernatants).

Procedure 2 (mersalyl protection). Mitochondria were preincubated with 15 nmol mersalyl/mg protein for 1 min at 0°C followed by the addition of 2 mM NEM. After further 2 min the reaction was stopped by 5 mM cysteine which also released mersalyl reversibly bound to the sulphydryl groups of the phosphate carrier. Cysteine and unbound NEM were washed (3 times) with STE medium containing 0.5 mM cysteine (in the first wash). Mitochondria were then incubated with MSL as described in procedure 1. All incubations were performed at 0°C.

ESR measurements. Electron paramagnetic resonance spectra were carried out using a JEOL JES PE-3X spectrometer at 22°C in quartz flat cells JES-LC-11. First derivative spectra were recorded at the following instrumental setting: field set, 3268 ± 100 G (diphenylpicrylhydrazyl EPR marker); frequency and intensity of X-band, 9.22 GHz and 20 mW, respectively; field modulation amplitude, 1.6 G; time constant 1s; chart speed, 16 min/360 mm.

Labeling of mitochondria with ^{14}C-N-ethylmaleimide (NEM). Non--protected mitochondria (procedure 1) or mersalyl-protected mitochondria (procedure 2) (5 mg protein/ml) were incubated with 15 nmoles ^{14}C-NEM/mg protein for 2 min at 0°C. The reaction was terminated by 2 mM cysteine and mitochondria were washed (three times) with STE medium. When indicated, mitochondria were incubated with MSL (see above) prior to their labeling with ^{14}C-NEM.

RESULTS

The effect of maleimide spin labels on phosphate transport. Fig.1 shows that the three paramagnetic analogues of NEM (MSL_a, MSL_b, MSL_c) react with the sulphydryl groups of the phosphate carrier since they inhibit the ammonium phosphate induced swelling of mitochondria. However, they differ in the extent of the inhibitory effect. The shortest analogue, MSL_a, is the most powerful and is almost as effective as NEM. The concentration required for a 50% inhibition is only 1.2 times higher than that of NEM. The other two analogues, MSL_b and MSL_c, are significantly less inhibitory. The concentrations required for a 50% inhibition are 1.7 and 3.6 times, respectively, higher than that of NEM. Nevertheless, both MSL_b and MSL_c may also induce the maximal (up to 90%) inhibition of phosphate transport when are used at higher concentrations and/or for longer incubation times (not shown).

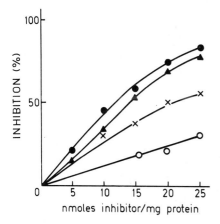

Fig.1 - The effect of MSL on phosphate transport.
Mitochondria (5 mg protein/ml) were incubated for 2 min at 0°C with the indicated concentrations of NEM (●), MSL_a (▲), MSL_b (✗) or MSL_c (○). Reaction was terminated by 2 mM cysteine and aliquots were transferred into 0.1 M ammonium phosphate, 5 mM Tris-HCl, 1 mM EDTA, pH 7.2. Pi transport was measured by recording the decrease in $O.D._{546}$ at 25°C.

The inhibition of phosphate transport induced by both NEM and MSL_a, MSL_b or MSL_c exhibits the same, biphasic time dependence indicating that most of the inhibitor which binds to the phosphate carrier is bound within the first 2 min (not shown). This rapid binding of inhibitors can be further accelerated by increasing the temperature. A change from 0 to 30°C increases the inhibitory effect induced by a 30 sec incubation more than two times.
The effect of maleimide spin labels on ^{14}C-NEM labeling of the phosphate translocator. The selectivity of labeling of the phosphate carrier by maleimide spin labels was controlled with the aid of ^{14}C-NEM. In control mitochondria (not preincubated with MSL), the most of the bound radioactivity is associated with a protein of approx. 33K Mr (Fig.2, lane A) with the same mobility of the isolated phosphate carrier (4), although several other mitochondrial proteins are also labelled to a much lesser extent. Preincubation

of mitochondria with MSL_a, MSL_b or MSL_c (at a concentration giving 50% inhibition) markedly decreases the intensity of the radioactive band corresponding to the isolate phosphate carrier (lanes B, C and D). The effect of similar preincubations with mersalyl or NEM

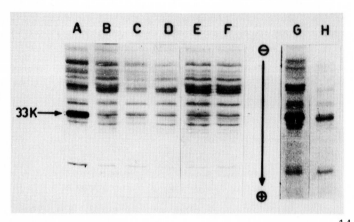

Fig.2 - The effect of MSL, NEM and mersalyl on ^{14}C-NEM binding to mitochondrial proteins. Mitochondria (5 mg protein/ml) were incubated for 2 min at 0°C without (A and G) or with 15 nmol MSL_a/mg protein (B), 20 nmol MSL_b/mg protein (C), 35 nmol MSL_c/mg protein (D), 15 nmol mersalyl/mg protein (E) and 15 nmol NEM/mg protein (F). In (H) mitochondria were incubated for 1 min with 15 nmol mersalyl/mg protein followed by 2 mM NEM (2 min). Reaction in all samples(except for E) was terminated by 2 mM cysteine and after washing mitochondria were labeled for 2 min with 15 nmol ^{14}C-NEM/mg protein at 0°C. Aliquots were precipitated with 90% acetone and subjected to electrophoresis (gradient 12-20%). Radioactivity in gels was detected by fluorography using impregnation with PPO and KODAK X-OMAT R films.

is also shown (lanes E and F) for comparison with those of MSL.

With mersalyl-protected mitochondria (Fig.2, lane H) the number of ^{14}C-NEM labeled proteins decreases several times as compared with non protected, control mitochondria. Thus, of the total bound radioactivity approx. 80% becomes localized in the phosphate carrier. The remaining ^{14}C-NEM label is almost equally distributed among the other minor radioactive bands. Preincubation with MSL_a, MSL_b or MSL_c (at a concentration giving 50% inhibition) drastically reduces the radioactivity in the 33K Mr band, without affecting the ^{14}C label associated with the other few proteins demonstrating that in mersalyl-protected mitochondria maleimide spin labels react selectively with the free sulphydryl groups of the Pi carrier.

These data show that the paramagnetic analogues of NEM can react
quite specifically with the sulphydryl groups of the phosphate
carrier if mersalyl-protected mitochondria are labeled by a 50%
inhibitory concentrations of MSL.

<u>In situ specific spin labeling of the phosphate carrier.</u> Fig. 3B
shows the ESR spectrum of non-mersalyl protected mitochondria
labeled with a concentration of MSL_a which inhibits 50% of phosphate
transport (15 nmol/mg protein, 2 min, 0°C). This spectrum is
composite and can be regarded as a superimposition of spectra from
two kinds of bound spin labels, the molecular motions of which are
restricted to different degrees. As compared with the spectrum of
MSL_a in buffer (Fig. 3A), the spectrum of Fig. 3B shows a new
component (outer hyperfine extreme 65 ± 0.5 G) in the low field
region which apparently corresponds to what is usually called a
strongly immobilized signal (see the arrow in Fig. 3B) with respect
to the mobile component representing a large fraction of the signal.
According to Hamilton and McConnel (5) the mobile (or weakly
immobilized) signal (W) reflects the spin probe bound to an external
site, whereas the strongly immobilized signal (S) reflects the spin
label restricted in its local motion.

To increase the specificity of the MSL-phosphate carrier interaction,
the maleimide spin labels were added to mersalyl-protected
mitochondria. As a result of this more selective labeling of the
phosphate translocator, somewhat lower concentrations of spin
labels were required for 50% inhibition of phosphate transport, and
the shapes and the amplitudes of the ESR spectra changed. The ESR
spectrum of mersalyl-protected mitochondria labeled with MSL_a
(Fig. 3C) shows a remarkable increase of the immobilized signal
(S) at the expense of the weakly immobilized one (W). This indicates
that the interaction of MSL with some easily accessible membrane
sulphydryl groups localized close to the surface of the membrane
has been abolished by the "mersalyl protection" procedure. In turn,
MSL has reacted to a greater extent with some of the sulphydryl
groups of the phosphate carrier which are less accessible to the
probe. Both the weakly and the strongly immobilized components of
the spectrum of Fig. 3C undoubtedly represent MSL which has reacted
with SH groups (and not free spin label), since a drastic decrease
in all lines amplitude is observed (Fig. 3D) after preincubation

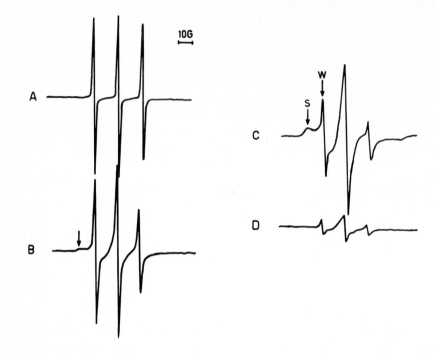

Fig.3 - The ESR spectra of mitochondria labeled with MSL$_a$. Mitochondria were incubated with 15 nmol MSL$_a$/mg protein following procedure 1 (B), procedure 2 (C) or procedure 2 without preincubation with mersalyl (D). (A) represents MSL$_a$ in STE medium. Amplification: spectrum B 3.2 x 10, spectra C and D 5 x 10^2.

of mitochondria with NEM, followed by cysteine and washing, i.e. as in the expt. of Fig. 3C but without mersalyl preincubation. When decreasing the incubation time from 2 min to 30 sec the amount of the bound spin label decreases similarly as the inhibition of phosphate transport measured after these time intervals (2 min - 53%, 30 sec - 25%). In these expts. a parallel decrease of both components of the ESR signal is observed (not shown), indicating that both components of the ESR spectra relate to sites involved in the biological activity of the carrier.

Differentiation between the W and S components of the MSL$_a$ spectra by means of ascorbate and chromium oxalate. The composite ESR spectrum of MSL$_a$ bound to mersalyl-protected mitochondria as observed in Fig. 3C might be due to a dual localization and character of the phosphate carrier sulphydryl groups. The ESR

signal of externally or internally membrane bound spin label (or
rather exposed and non-exposed to the aqueous medium) may be
distinguished using ascorbate or chromium oxalate since the signal
of externally localized spin probe is either reduced by ascorbate
(6) or broadened by chromium oxalate (exchange and dipole-dipole
interactions, ref.7). The results presented in Fig. 4A clearly
show that the weakly immobilized component W gradually decreases

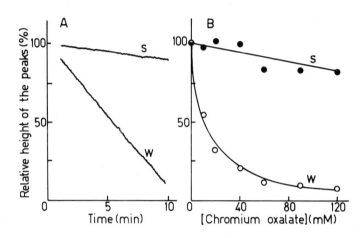

Fig.4 - The effect of ascorbate and chromium oxalate on the W and S components
of the ESR spectrum from mersalyl-protected mitochondria labeled with MSL_a
(15 nmol/mg protein) following procedure 2. 0.5 mM ascorbate in A and
increasing concentrations of chromium oxalate in B were added at time zero. The
S and W components were measured at 3256 G and at 3267 G, respectively, as a
function of time in A and after 2 min in B.

after the addition of ascorbate, whereas the strongly immobilized
component S remains nearly unchanged. Similarly, chromium oxalate
(Fig. 4B), although having a different mode of action, affects
preferentially the W component of the spectrum. These results
clearly indicate the existence of two distinct types of sulphydryl
groups of the native membrane-bound phosphate carrier with different
localization within the membrane.

Localization of sulphydryl groups of the phosphate carrier by
maleimide spin labels with different chain length. In order to
further characterize the localization in the membrane of the
phosphate carrier sulphydryl groups, the mersalyl-protected mitochondria
were labeled with MSL of different chain length.

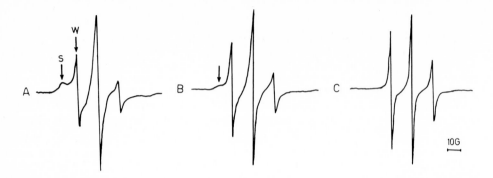

Fig.5 - The ESR spectra of mersalyl-protected mitochondria labeled with 15 nmol MSL_a/mg protein (A), 25 nmol MSL_b/mg protein (B) or 40 nmol MSL_c/mg protein (C) following procedure 2. Amplification of the spectra: 5×10^2 (A), 3.2×10^2 (B) and 1.4×10^2 (C).

As demonstrated in Fig.5, the spectra obtained are different as the length of the label molecule varies, in spite of the fact that in all cases a 50% inhibitory concentration of MSL was used. In the case of the short chain MSL_a (~ 6.8 Å) a considerable number of spin label molecules is in a strongly immobilized position. The outer hyperfine extreme (~ 65 G) indicates a very long correlation time ($\tau_c > 10^{-7}$ s) which cannot be measured by conventional ESR. When the distance of the probe from the reacting groups increases (~ 11.6 Å for MSL_b and ~ 15.3 Å for MSL_c) the motion of the probe becomes less restricted. Thus the spectrum of MSL_b (Fig. 5B) shows still the immobilized component but the major part of the signal is represented by the mobile one. In the case of MSL_c (Fig. 5C) the ESR spectrum shows mainly one mobile component. The observed correlation time ($\tau_c \sim 6.1 \times 10^{-10}$ s) is markedly shorter than that of MSL_b ($\tau_c \sim 1.9 \times 10^{-9}$ s) and especially than that of MSL_a (see above). Since the hyperfine coupling constant (15.0 ± 0.2 G) for MSL_c is equal to that observed with the buffer solution of the label, it appears that the nitroxide groups of the long chain MSL are located towards the aqueous environment and are not trapped into the hydrophobic pocket of the membrane. Considering the differences in the length of the chains of individual MSL analogues and assuming their equal perpendicular location in the membrane, the distance of the deep sites should be equal or not greater than 12-15 Å .

DISCUSSION

The present attempt to use maleimide spin labels for a characterization of the sulphydryl groups of the mitochondrial phosphate carrier are based on the finding that, similarly to NEM, the maleimide spin labels inhibit phosphate transport in mitochondria (Fig.1).

It was shown before that sulphydryl groups of the phosphate translocator react with NEM (low concentration, low temperature and short time) more easily than sulphydryl groups of most of the other inner mitochondrial membrane proteins (4,8). Furthermore, a highly specific interaction of NEM with sulphydryl groups of the phosphate carrier can be achieved by mersalyl protection (9). The conditions used in this study which combine protection by mersalyl and labeling with low concentrations of MSL giving 50% inhibition of phosphate transport are characteristic of a predominant labeling of the phosphate carrier sulphydryl groups by MSL (Fig.2).

The ESR data presented above show that the sulphydryl groups of the phosphate carrier are at least of two types. One type of the groups (site I) giving rise to a mobile ESR signal is easily accessible from the aqueous phase and its environment does not considerably influence the motion of the bound probe. The ESR signal corresponding to site I is strongly affected by non-permeant agents like ascorbate and chromium oxalate. Therefore, the site I sulphydryl groups are apparently localized close to the surface of the membrane and are exposed to the aqueous phase. The second type of sulphydryl groups as sensed by maleimide derivatives (site II) is completely different. The close environment of site II as analysed by the short chain MSL_a strongly restricts the motion of the bound probe, which results in the strongly immobilized ESR signal. The accessibility of the site II sulphydryl groups to MSL_a is significantly smaller than that of site I and of some nonspecific sulphydryl groups also located at the surface of the membrane, since the immobilized component S of the ESR spectra is markedly increased by the mersalyl protection of the phosphate carrier sulphydryl groups. As the short chain MSL_a bound to site II is not accessible to ascorbate and chromium oxalate, it may be assumed to be localized in the hydrophobic membrane interior. Furthermore, from comparative experiments using various MSL analogues with different chain length, it may be inferred that site II is localized not deeper than 12-15 Å. Interestingly, the same conclusion of two types of phosphate carrier sulphydryl groups similarly located at different depth of the membrane was made using carboxylated maleimides with varying length of the chain linking the carboxyl group to the maleimide molecule (10). It is also possible that sites I and II are identical with the two classes of the phosphate carrier sulphydryl groups differentiated by a very low concentration (4-5 nmol/mg protein) of mersalyl (11-12).

In conclusion, the data reported here are in favour of the existence of two types of sulphydryl groups of the phosphate translocator (site I and II) which differ in the surrounding microenvironment, accessibility and localization in the membrane. However, it cannot be decided yet whether these non-equivalent SH groups are re-orienting or stable positions.

REFERENCES

1. Fonyò, A. and Bessman, S.P. (1968) Biochem. Med. 2, 145-163.

2. Tyler, D.D. (1969) Biochem. J. 111, 665-678.

3. Delmelle, M. and Virmaux, M. (1977) Biochim. Biophys. Acta 464, 370-377.

4. Kolbe, H.V.J., Böttrich, J., Genchi, G., Palmieri, F. and Kadenbach, B. (1981) FEBS Lett. 124, 265-269.

5. Hamilton, C.L. and McConnell, H.M. (1968) in "Structural Chemistry and Molecular Biology" (H.M. Rich and N. Davidson, eds.), p.115, W.H. Freeman, San Francisco.

6. Kornberg, R.D. and McConnell, H.M. (1971) Biochemistry 10, 1111-1120.

7. Vistnes, A.I. and Puskin, J.S. (1981) Biochim. Biophys. Acta 644, 244-250.

8. Wohlrab, H. (1979) Biochemistry 18, 2098-2102.

9. Hadvary, P. and Kadenbach, B. (1976) Eur. J. Biochem. 67, 573-581.

10. Griffith, D.G., Partis, M.D., Sharp, R.N. and Beechey, R.B. (1981) FEBS Lett. 134, 261-263.

11. Fonyò, A. (1974) Biochem. Biophys. Res. Commun. 57, 1069-1073.

12. Fonyò, A., Palmieri, F., Ritvay, J. and Quagliariello, E. (1974) in "Membrane Proteins in Transport and Phosphorylation" (G.F. Azzone et al., eds.), pp.283-286, North Holland, Amsterdam.

© 1983 Elsevier Science Publishers B.V.
Structure and Function of Membrane Proteins,
E. Quagliariello and F. Palmieri editors.

CONFORMATIONAL TRANSITIONS IN DETERGENT TREATED AND MEMBRANE BOUND Ca^{2+}-ATPase FROM SARCOPLASMIC RETICULUM.

JENS PETER ANDERSEN[1], JESPER VUUST MØLLER[1] AND PETER LETH JØRGENSEN[2]
[1]Institute of Medical Biochemistry and [2]Institute of Physiology, University of Aarhus, 8000 Aarhus C (Denmark)

INTRODUCTION

The sarcoplasmic reticulum Ca^{2+}-ATPase is an intrinsic membrane protein, which can be solubilized in monomeric form retaining full enzymatic activity (1). The non-ionic detergent octaethyleneglycol monododecyl ether (C$_{12}$E$_8$) is well suited for this purpose (1-3). However, several reports suggest that the functional unit of Ca^{2+}-transport may be a dimer (1,4,5). The present study has been undertaken to investigate the capability of the C$_{12}$E$_8$ solubilized monomer for undergoing the complex conformational transitions involved in Ca^{2+}-transport (6). In order to examine a possible effect of detergent per se we have also studied the properties of a membranous Ca^{2+}-ATPase preparation perturbed by incorporation of non-solubilizing amounts of detergent into the bilayer.

MATERIALS AND METHODS

Purified Ca^{2+}-ATPase membranes were prepared and solubilized in C$_{12}$E$_8$ as described in (3). Sedimentation velocity studies in the analytical ultracentrifuge indicated that more than 70% of the soluble preparation was monomeric, both in presence and in absence of vanadate. Incorporation of small non-solubilizing amounts of C$_{12}$E$_8$ into the membrane was performed by mixing detergent and purified Ca^{2+}-ATPase in such a proportion that the concentration of free detergent did not exceed 0.04 mg/ml (the critical micellar concentration is 0.05 mg/ml). Gel chromatographic experiments as well as freeze fracture electron microscopic examinations confirmed that no solubilization had occurred by this treatment.

Conformational changes in the Ca^{2+}-ATPase protein were studied by recording changes in intrinsic tryptophan fluorescence (7) (λ_{ex} = 290 nm, λ_{em} = 320 nm) on a Perkin Elmer MPF 44A spectrofluorimeter fitted with a continously stirred cuvette. Ca^{2+}-ATPase activity in presence and in absence of vanadate was measured spectrophotometrically by a NADH-coupled assay (3).

RESULTS AND DISCUSSION

Fig. 1 shows the fluorescence responses induced by changing the Ca^{2+}-concentration or adding C$_{12}$E$_8$ to Ca^{2+}ATPase. The upper trace is the response obtained in unperturbed Ca^{2+}-ATPase membranes. This response decreases in amplitude after addition of non-

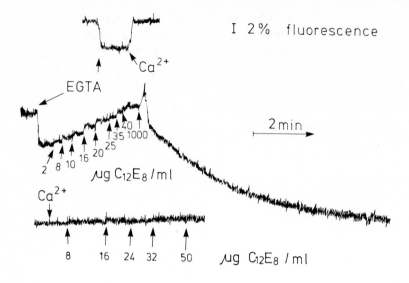

Fig. 1. Effect of Ca^{2+} and $C_{12}E_8$ on intrinsic fluorescence of Ca^{2+}-ATPase. Addition EGTA (0.2 mM) lowers the Ca^{2+} concentration to less than 10^{-8}M. Medium: 0.01 M Tes (pH 7.5), 0.1 M KCl, 1 mM $MgCl_2$, 0.05 mg/ml Ca^{2+}-ATPase, 0.3 mM Ca^{2+} was added, 20°C.

solubilizing amounts of $C_{12}E_8$, but is fully regained on solubilization in monomeric form. As revealed by the middle and lower traces of fig. 1 the decrease in amplitude induced by $C_{12}E_8$ is mainly due to a rise in fluorescence of the Ca^{2+} free state of the enzyme towards the level of the Ca^{2+} bound form. This is in accordance with a shift in the equilibrium between the two main Ca^{2+}-ATPase conformations (E_1 and E_2 cf. ref (6)) in favour of the species normally observed only in the presence of micromolar Ca^{2+} (E_1).

The middle trace of fig. 1 also shows the biphasic response obtained by solubilization (addition of 1000 µg $C_{12}E_8$/ml) in absence of Ca^{2+}. A rapid fluorescence change similar to the usual response induced by removal of Ca^{2+}, is followed by a much slower, but larger change in fluorescence. We have found that the slow fluorescence change is mono-exponential and reflects irreversible inactivation of the soluble enzyme. This inactivation is not a photolysis process, since it occurs at an unchanged rate when the exitation light is shut off for several minutes, or when detected with another fluorescent probe, which is not exited in the UV region (3). The slow inactivation of soluble Ca^{2+}-ATPase does not occur in the presence of Ca^{2+} to saturate the high affinity sites and thus the inactivation response can be used to study the equilibrium between the Ca^{2+}-bound (E_1) and Ca^{2+}-

depleted (E_2) conformational states. The dependence of the inactivation rate constants on pCa is depicted in fig. 2 together with the fluorescence responses of unperturbed membranous Ca^{2+}-ATPase. The data indicate that the soluble monomer has retained the ability to bind Ca^{2+} with high affinity in a cooperative process (Hill coefficient approx. 2) similar to that characteristic of membranous Ca^{2+}-ATPase. Thus, the cooperative interaction of Ca^{2+} with the ATPase can not be caused by coupling between different peptides but rather there are interacting Ca^{2+}-sites on the same polypeptide.

Fig. 2. Effect of the final Ca^{2+} concentration on fluorescence responses of membranous (□) and soluble, monomeric (●) Ca^{2+}-ATPase. The Ca^{2+} concentration before addition of various concentrations of EGTA was 0.1 mM.

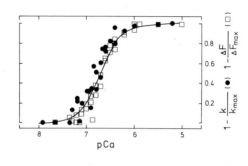

The rate constants were calculated from semilog plots of the slow inactivation response. Medium composition was as described for fig. 1 $\pm$ 1 mg/ml $C_{12}E_8$.

In order to examine the enzymatic cycle of the detergent treated Ca^{2+}-ATPase preparations we have studied the effect of vanadate on ATP-hydrolysis. Vanadate is a specific inhibitor which interacts with the E_2 state only. As can be seen from fig. 3 addition of only 6 µM vanadate to the unperturbed Ca^{2+}-ATPase membranes leads to a strong inhibition of ATP hydrolysis, which evolves within approx. ten minutes. In contrast, the membranous Ca^{2+}-ATPase perturbed by 0.025 mg/ml $C_{12}E_8$ is almost unaffected by 25 µM vanadate. However, after solubilization in monomeric form with 1 mg/ml $C_{12}E_8$ a high degree of inhibition by vanadate is detected. This indicates that in the case of the soluble monomer, the enzymatic cycle does contain the E_2 state, whereas the concentration of this species is negligible during ATP hydrolysis by the detergent perturbed membranous enzyme.

CONCLUSIONS

Our results suggest that incorporation of non-solubilizing amounts of $C_{12}E_8$ into Ca^{2+}-ATPase membranes shifts the E_1-E_2 conformational equilibrium in favour of E_1 both in the absence of ATP and during ATP hydrolysis. This is probably not caused

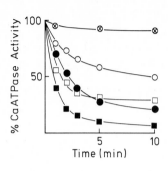

Fig. 3. Effect of vanadate on ATP hydrolysis. Ca^{2+}-ATPase activity is shown as percentage of that measured after the same times of incubation in absence of vanadate. At time zero vanadate (6 µM (O, ●), 12 µM (□,■) or 25 µM (⊗)) is added together with 10 µM ATP; (●,■) unperturbed membranous Ca^{2+}-ATPase, (O,□) soluble monomeric Ca^{2+}-ATPase at 1 mg/ml $C_{12}E_8$, (⊛) membranous Ca^{2+}-ATPase in presence of 0.025 mg/ml $C_{12}E_8$. Medium: 0.01 M Tes (pH 7.5), 0.1 M KCl, 10 mM $MgCl_2$, 0.2 mM EGTA, 0.196 mM $CaCl_2$, 0.005 mg/ml Ca^{2+}-ATPase, 5 mM phosphoenolpyruvate, 0.15 mM NADH, 0.10 mg/ml pyruvate kinase, 20°C.

by dissociation of functional oligomers, since the conformational equilibrium is much less disturbed after solubilization in monomeric form. The monomer has retained the ability to perform a full enzymatic cycle, including the E_1-E_2 conformational transition considered essential for the Ca^{2+}-transport function (6).

ACKNOWLEDGEMENTS

This investigation has been supported by The Danish Medical Research Council to whom we express thanks.

REFERENCES

1. Møller, J.V., Andersen, J.P. and leMaire, M. (1982) Mol. Cell. Biochem., 42, 83-107.
2. Dean, W.L. and Tanford, C. (1978) Biochemistry, 17, 1683-1690.
3. Andersen, J.P., Møller, J.V. and Jørgensen, P.L. (1982) J. Biol. Chem., 257, 8300-8307.
4. Ikemoto, N., Miyao, A. and Kurobe, Y. (1981) J. Biol. Chem., 256, 10809-10814.
5. Silva, J.L. and Verjovski-Almeida, S. (1983) Biochemistry, 22, 707-716.
6. De Meis, L. and Vianna, A.L. (1979) Annu. Rev. Biochem., 48, 275-292.
7. Dupont, Y. (1976) Biochem. Biophys. Res. Commun., 71, 544-550.

ATP INDUCED Ca^{2+} RELEASE FROM MONOMERIC SARCOPLASMIC RETICULUM Ca^{2+}-ATPase

JENS PETER ANDERSEN, ULRIK GERDES AND JESPER VUUST MØLLER
Institute of Medical Biochemistry, University of Aarhus,
8000 Aarhus C (Denmark)

INTRODUCTION

The structural basis for active transport of Ca^{2+} through the membrane of sarcoplasmic reticulum is presumably a conformational change in the Ca^{2+}-pump protein, which changes the affinity of the binding site for Ca^{2+} (1). Consistent with this model it has been found that addition of ATP to purified Ca^{2+}-ATPase may lead to release of Ca^{2+} from the enzyme (2,3). If only one Ca^{2+}-ATPase polypeptide constitutes the functional unit of Ca^{2+}-transport as suggested by our fluorescence and inhibition studies (4,5) it is expected that ATP induced Ca^{2+}-release can be detected also in the detergent solubilized preparation of enzymatically active monomeric Ca^{2+}-ATPase. However, in a recent study an apparent uncoupling between ATP-hydrolysis and Ca^{2+}-release was reported after membrane bilayer disruption with octaethyleneglycol monododecyl ether ($C_{12}E_8$) (6). Concurrently with their investigation we have been engaged in similar experiments, but have reached the opposite conclusion. Using the same dye (Arsenazo III) to monitor changes in free Ca^{2+}-concentration we have found it possible under appropriate conditions to demonstrate ATP induced Ca^{2+}-release in presence of $C_{12}E_8$, even when the enzyme is present in solution in predominantly monomeric form.

MATERIALS AND METHODS

The preparation of purified leaky Ca^{2+}-ATPase membranes was the same as in the fluorescence studies (3,4). The buffer used in the Ca^{2+}-release experiments contained 50 mM maleic acid titrated to pH 6.70 with Tris, 100 mM KCl, 10 mM $MgCl_2$ and 30 µM Arsenazo III. The free concentration of Ca^{2+} was monitored by the difference in dye absorbance between 598 nm and 572 nm, read on an Aminco DW-2A spectrophotometer. In order to obtain a calibration of the absorbance change $CaCl_2$ was added in 4 µM steps. The final concentration of Ca^{2+} in the medium (including Ca^{2+} present in the Ca^{2+}-ATPase preparation) was approx. 75 μM. ATP was added in small volumes (1-5 μl) with a complete mixing time of less than 0.5 seconds in the continuously stirred cuvette. Control experiments in the absence of protein were performed to ensure that the effect of ATP per se was negligible.

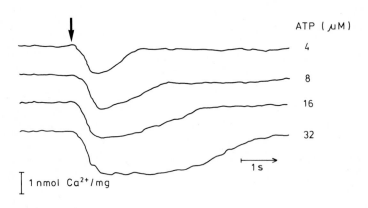

Fig. 1. ATP induced Ca^{2+} release from purified Ca^{2+}-ATPase vesicles, monitored with Arsenazo III. The medium was as described in "Materials and Methods", protein concentration 1.0 mg/ml. At the arrow ATP (less than 5 μl) was added from Hamilton syringes to give the final concentrations indicated in the figure. A decrease in absorption corresponds to release of Ca^{2+} into the medium.

RESULTS

Fig. 1 shows the absorption change (transformed to nmol Ca^{2+} per mg protein) associated with addition of various concentrations of ATP to leaky Ca^{2+}-ATPase vesicles. The magnitude of the response corresponds to release and rebinding of approx. 2 nmol Ca^{2+} per mg protein. The time integral of the Ca^{2+} release is approximately proportional to the amount of added ATP suggesting that all ATP has to be hydrolyzed before complete rebinding takes place. Probably, at the higher ATP concentrations it is not a single cycle, which is observed, but a steady state distribution of low and high Ca^{2+}-affinity forms of the enzyme.

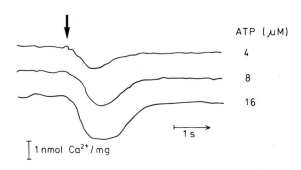

Fig. 2. ATP induced Ca^{2+}-release from Ca^{2+}-ATPase (1.5 mg/ml) solubilized in $C_{12}E_8$ (4 mg/ml). Other conditions were as described in Fig. 1.

When Ca^{2+}-release experiments were performed with Ca^{2+}-ATPase solubilized in $C_{12}E_8$ the results shown in Fig. 2 were obtained. It is seen that the response of the soluble Ca^{2+}-ATPase is similar to that of the membranous enzyme except for a somewhat shorter duration, which may be due to a higher rate of ATP hydrolysis by the soluble Ca^{2+}-ATPase at these low ATP concentrations (7).

We also studied the effect of ATP on Ca^{2+}-binding at pH 7.5 in Tes buffer. In these conditions we found an apparent increase in Ca^{2+}-binding instead of release of Ca^{2+}, both in the membranous and soluble preparation (not shown).

At the high protein concentration (1.5 mg Ca^{2+}-ATPase/ml) used in the present study the soluble enzyme is not entirely monomeric. Sedimentation velocity studies indicate that as much as 30-35 % of the protein may be aggregated. However, since the magnitude of the response of the soluble Ca^{2+}-ATPase is as high as that of the membranous enzyme, it seems safe to conclude that the monomers do contribute significantly to the ATP induced Ca^{2+}-release.

DISCUSSION

The ATP induced Ca^{2+}-release is conveniently considered in terms of the following minimal scheme for the enzymatic cycle. E_1 and E_2 are the two main conformational states, having high affinity and low affinity Ca^{2+} sites, and $E_1{\sim}P$ and E_2-P are high energy and low energy forms of covalently bound phosphate, respectively (8). The $E_1{\sim}P \rightarrow E_2$-P transition is considered to be the essential step involved in translocation of Ca^{2+} across the membrane, and is tentatively associated with the release of Ca^{2+}, under conditions where the cation is not accumulated inside a vesicular compartment (3).

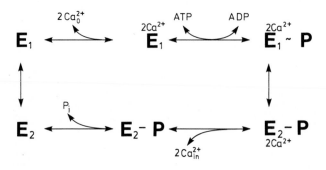

Our data indicate that it is possible to observe ATP induced Ca^{2+}-release from the soluble monomeric Ca^{2+}-ATPase, and thus it is suggested that the monomer can perform the $E_1{\sim}P \rightarrow E_2$-P conformational transition. However, some caution is needed in the evaluation of the data. In effect changes in Ca^{2+}-binding by addition of ATP may represent a balance between processes that lead to enhanced binding and to release of Ca^{2+}. In particular increased binding of Ca^{2+} has been described under conditions where the $E_1{\sim}P$ form is stabilized in preference to E_2-P, and Ca^{2+} may be present on the enzyme in an occluded state (9-11). We find that at pH 7.5 in Tes buffer, Arsenazo III seems to indicate an increased binding of Ca^{2+} after addition of ATP. By an appropriate change in medium composition, as reported in the present paper, we are able to demonstrate the reverse process, i.e. release of Ca^{2+}, both in the membranous and in the $C_{12}E_8$

solubilized state. Dean and Gray (6), after exposure of Ca^{2+}-ATPase to $C_{12}E_8$, found no ATP induced change in Ca^{2+}-binding, even in conditions (pH 6.1 and presence of dimethylsulfoxide) which should be optimal for detection of Ca^{2+}-release by the membranous enzyme (3). The reason for this discrepancy with the present results is not entirely clear and requires further experimentation, including a critical assessment of the reliability of Arsenazo III for the present purpose. However, it is also of notice that in the detergent solubilized state the E_2 conformation is highly unstable (4,5), particularly at low pH. This is a significant factor to be taken into account in studies on the transport capability of detergent solubilized preparations.

ACKNOWLEDGEMENTS

This investigation has been supported by The Danish Medical Research Council to whom we express thanks.

REFERENCES

1. Tanford, C. (1982) Proc. Natl. Acad. Sci. USA, 79, 2882-2884.

2. Ikemoto, N. (1976) J. Biol. Chem., 251, 7275-7277.

3. Watanabe, T., Lewis, D., Nakamoto, R., Kurzmark, M., Fronticelli, C. and Inesi, G. (1981) Biochemistry, 20, 6617-6625.

4. Andersen, J.P., Møller, J.V. and Jørgensen, P.L., preceeding paper in this volume.

5. Andersen, J.P., Møller, J.V. and Jørgensen, P.L. (1982) J. Biol. Chem., 257, 8300-8307.

6. Dean, W.L. and Gray, R.D. (1983) Biochemistry, 22, 515-519.

7. Møller, J.V., Lind, K.E. and Andersen, J.P. (1980) J. Biol. Chem., 255, 1912-1920.

8. de Meis, L. and Vianna, A.L. (1979) Annu. Rev. Biochem., 48, 275-292.

9. Dupont, Y. (1980) Eur. J. Biochem., 109, 231-238.

10. Takisawa, H. and Makinose, M. (1981) Nature, 290, 271-273.

11. Nakamura, Y. and Tonomura, Y. (1982) J. Biochem., 91, 449-461.

STUDIES OF THE ACTIVE SITE STRUCTURE OF KIDNEY (Na^+ + K^+)-ATPASE WITH ATP ANALOGUES AND PARAMAGNETIC PROBES.

CHARLES M. GRISHAM

Department of Chemistry, University of Virginia, Charlottesville, Virginia, 22901 (U.S.A.)

INTRODUCTION

A thorough understanding of the mechanism of membrane-bound (Na^+ + K^+)-ATPase and its role in monovalent cation transport in mammalian systems will require the determination of the three-dimensional structure of the enzyme. A structural model should describe (at a minimum) the primary sequence of the polypeptide chains, the folding of these peptides in (and across) the plasma membrane, identification of amino acid residues involved in binding of substrates, activators and inhibitors, and the arrangement and conformation of the substrates and activators themselves on the enzyme. While many other groups have pursued the first three of these issues with notable success, our own laboratory has been primarily concerned with the fourth. Recently we have examined the interactions of ATP, inorganic phosphate (P_i) and divalent metals using Co(III) and Cr(III) analogues of ATP and inorganic phosphate. Our results demonstrate that (a) there are two divalent cation sites at the active site of the ATPase, (b) these two sites are 8.1 Å apart, (c) inorganic phosphate and ATP can interact with both these divalent cation sites and (d) ATP and P_i can bind simultaneously to the ATPase.

MATERIALS AND METHODS

Enzyme. The (Na^+ + K^+)-ATPase was purified from the outer medulla of sheep kidney as previously described (1).

Substrate Analogues. The β,γ-bidentate complex of Co(NH3)4ATP was prepared as described by Cornelius et al. (2). The β,γ-bidentate complex of Cr(H2O)4ATP was prepared according to Cleland and Mildvan (3). Bidentate tetraammine phosphatocobalt(III) [Co(NH3)4(PO4)] was synthesized according to Siebert (4).

NMR Measurements. Phosphorus-31 NMR spectra and relaxation times were obtained at 145.9 MHz on a Nicolet Magnetics Corp. NT-360/Oxford spectrometer equipped with a 1280/293B data system. Spectra were measured at 4°C (to prevent breakdown of Co(NH$_3$)$_4$ATP) with a 10 mm broad-band probe with an internal ^{2}H lock (20% D$_2$0, total volume of 2.4 ml).

RESULTS AND DISCUSSION

Our studies depend in part on the identification of unique sites on the ATPase for paramagnetic probes. The first such probe site characterized by us was a single, high affinity Mn^{2+} binding site which we have since shown to be a catalytically efficacious site for Mg^{2+} which exists on the ATPase both in the presence and absence of ATP or P$_i$ (1,5). The dissociation constant measured for Mn^{2+} at this site by EPR (0.2 µM) agreed well with the K$_m$ values for Mn^{2+} in the ATPase reaction as measured by us (5) and by Robinson (6).

One of the most interesting questions about (Na$^+$ + K$^+$)-ATPase concerns whether Mg^{2+} and ATP bind at the active site as the free species or as the MgATP complex. The initial observation by us of the single, catalytic site for Mg^{2+} (or Mn^{2+}) on the ATPase was followed by a series of important observations by ourselves and others which now permit a new answer to this old question of the nature of metal and nucleotide binding to the ATPase. The first of these was our demonstration by Mn^{2+}-EPR measurements that binding of ATP to the enzyme caused major changes in the coordination geometry of enzyme-bound Mn^{2+} (consistent with binding of ATP in the vicinity of the Mn^{2+} site) bud *did not* give evidence for formation of a typical inner sphere MnATP complex on the enzyme (1,7). Our NMR studies of water protons on the enzyme-bound Mn^{2+} were consistent with the above observations. We demonstrated (5) that Mn^{2+} at the high affinity site coordinates four rapidly exchanging water protons (i.e., two water molecules), one proton of which is lost (either displaced or "frozen" in place) upon addition of P$_i$. ^{31}P NMR studies subsequently established the existence of a phosphate site 6.9 Å from the enzyme-bound Mn^{2+} (8). These results were interpreted in terms of a second sphere complex of Mn^{2+} and phosphate with an intervening water molecule.

Cantley and co-workers (9,10) later observed that vanadate ion, VO$_4^{3-}$, was a potent inhibitor of the ATPase in the presence of divalent cations. They suggested that VO$_4^{3-}$ represented a transition state analogue for the ATPase, mimicking the putative trigonal bipyramidal structure appropriate to an S$_N$2-type phosphoryl transfer. They further showed that inhibition by VO$_4^{3-}$ requires the presence of divalent cations (e.g., Mg^{2+} or Mn^{2+}) and that binding of VO$_4^{3-}$ creates a divalent cation binding site on the enzyme (9,10). At about the same time, we showed that Cr(III) and Co(III) complexes of ATP could be useful structural and kinetic probes of the ATP site of the ATPase. We demonstrated (11,12) that β,γ-bidentate Cr(H$_2$O)$_4$ATP could bind to the (Na$^+$ + K$^+$)-ATPase simultaneously with the binding of Mn^{2+} to the single, high affinity site on the enzyme. Pauls et al. (13) showed that slow and irreversible inhibition of the ATPase by Cr(H$_2$O)$_4$ATP was enhanced in the presence of free Mg^{2+}, and, in a series of Mn^{2+} EPR studies, we showed (12) that the Mn^{2+} site on the ATPase and the Cr^{3+} ion of bound Cr(H$_2$O)$_4$ATP were separated by approximately 8.1 Å at the active site. It has occurred to us that the divalent metal site created by the binding of VO$_4{}^{3-}$ could be related to the metal site on enzyme-bound ATP. These studies with Mg-VO$_4$ and CrATP or CoATP are consistent with the involvement in the ATPase mechanism of a second divalent metal ion which interacts directly with the substrate (ATP) and product (P$_i$) of the ATPase reaction.

Recently, in order to further examine the roles of these two different divalent cation sites, we have used ^{31}P NMR to study complexes of the ATPase, Mn^{2+} and Co(NH$_3$)$_4$ATP. Our kinetic studies have shown that Co(NH$_3$)$_4$ATP (shown in Fig. 1) is a competitive inhibitor with respect to MnATP for the (Na$^+$ + K$^+$)-ATPase. The K$_i$ values under both high- and low-

Figure 1.

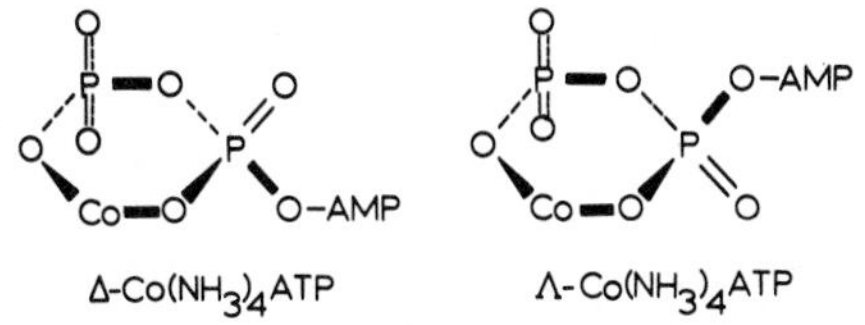

Δ-Co(NH$_3$)$_4$ATP Λ-Co(NH$_3$)$_4$ATP

affinity conditions (K$_i$ = 10 µM and K$_i$ = 1.6 mM, respectively) are similar to the K$_m$ values for MnATP under the same conditions (2.88 µM and 0.902 mM). In the NMR studies, the presence of the (Na$^+$ + K$^+$)-ATPase causes a 6-10-fold enhancement of the effect of Mn^{2+} on the ^{31}P relaxation rates of Co(NH$_3$)$_4$ATP, establishing the formation of a ternary Mn^{2+}-ATPase-Co(NH$_3$)$_4$ATP complex. Addition of saturating levels of diamagnetic Mg^{2+} ion produced no measurable increase in the relaxation rates of the phosphorus nuclei of Co(NH$_3$)$_4$ATP, indicating that the effects of added Mn^{2+} are due entirely to electron-nuclear dipolar interactions. Under these conditions, we can use the paramagnetic effects of Mn^{2+} on enzyme-bound Co(NH$_3$)$_4$ATP to calculate the internuclear separations between the single Mn^{2+} site and the three phosphorus nuclei of the bound analogue (14). As shown in Figure 2, the Mn-P distances are the same for the γ and β phosphorus atoms of Co(NH$_3$)$_4$ ATP on the enzyme. On the other hand, the distance from Mn^{2+} to the α-P is

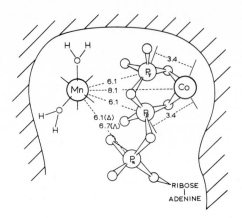

Figure 2. Conformation of triphosphate chain of Co(NH$_3$)$_4$ATP with respect to bound Mn^{2+} ion on (Na$^+$ + K$^+$)-ATPase. Distances are in angstroms.

slightly larger than for the β-P and α-P. Moreover, the M-α-P distances for the Λ- and Δ-diastereomers of Co(NH$_3$)$_4$ATP are slightly different. The distances from Mn^{2+} to the γ-P and β-P are consistent with the formation of a second sphere coordination complex on the enzyme between Mn^{2+} and the phosphates of Co(NH$_3$)$_4$ATP.

The roles of the two divalent metals in effecting the hydrolysis of ATP are not entirely clear, but certain inferences may be made. The nucleotide-bound metal is probably not involved in adjusting the protein conformation, since its role can be filled by Cr^{3+}, which is substitution-inert and therefore cannot acquire ligands from the enzyme. (It should be noted here that, despite its substitution-inert nature, Cr(H$_2$O)$_4$ATP is a substrate for the (Na$^+$ + K$^+$)-ATPase (15).) A more likely role for the nucleotide-bound metal would be to adjust the conformation of the polyphosphate backbone and to polarize and thereby increase the electrophilicity of the γ-phosphoryl group of ATP. The role of the enzyme-bound metal may be to adjust the protein conformation and/or to orient water ligands near the phosphoryl groups of ATP and phosphate. Such interactions could serve to stabilize the phosphorane transition state in the S$_N$2 displacements occurring during formation and breakdown of the aspartyl phosphate intermediate, which is a requisite intermediate in the hydrolytic pathway of this enzyme.

Mechanistic details of the phosphoryl transfer steps of the (Na$^+$ + K$^+$)-ATPase reaction can also be obtained from studies of the interaction of the enzyme with P$_i$ and with phosphate analogues. It has long been known that the covalent phosphoenzyme intermediate, which is part of the catalytic pathway of this enzyme, can be formed from P_i and Mg^{2+}, as well as from ATP, Na^+ and Mg^{2+}. On the other hand, the existence in this same pathway of a noncovalent complex between the ATPase and P_i has been appreciated only recently. Our own ^{31}P NMR studies (8) first demonstrated the existence of such a noncovalent $E \cdot P_i$ complex, and Froehlich et al. (16) later reported kinetic evidence for the involvement of $E \cdot P_i$ in the ATPase pathway.

Critical to a full understanding of the role of phosphate in the mechanism of the $(Na^+ + K^+)$-ATPase are the questions of how phosphate at the active site is related to 1) the ATP substrate and 2) the two divalent cations at the active site. For example, is the phosphate binding site on the enzyme identical with (or overlapping with) the γ-PO_4 of bound ATP, or can ATP and P_i bind simultaneously? An argument for the latter possibility can be made in two ways. We have compared the effects of ATP and P_i on the inhibition of the ATPase by 2,3-butanedione, an arginine-directed protein modification reagent (17). Inhibition of the ATPase by 2,3-butanedione is a biphasic process, in which a rapid loss of roughly 40-45% of the enzyme activity is followed by a much slower loss of the balance of the activity. ATP alone completely abolishes the slow phase of the inhibition. ATP plus P_i can effectively reverse the rapid phase of the inhibition as well, consistent with the simultaneous binding of ATP and P_i to the enzyme (17). Recently Karlish et al. have obtained additional evidence for the simultaneous binding of ATP and P_i from studies of Rb-Rb exchange in phospholipid vesicles reconstituted with $(Na^+ + K^+)$-ATPase (18). They observe that the presence of both ATP and P_i at optimal levels produces a seven-fold stimulation of Rb-Rb exchange over that observed with optimal concentrations of either ATP or P_i alone.

The matter of how phosphate interacts with the two divalent cation sites of the ATPase active site is potentially a more complicated issue. As we described above, phosphate analogues form a second sphere complex with the single high affinity Mn^{2+} site which exists on the ATPase even in the absence of ATP and P_i. The divalent cation site created by VO_4^{3-} binding may be analogous to a second metal site involved in the interaction of P_i with the enzyme, but it is not clear whether this second site is related to the metal of the metal-nucleotide analogues we have studied (11,12,14,15) and how this second metal site may be involved in catalysis. In order to study the latter question we have prepared a complex of tetraammine Co(III) and inorganic phosphate. As Dunaway-Mariano and coworkers have shown (19), this complex can exist as either a monodentate or bidentate complex:

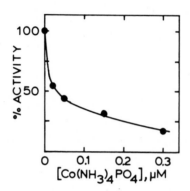

I

II

The equilibrium is pH dependent and at pH 7, the complex is approximately equally distributed between the forms I and II. We have found this metal-phosphate analogue to be strongly inhibitory to the $(Na^+ + K^+)$-ATPase. As shown in Figure 3, in the presence of 3 mM ATP and 3 mM $MgCl_2$, an apparent

Figure 3. Inhibition of $(Na^+ + K^+)$-ATPase by $Co(NH_3)_4ATP$. pH = 7.0, T = 23°C.

K_i of 30 μM is measured. However, if only one of the forms I and II are responsible for the inhibition, the K_i would be lower. In addition, if the binding of $Co(NH_3)_4PO_4$ is competitive with respect to ATP or MgATP, the true K_i could be much lower.

Since $MgHPO_4$ is not inhibitory to the $(Na^+ + K^+)$-ATPase under these conditions, it is not immediately obvious why $Co(NH_3)_4PO_4$ is so strongly inhibitory. One possible explanation may lie in the fact that, while substantial amounts of $(NH_3)_4Co\overset{O}{\underset{O}{<}}PO_2$ exist at physiological pH, complexes of Mg^{2+} and phosphate are clearly monodentate except at pH values greater than 10 (19). Thus if bidentate $MgHPO_4$ is a substrate for the ATPase in the direction of ATP synthesis, it must be stabilized in the active site and therefore would be an activated form of orthophosphate. As shown in Figure 4, Mg^{2+} could activate the phosphorus atom via electron withdrawal, stabilize the departing hydroxide ion via coordination, and, by virtue of

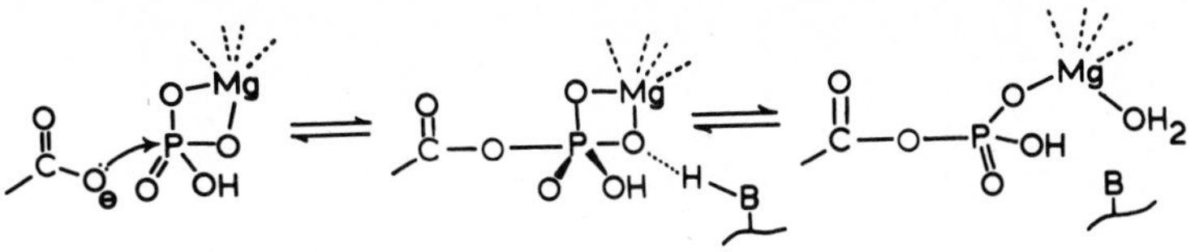

Figure 4. A mechanism for phosphorylation of the ATPase via a bidentate metal-phosphate transition state.

the chelate ring, provide a mechanism for strain catalysis (20). The strong inhibition by $Co(NH_3)_4PO_4$ could reflect the ability of this complex to mimic the transition state for phosphoryl transfer.

ACKNOWLEDGMENTS

This research was generously supported by the Muscular Dystrophy Association of America, and by National Institutes of Health Research Grants AM19419 and AM00613. The NMR instrumentation used in these studies was provided by grants from the National Science Foundation and the University of Virginia. C.M.G. is a Research Career Development Awardee of the U.S. Public Health Service. The author wishes to acknowledge the technical assistance of Cindy Klevickis and Jennifer Jefferies and the expert assistance of Mrs. Teresa Napier in the preparation of the manuscript.

REFERENCES

1. O'Connor, S.E. and Grisham, C.M. (1979) Biochemistry 18, 2315.
2. Cornelius, R., Hart, P. and Cleland, W. (1977) Inorg. Chem. 16, 2799.
3. Cleland, W. and Mildvan, A. (1979) in Advances in Inorganic Biochemistry (Eichhorn, G. and Marzilli, L., Eds.) Vol. 1, p. 163, Elsevier/North-Holland, New York.
4. Siebert, H. (1958) Z. Anorganische und Allgemeine Chemie 296, 280.
5. Grisham, C. and Mildvan, A. (1974) J. Biol. Chem. 249, 3187.
6. Robinson, J. (1981) Biochim. Biophys. Acta 642, 405.
7. O'Connor, S.E. and Grisham, C.M. (1980) Biochem. Biophys. Res. Commun. 93, 1146.
8. Grisham, C. and Mildvan, A. (1975) J. Supramol. Structure 3, 304.
9. Cantley, L.C., Cantley, L.G. and Josephson, L. (1978) J. Biol. Chem. 253, 7361.
10. Smith, R., Zinn, K. and Cantley, L. (1980) J. Biol. Chem. 255, 9852.
11. Grisham, C. (1981) J. Inorg. Biochem. 14, 45.

12. O'Connor, S.E. and Grisham, C. (1980) FEBS Letters 118, 303.

13. Pauls, H., Bredenbrocker, B., and Schoner, W. (1980) Eur. J. Biochem. 109, 523.

14. Klevickis, C. and Grisham, C. (1982) Biochemistry 21, 6979.

15. Gantzer, M.L., Klevickis, C. and Grisham, C. (1982) Biochemistry 21, 4083.

16. Froehlich, J., Albers, R.W., Koval, G., Goebel, R. and Berman, M. (1976) J. Biol. Chem. 251, 2186.

17. Grisham, C. (1979) Biochem. Biophys. Res. Commun. 88, 229.

18. Karlish, S.J.D., Lieb, W. and Stein, W.D. (1982) J. Physiol. 328, 333.

19. Haromy, T.P., Knight, W.B., Dunaway-Mariano, D. and Sundaralingam, M. (1983) Biochemistry, in press.

20. Farrell, F.J., Kjellstrom, W. and Spiro, T.G. (1969) Science 164, 320.

IDENTIFICATION AND MAPPING OF SITES IN MEMBRANE PROTEINS

CHARACTERIZATION OF A PLASMA MEMBRANE KINASE WHICH SPECIFICALLY PHOS-
PHORYLATES THE (NA,K)PUMP

LEWIS CANTLEY, LI-AN YEH, LEONA LING, JOHN SCHULZ AND LEIGH ENGLISH
Department of Biochemistry and Molecular Biology, Harvard University,
Cambridge, MA   02138

INTRODUCTION

(Na,K)ATPase has been purified from a variety of tissues and shown to
be the enzyme which maintains Na^+ and K^+ gradients in most animal cells
(For review see reference 1). The kinetics, transport activities and
structure of this enzyme have been extensively studied in erythrocyte
membranes, purified enzyme, and enzyme reconstituted in lipid vesicles.
These studies have been aided by the ability of cardiac glycosides (such
as ouabain) to specifically inhibit this enzyme from the extracellular
side. These studies all suggest that the (Na,K)ATPase pumps 3 Na^+ ions
out of the cell and 2 K^+ ions into the cell per ATP molecule hydrolyzed.

Despite extensive knowledge about the structure and kinetics of this
enzyme, very little is known about its regulation *in vivo*. A number of
hormones are known to cause rapid changes in ouabain sensitive K^+ pumping
into cells (2-6). In some cases these changes are a consequence of a
change in intracellular Na^+ (i.e. substrate level control) (6). However,
in the case of insulin and cAMP stimulation of the (Na,K) Pump, neither
a change in copy number nor increase in substrate occurs (4,5,7). The
molecular mechanism by which such changes occur is unknown partly because
it has been difficult to purify and characterize the enzyme from tissues
which show hormonal regulation and because hormonal regulation is
generally lost upon cell lysis.

Recently we have been studying regulation of the (Na,K)Pump in Friend
virus transformed murine erythroleukemia cells. These cells can be
induced to undergo erythropoesis when treated with dimethyl sulfoxide
or a variety of other agents. All the inducers examined have in common
the ability to decrease ouabain sensitive K^+ uptake 6 to 10 hours after
addition to the cells (8). Addition of ouabain to these cells along with
DMSO accelerates commitment to differentiation (9) and in some cell lines,
ouabain alone causes differentiation (10). These results suggest that
regulation of the (Na,K) Pump is essential for terminal differentiation of
Friend cells. We purified plasma membranes from Friend cells and discovered

a protein kinase which phosphorylates the catalytic subunit of the (Na,K)-
ATPase both *in vitro* and *in vivo* (11). Here we further characterize this
kinase and show that it is highly specific for the (Na,K)ATPase.

MATERIALS AND METHODS
Cell Culture Conditions
   Friend virus transformed murine erythroleukemia cell line 745-PC4 was
obtained from David Housman's laboratory and was maintained in α medium
lacking nucleosides and supplemented with 13% fetal calf serum (Grand
Island Biological Co., Grand Island, NY).
Plasma Membrane Preparation
   Plasma membranes were prepared from Friend cells using a modification of
the Brunette and Till (12) procedure as previously described (11). Membranes
were stored frozen at - 70° and used immediately after thawing.
Gel Electrophoresis
   Sodium dodecyl sulfate polyacrylamide gel electrophoresis was performed
by the method of Laemmli (13).

RESULTS
   We previously showed that the 100,000 dalton catalytic subunit of the
(Na,K)ATPase is phosphorylated by a plasma membrane bound protein kinase
in Friend cells (11). Both the *in vivo* and *in vitro* labeled 100,000 dalton
phosphopeptides were retained on a ouabain affinity column and were specifically eluted by ouabain. These phosphopeptides also migrated at the same
molecular weight but slightly more acidic pH than dog kidney (Na,K)ATPase
α subunit on two dimmensional gel electrophoresis. One dimmensional peptide
maps of the 100,000 dalton phosphopeptides labeled *in vivo* or *in vitro*
were similar to peptide maps of dog kidney (Na,K)ATPase *a* subunit confirming
the identification. Phosphothreonine was detected in both the *in vivo*
in vitro labeled peptides.

   Although initial *in vitro* labeling experiments with purified Friend
cell plasma membranes revealed labeling of numerous phosphopeptides (11)
we later found conditions in which the (Na,K)ATPase α subunit is specifically
labeled. Figure 1 shows a sodium dodecyl sulfate polyacrylamide gel of
plasma membranes labeled for 5 min with 50 μM [γ - ^{32}P] - ATP. Greater
than 90% of the phosphate is incorporated into the 100,000 dalton peptide
previously identified as the α subunit (11). A 40 min chase with excess
unlabelled ATP released very little of the ^{32}P label from this peptide
showing that this label is not due to an intermediate of ATP hydrolysis

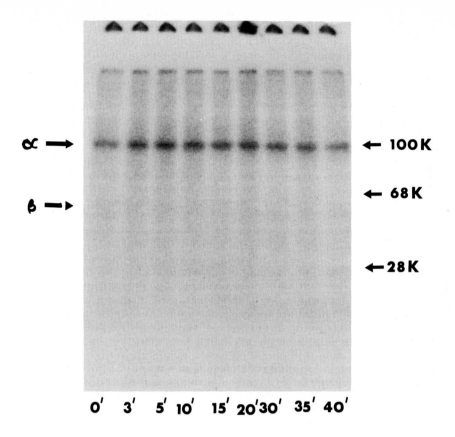

Fig. 1. _In Vitro_ Phosphorylation of Friend cell plasma membranes. 50 μM of [γ - ^{32}P]ATP was added to 40 μg of Friend cell membranes in the presence of 20mM Tris-Cl, 5mM sodium phosphate, 10 mM $MgCl_2$ pH 7.4, 30° C (200 μl total volume). After five minutes, unlabeled ATP was added to give a final concentration of 50 mM and incubation was continued at 30° C. At the times indicated in the figure, a 20 μl fraction was extracted and to it 4 μl of 20% sodium dodecyl sulfate was added. The fractions were analyzed by sodium dodecyl sulfate polyacrylamide gel electrophoresis (10% acrylamide). An autoradiograph of the gel is presented. The migration positions of the α and β subunits of dog kidney (Na,K)ATPase are indicated on the left. Times are in minutes after the chase with unlabeled ATP.

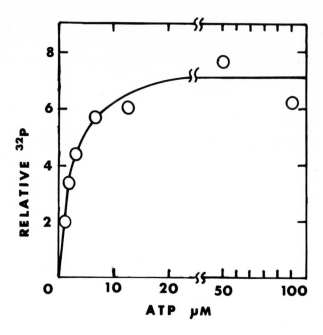

Fig. 2. The ATP dependence of phosphorylation of the α subunit of (Na,K)-ATPase in Friend cell plasma membranes. The phosphorylation conditions were the same as in Fig. 1 except that 1 mM ouabain was added to decrease the rate of ATP hydrolysis and the ATP concentration was varied from 0.78 μM to 100 μM. The reaction was stopped after 10 min and samples were analyzed by gel electrophoresis as in Fig. 1. Relative intensities were determined by scanning the autoradiograph of the gel.

and that very little phosphatase actvity is present in the plasma membrane.

Since more specific lableing of the α subunit was detected at relatively low ATP concentrations, we investigated the ATP dependence of phosphorylation. Fig. 2 shows that the K_m for ATP is less than 3μM. Since some of the ATP is hydrolysed by Mg^{2+}-ATPase activity in the membranes during the 10 min incubation, the true K_m may be somewhat lower.

The time dependence of the phosphorylation is described in Fig. 3. Sufficient ATP was added to insure that the kinase was saturated throughout the assay. The $t_{1/2}$ for saturation was 5-10 min at 37° C. The maximum stoichiometry of *in vitro* phosphorylation at saturation was previously estimated to be approximately 3 mol per mol of α subunit (11).

Two dimmensional gels of lysed Friend cells indicate that the (Na,K)-ATPase is only partially phosphorylated *in vivo*. We previously showed that the *in vivo* phosphorylated α subunit focuses between pH 5.1 - 5.3 (11).

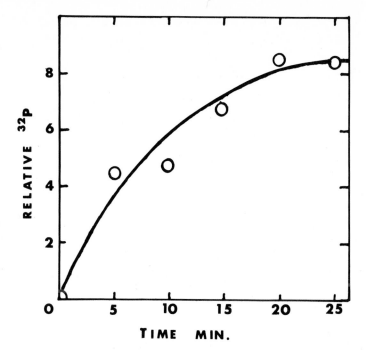

Fig. 3. The time dependence of phosphorylation of the α subunit of (Na.K)-
ATPase in Friend cell plasma membranes. The phosphorylation and analysis
were performed as described in Fig. 2 except that the reaction contained
400 μg/ml plasma membranes, 20 mM Tris. Cl, 100 mM NaCl, 20 mM KCl, 5mM
[γ-^{32}P]ATP, 5mM $MgCl_2$, 1mM EGTA pH 7.4. The reaction was carried out
at 37° and fractions were removed and precipitated with trichloroacetic
acid at the indicated times. The precipitates were analyzed by sodium
dodecyl sulfate polyacrylamide gel electrophoresis as in Fig. 2.

Western blots of two dimmensional gels of Friend cells were performed using
antibody raised against the α subunit of dog kidney (Na,K)ATPase. These
antibodies were previously shown to react specifically with the 100,000
dalton Friend cell α subunit in Western blots of sodium dodecyl sulfate
polyacrylamide gels (11). In the two dimmensional gels the antibodies
reacted with a doublet of spots migrating at the position of the phospho-
peptide (5.1 - 5.3). However, they also reacted with an equally intense
doublet of 100,000 daltons which focused at pH 5.5 - 5.6 (data not presented).
Although it is difficult to accurately determine protein concentrations
from the Western blot procedure, these results suggest that approximately
half the enzyme is unphosphorylated *in vivo*. This result is consistent
with our observation that additional phosphorylation can occur *in vitro*

and that pretreatment of membranes with alkaline phosphatase approximately doubles the amount of phosphate incorporated into α (data not presented).

SUMMARY

   When low concentrations of ATP are added to the Friend cell plasma membranes, the endogenous kinase is highly specific for the α subunit of (Na,K)ATPase. The K_m (ATP) is 3 μM and the $t_{1/2}$ for maximum phosphorylation is ∿ 5 min at 37°. The high specificity and rapid phosphorylation by this kinase strongly suggest that it plays an important role in regulating the (Na,K)ATPase. Such a role is also suggested by the observation that approximately 50% of the (Na,K)ATPase molecules in living Friend cells are phosphorylated. The number of residues phosphorylated per peptide chain *in vivo* is unknown but Cleveland Maps indicate that the same region of the peptide is phosphorylated *in vivo* as *in vitro* (11). The mechanism by which the kinase is regulated is also unknown. It is not affected by cAMP, cAMP dependent protein kinase inhibitor, Ca^{2+}, EGTA, or trifluoperazine in *in vitro* experiments. We are investigating the effect of the kinase on (Na,K)ATPase acitvity and ATP dependent cation transport in resealed plasma membranes.

ACKNOWLEDGEMENTS

   This research was supported by NIH Grant GM 28538. L.C. is an Established Investigator of the American Heart Association. L.-A. Y. is a post-doctoral fellow of the Massachusetts Affiliate of the American Heart Association.

REFERENCES

1. Cantley, L.C. (1981) Current Topics in Bioenergetics 11, 201-237.

2. Clausen, T. and Kohn, P.G. (1977) J. Physiol. 265, 19-42.

3. Rozengurt, E. and Heppel, L.A. (1975) Proc. Natl. Acad. Sci. USA 72, 4492-4495.

4. Resh, M.D., Nemenoff, R.A. and Guidotti, G. (1980) J. Biol. Chem. 255, 10938-10945.

5. Flatman, J.A. and Clausen, T. (1979) Nature 281, 580-581.

6. Smith, J.B. and Rozengurt, E. (1978) J. Cell Physiol. 97, 441-450.

7. Resh, M.D. (1982) J. Biol. Chem. 257, 11946-11952.

8. Mager, D. and Bernstein, A. (1978) J. Cell Physiol. 94, 275-285.

9. Smith, R.L., Macara, I.G., Levenson, R., Housman, D. and Cantley, L. (1982) J. Biol. Chem. 257, 773-780.

10. Bernstein, A., Hunt, V., Crichley, V. and Mak, T.W. (1976) Cell 9, 375-381.

11. Yeh, L.-A., Ling, L., English, L. and Cantley, L. (1983) J. Biol. Chem. 258 (in press).

12. Brunette, D.M. and Till, J.E. (1971) J. Membr. Biol. 5, 215-224.

13. Laemmli, U.K. (1970) Nature (London) 227, 680-685.

PHOTOACTIVATABLE REAGENTS IN PROBING THE PROTEIN-LIPID INTERFACE OF MEMBRANES

JOSEF BRUNNER[1], GIORGIO SEMENZA[1] AND JUERGEN HOPPE[2]

[1]Laboratorium für Biochemie der ETH Zürich, Universitätstrasse 16,
CH-8092 Zürich (Switzerland) and [2]Gesellschaft für Biotechnologische Forschung,
Abteilung Stoffwechselregulation, Mascheroder Weg 1, D-3300 Braunschweig (FRG)

STATEMENT OF PROBLEMS

Several reagents have now been described that selectively label those portions
of intrinsic proteins that are embedded in the lipid bilayer. Generally, these
reagents contain a photosensitive group which upon activation generates one or
several reactive intermediates. These species may differ considerably in their
chemical properties and, hence, in the way they react within a biological membrane. It explains why with reagents,differing mainly in the nature of the
photoreactive unit, often different labeling patterns are obtained. One striking
example is the labeling of the Na,K-ATPase by various hydrophobic reagents which
has led to a controversal view as to the membrane assembly of the two subunits
(1-5). These difficulties in reliably interpreting labeling patterns result from
the lack of adequate knowledge of the principal reaction paths existing for each
intermediate.

Considering the high abundance of amino acid residues with chemically inert
side chains present in membraneous polypeptide segments and the putative structural organization of these segments, a general labeling reagent of the lipid
core of membranes must react even with saturated hydrocarbon chains. This argument has been used to question the general suitability of arylazide-based
reagents and to propose instead generators of very reactive carbenes (6-8).
That arylazides are still used as photolabeling reagents is - among other
reasons - due to the fact that protein labeling yields are often much higher
than with carbene generators. However, several lines of evidence suggest that
this apparent advantage is at the expense of an increased likelyhood that compa-
ratively low labeling of other, more inert proteins, may escape detection. The
reason for this is the occurence of long-lived reactive intermediates which
react in a specific manner and with high efficiency with nucleophiles at the
protein surface.

The apparence of long-lived intermediates is not confined to arylazides but
must be considered as a general feature of the photochemistry of precursors

of nitrenes and carbenes (9-11). 3-Trifluoromethyl-3-phenyldiazirine (TPD)*
(12) is just one example of a precursor of a carbene that had been designed to
eliminate potential problems related to the diazirine-diazo photoisomerization
(7,13,14). Although photolysis of TPD results in the formation of both the carbene and the diazo isomer of TPD, the latter is chemically unreactive under the
usual labeling conditions and can be removed from the photolysis medium. In contrast, there is recent evidence that with derivatives of 3H,3-phenyldiazirine,
alkylation of carboxyl groups by the diazo isomers can occur at physiological
pH. One major goal of current research in this area is the design and synthesis
of novel photoactivatable units which are (i) highly sensitive to photolysis
(ii) chemically appropriate to be built in into other molecules and (iii) which
give rise to a single, short-lived reactive intermediate.

Hydrophobic labeling reagents must react exclusively from within the lipid
bilayer. Generally, this is achieved by using very hydrophobic compounds presumed to dissolve quantitatively within the lipid phase of the membrane, or amphipathic molecules which within the bilayer adopt a preferential orientation
(with respect to the plane of the bilayer) and of which transversal diffusion is
highly restricted. The aim in the use of such reagents is to label the lipid
core within spatially definable sub-compartments.

A problem not directly related to the labeling reagents is the analysis of a
labeled membrane or protein. So far most analyses have been restricted to determinations of the distribution of (radio)label among individual proteins or
protein fragments. As shown by a few studies, Edman degradation provides a means
of increasing the resolution to the level of single amino acid residues. However,
unless the properties of a reagent have been adequately defined, interpretation
of Edman degradation data in terms of detailed structural (topological) features
is not possible and may lead to erroneous conclusions.

REACTION OF [^{125}I]TID WITH THE PROTEIN SURFACE INCLUDES ALL TYPES OF AMINO ACID
RESIDUES

Although C-H insertion is a common reaction of various carbenes, its significance for hydrophobic photolabeling of membranes has not been fully established.
However, successful labeling of various polypeptide segments that are

*The abbreviations used are:TPD, 3-Trifluoromethyl-3-phenyldiazirine; [^{125}I]TID,
3-Trifluoromethyl-3-(m-[^{125}I]iodophenyl)diazirine; CNBr, Cyanogen bromide; PTH,
Phenylthiohydantoin;

extraordinarily rich in aliphatic amino acid side chains followed by Edman degradation suggest that C-H insertion occurs to extents that are readily detectable by the usual techniques. This is shown here by recent results of a $[^{125}I]$TID
labeling study of F_1F_0 ATP synthase from *E. coli* (15).

3-Trifluoromethyl-3-phenyl-
diazirine

3-Trifluoromethyl-3-(m-$[^{125}I]$iodo-
phenyl)diazirine

Fig. 1. Formulas of the two reagents TPD and $[^{125}I]$TID

All three subunits (a,b and c) of the membrane inserted F_0 part were labeled
by $[^{125}I]$TID, two of them (b and c) were subjected to further analyses. More
than 98% of the radioactivity bound to polypeptide b was confined to an N-terminal portion of 30 residues as determined by CNBr fragmentation. Using Edman
degradation, this radiolabel was traced back to individual residues. These data
are reported by the radioactivity profile shown in the upper half of Fig. 2.
A cysteine residue (Cys_{21}) was most heavily labeled but also chemically inert
residues (notably Leu_8, Ile_{12} and essentially all residues between Phe_{15} and
Trp_{25}) received detectable label. Thiazolinones were converted to PTH derivatives and these chromatographed on silica gel thin-layer plates. Fig. 2 (lower
half) shows an autoradiogram obtained from such plates. Undoubtedly, the autoradiogram provides valuable information concerning the likely nature and composition of the radioactive products released upon each degradation cycle. Thus,
although the structure of the numerous radioactive products has first to be determined, it is reasonable to assume that $[^{125}I]$TID-labeled amino acid residues
(PTH derivatives) give rise to characteristic radioactivity patterns on the
thin-layer plate. This should, in principle, allow one to distinguish clearly
between radioactivity truly bound to amino acid side chains and other forms
of radiolabel (e.g. associated with lipids or with peptides).

Edman degradation of b and c subunits, labeled by $[^{125}I]$TID in various states
(see below), led to the main conclusion that all types of amino acid residues
are potentially reactive toward the carbene. Radioactive photolysis products
were found with Cys, Lys, Ser, Met, Arg and Tyr. These residues yielded rather

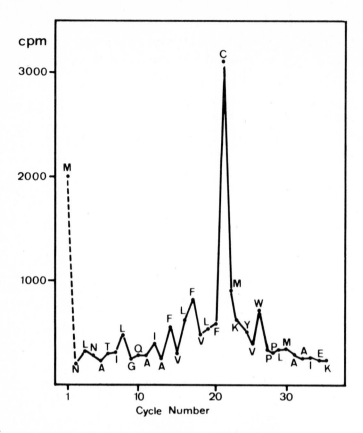

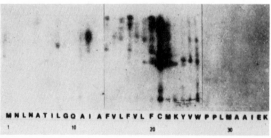

Fig. 2. Distribution of radioactivity among amino acid residues of the N-terminal region of subunit b, labeled with $[^{125}I]TID$ in intact F_1F_0 in Aminoxid WS 35. The upper part of the figure shows the radioactivity profile as obtained by Edman degradation of the labeled protein immobilized to porous glass. The lower half shows the autoradiogram of a thin-layer chromatogram of respective PTH derivatives.

uniform products and were formed presumably upon electrophilic attack by the carbene of a nonbonding electron pair in the heteroatom(s) of the respective side chains. Phe and Trp gave various (heterogeneous) products indicating that insertion of the carbene into the aromatic rings occured at different positions. Although the products formed with amino acids containing aliphatic side chains (Leu, Ile, Val) were generally present in low yields, distinctly different patterns for each amino acid residue were observed. Presumably, the dominant component in each case resulted from insertion of the carbene into the tertiary C-H bond present in these residues. A similar reaction might have occured at the α-carbon C-H bond of glycyl residues. Labeling of alanine has not been indicated so far and products with acidic amino acids (Asp and Glu) may exhibit limited stability in strong acid and may, therefore, not have been detected.

Clearly, more data with other proteins will be required to characterize in more detail the reaction of the carbene with the various types of amino acid residues. However, there is suggestive evidence that labeling is *not* confined to the more reactive amino acid residues but includes even chemically "inert" aliphtic chains.

SCOPE OF PHOTOACTIVATABLE REAGENTS IN PROBING THE LIPID CORE OF MEMBRANES

Interpretation of labeling patterns in terms of possible structural features is hampered by the facts that intrinsic reactivities of amino acid side chains toward a carbene are not comparable, that no data exist as to the precise distribution of reagent molecules within a lipid bilayer and that nothing is known about the existence of specific or nonspecific interactions between reagent and protein. It is necessary, therefore, to develop different labeling strategies to ascertain conclusions drawn from one type of experiments and to obtain a view of increased structural resolution.

One promising approach along this line is the labeling of a protein in different structural (conformational) states. In the intact F_1F_0 ATP synthase, subunit c forms an oligomer containing 10 ± 1 copies per F_0 (16). In SDS, this complex dissociates to yield monomers of subunit c, a process which is likely to be accompanied by major structural changes and, hence, changes in the residues available at the protein surface. Even though the SDS-denatured protein may still be highly structured, labeling in this state and in the native complex by $[^{125}I]$TID resulted in pronounced differences in the label distribution among individual amino acid residues. Thus, when labeling was performed with intact

F_1F_0, only a few residues received clearly detectable amounts of label (Fig.3A). One stretch of such residues is located at the N-terminus (Tyr_{10}, Met_{11}, Val_{15}, Leu_{19}). Interestingly, in an α-helix, these residues would lie on the same side of the cylindrical envelope. Such a pattern might therefore be indicative of an α-helix accessible only from one side. Arrangements of densely packed helical rods seems to be a very common structural element of integral membrane proteins. We are currently investigating whether a periodicity in the labeling pattern is also obtained with bacteriorhodopsin, a protein excellently suited to examine this point in more detail.

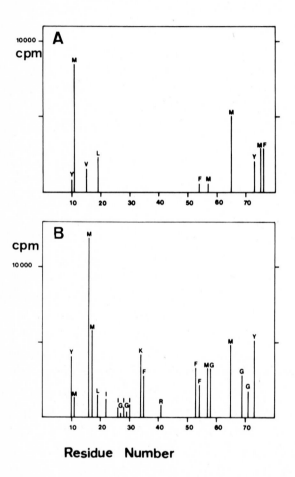

Fig. 3. Histograms of the distribution of radiolabel among residues of subunit c labeled with [^{125}I]TID in intact ATP synthase (in Aminoxid WS 35) (Fig.3 A) and in SDS solution (Fig. 3 B).

In the SDS-denatured state, most of the polypeptide chain was accessible to $[^{125}I]$TID. In the putative helical segment mentioned above, two adjacent methionyl residues (Met_{16} and Met_{17}) are heavily labeled while they received very little or no label in the intact F_1F_0 despite their being among the most reactive residues (Fig. 3B).

Alternatively to labeling proteins in different states, reagents of distinct molecular shape and physical properties may be utilized. In addition to lipid-soluble reagents such as $[^{125}I]$TID, the use of phospholipids carrying a photosensitive group appears particularly promising. As discussed in detail by Khorana (17), in whose laboratory this class of reagents was introduced, they allow one to study protein-lipid interactions in a way and at a level different to that with more simple, hydrophobic reagents. It must be emphasized that in choosing the photosensitive group, the same or even more stringent criteria apply as for other reagents. A striking example of the decisive role of the photoreactive unit has been obtained recently by labeling the transmembrane segment of glycophorin A with two phospholipids which mainly differed in the photosensitive part of the molecules (14,18). One probe (PL I; Fig. 4) contained the diazirinophenoxy group known to photoisomerize to a reactive diazo isomer (14) whereas the other lipid carried the trifluoromethyl-diazirinophenyl group which yields a diazo compound unreactive under labeling conditions. Labeling of glycophorin A was examined by cyanogen bromide fragmentation and determination of the label associated with individual fragments. According to the model of the transmembrane domain of glycophorin A, Met_{81} must be located near the middle of the membrane-spanning segment implying that the two major CNBr fragments (residues 9-81 and residues 82-131, respectively) contain those parts of the helical segment which traverses the outer half of the lipid bilayer (fragment 9-81) and the inner half (fragment 82-131). While the phospholipid which generates a reactive diazo isomer apparently labeled only one half of the transmembrane segment (14), both were labeled by the other phospholipid, PTPC (18).

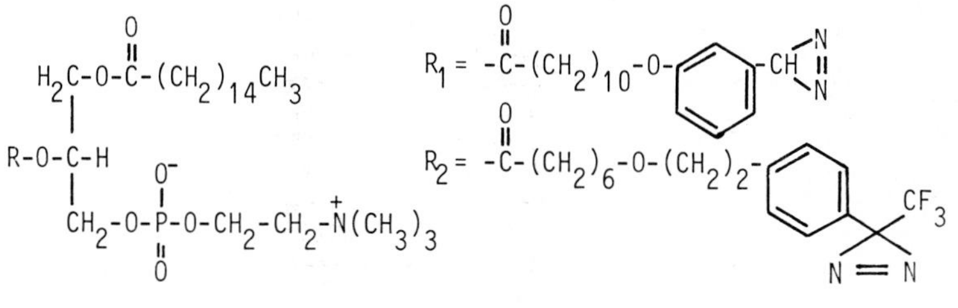

Fig. 4. Structure formula of PL I ($R=R_1$) and PTPC ($R=R_2$).

The apparent failure of PL I to label segment 82-131 has been interpreted to reflect steric exclusion of the probe and/or a low reactivity of the membraneous part of this segment. However, the additional finding that PL I crosslinked essentially to a single residue (Glu_{70}) of glycophorin raises the question of the nature of the reactive intermediate responsible for that labeling. As there is also evidence that the diazo isomer did contribute substantially to the overall labeling and as such labeling selectivity is not compatible with a highly reactive carbene, it is very possible that labeling by PL I of glycophorin A was mainly due to the diazo isomer. This does not rule out that as in the case of PTPC, the carbene derived from PL I did attack both halves of the transmembraneous segment, but these reactions may have been much less efficient than esterification of Glu_{70} by the diazo isomer and may have escaped detection by the analytical procedure used.

Provided that membrane preparations can be obtained that consist of sealed vesicles of unique orientation, it should, in general, be possible to selectively label the outer leaflet of these vesicles using phospholipid analogues. Spontaneous or phospholipid exchange protein mediated transfer of the probe from liposomes to the target membrane provides a means of incorporating the reagent into the outer monolayer of the sealed vesicle. As asymmetric labeling rests decisively upon the formation of a nonequilibrium distribution of the photosensitive lipid in the target membrane, limitations of the applicability of this approach may exist in systems which exhibit high phospholipid flip-flop rates (18).

CONCLUDING REMARKS

This article summarizes our recent experience with photosensitive reagents designed to label the lipid core of membranes. Emerging evidence that labeling patterns can provide meaningful structural information at the level of single amino acid residues should make this technique to a valuable instrument in studies of membrane protein structure. However, it should be pointed out that with successful developments of highly photosensitive units which give rise only to short-lived intermediates, hydrophobic labeling may become a powerful tool to tackle questions of membrane dynamics such as, for example, the time course of protein insertion into membranes.

ACKNOWLEDGEMENT
This work was supported by SNSF, Berne.

REFERENCES

1. Karlish, S.J.D., Jørgensen, P.L. and Gitler, C. (1977) Nature, 269, 715.

2. Farley, R.A., Goldman, D.W. and Bayley, H. (1980) J. Biol. Chem. 255, 860.

3. Jørgensen, P.L., Karlish, S.J.D. and Gitler, C. (1982) J. Biol. Chem. 257, 7435.

4. Montecucco, C., Bisson, R., Gache, C. and Johannsson, A (1981) FEBS Lett. 128, 17.

5. Jørgensen, P.L. and Brunner, J., submitted

6. Bayley, H. and Knowles, J.R. (1978) Biochemistry 17, 2414.

7. Bayley, H. and Knowles, J.R. (1978) Biochemistry 17, 2420.

8. Gupta, C.M., Radhakrishnan, R., Gerber, G., Olsen, W.L., Quay, S.C. and Khorana, H.G. (1979) Proc. Natl. Acad. Sci. USA, 76, 2595.

9. Bayley, H. and Knowles, J.R. (1977) Methods Enzymol. 46, 69.

10. Chowdhry, V. and Westheimer, K.H. (1979) Annu. Rev. Biochem. 48, 293.

11. Staros, J.V. (1980) Trends Biochem. Sci. 6, 44.

12. Brunner, J., Senn, H. and Richards, F.M. (1980) J. Biol. Chem. 255, 3313

13. Bayley, H. and Knowles, J.R. (1980) Biochemistry 19, 3883.

14. Ross, A.H., Radhakrishnan, R., Robson, R.J. and Khorana, H.G. (1982) J. Biol. Chem. 257, 4152.

15. Hoppe, J., Brunner, J. and Jørgensen, B.B., submitted

16. Foster, D.L. and Fillingame, R.H. (1982) J. Biol. Chem. 257, 2009.

17. Khorana, H.G. (1980) Bioorg. Chem. 9, 363.

18. Brunner, J., Spiess, M., Aggeler, R., Huber, P. and Semenza, G. (1982) Biochemistry, in press.

A HYDROPHOBIC PHOTOLABELLING STUDY OF THE ATP-SYNTHASE FROM
ESCHERICHIA COLI

Cesare MONTECUCCO[1], Peter FRIEDL[2] and Jurgen HOPPE[2]
[1]Centro C.N.R. per la Fisiologia dei Mitocondri e Laboratorio
di Biologia e Patologia Molecolare, Istituto di Patologia
Generale, Università di Padova, ITALY and [2]Dpt of Stoffwechselregulation, G.B.F.-Gesellschaft fur Biotechnologische Forschung,
Braunschweig-Stockeim, WEST GERMANY

The ATP-synthase is a multi-subunit enzymic complex able to
form ATP using the energy derived from a transmembrane proton
gradient (1-3). Its activity is strictly lipid dependent. This
complex is formed by two distinct and separable entities: the
hydrophilic F_1 portion, provided with ATPase activity, which is
believed to project into the acqueous phase and F_o, a hydrophobic
domain, which forms a transmembrane proton channel. Bacterial, as
well as chloroplast and mitochondrial F_1-ATPases, are formed by
five different subunits termed α, β, γ, δ and ε in order of
decreasing molecular weight. E. coli F_o is formed by three different
subunits termed a, b and c, while those from mitochondria and
chloroplast appear to be more complex (1-3). Only for the E. coli
enzyme has the sequence of all subunits been determined as well
as their stoechiometry (4-7). This makes the E. coli enzyme the
best ATP-synthase for the study of its quaternary structure, organization and arrangement of its polypeptide components with respect
to the membrane.

Recently relevant information on lipid-protein interaction and
on polypeptide folding in the lipid bilayer have been obtained with
the use of photoreactive phospholipids (8-12). These radioactive
probes contain a photoactivatable group attached at a selected
position of one fatty acid chain. On illumination a highly reactive
intermediate is formed, which is able to react with neighbour
molecules forming a covalent adduct (13). In the probe used here
the photoreactive nitroarylazido group is localized at the polar
head group level (see inset of Fig. 1) and hence it is believed

to label only those polypeptide regions intercalated in the super-
ficial part of the lipid bilayer.

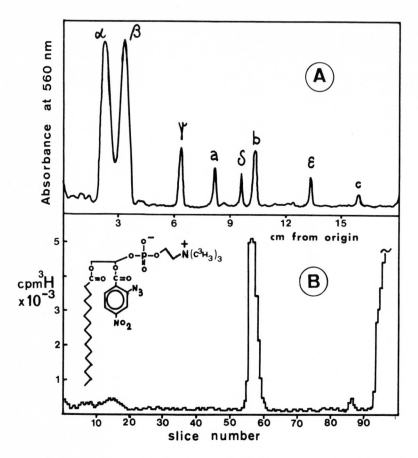

Fig. 1. A) densitometric trace of a Coomassie Blue-stained polyacrylamide gel run as in (9). B) pattern of labelling determined
by slicing and counting the gel as in (13).

Fig. 1 shows the pattern of labelling of the E. coli ATP-synthase obtained after incorporation of the enzyme in liposomes of
egg lecithin tagged with 0.8% of the azidolecithin probe and
illumination for 10' with a long wave U.V. lamp at 0°C. Most of
the label is associated with polypeptide b, minor amounts with
subunit c and β, while no radioactivity is associated with the

other F_1 subunits. This result supports the notion that F_1 is localized mainly outside the membrane. While we wannot offer a simple interpretation for the low labelling of subunit c, which is present in the membrane in ten copies (7), the high labelling of subunit b clearly indicates that it is in contact with lipids at the polar head group region of the membrane. To identify the site(s) of labelling, polypeptide b was isolated from the complex by HPLC and digested with S. aureus protease V8. Radioactivity was associated only with a large fragment containing the hydrophobic N-terminus. This peptide was attached to a solid support, sequenced and the radioactivity released at each step determined.

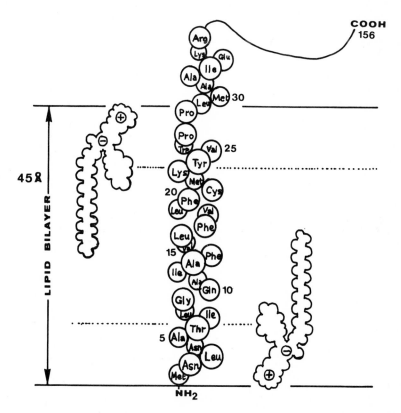

Fig. 2. Schematical model for the organization in the membrane of the N-terminal part of subunit b of the E. coli ATP-synthase.

Labelling was restricted to two short sequences (residues 2-5 and 20-25) intercalated by an unlabelled hydrophobic region of fifteen amino acid residues (12). This region is predicted to be arranged in an α-helical configuration (14,15). These results suggest that the N-terminal part of subunit b transverses the membrane adopting an α-helical structure as shown in Fig. 2. Given a lipid bilayer thickness of 45 Å, this organization brings the two labelled regions in contact with the two lipid polar head group regions of the membrane, where the photoreactive groups reside. The possibility of a folding back, with the two labelled segments on the same side of the membrane, is less likely on the base of the energetics of protein folding in a hydrophobic phase(16).

REFERENCES

1. Sebald. W. (1977) Biochim. Biophys. Acta 463, 1-27.
2. Fillingame, R.H. (1980) Annu. Rev. Biochem. 49, 1079-1113.
3. Capaldi, R.A. (1982) Biochim. Biophys. Acta 694, 291-306.
4. Gay,J.N. and Walker,J.E.(1981) Nucl. Acid Res. 9, 3919-3926.
5. Nielsen, J., Hansen, P.G., Hoppe, J., Friedl, P. and v.Meyenburg, K. (1981) Mol. Gen. genet. 184, 33-39.
6. Kanazawa, A.et al.(1981)Biochem.Biophys.Res.Commun. 103,613-620.
7. Foster,D.L. and Fillingame,R.H.(1982) J.Biol.Chem.257,2009-2015.
8. Montecucco, C., Bisson, R., Gache, C. and Johannsson, A. (1981) F.E.B.S. Lett. 128, 17-21.
9. Montecucco, C., Dabbeni-Sala, F., Friedl, P. and Galante, Y.M. (1983) Eur. J. Biochem. 132, 189-194.
10. Bisson, R., Steffens, G.C.M. and Buse, G. (1982) J. Biol. Chem. 257, 6716-6720.
11. Ross, A.H., Radhakrishnan, R., Robson, R.J. and Khorana, H.G. (1982) J. Biol. Chem. 257, 4152-4161.
12. Hoppe, J., Montecucco, C. and Friedl, P. (1983) J. Biol. Chem. 258, 2882-2885.
13. Bisson, R. and Montecucco,C. (1981) Biochem.J. 193, 757-763.
14. Walker, J.E., Saraste, M. and Gay,N.J.(1982) Nature 298, 867-869.
15. Hoppe, J. et al. (1982) 2nd Europ.Bioenerg.Conf. 85-86.
16. Henderson, R. (1979) in "Membrane Transduction Mechanisms" Cone, R.A. and Dowling, J.E.(Eds), Raven Press, New York,pp.3-15.

© 1983 Elsevier Science Publishers B.V.
Structure and Function of Membrane Proteins,
E. Quagliariello and F. Palmieri editors.

TRANSPORT AND MODIFIER SITES IN CAPNOPHORIN, THE ANION TRANSPORT
PROTEIN OF THE ERYTHROCYTE MEMBRANE

JENS OTTO WIETH AND POUL J. BJERRUM
Department of Biophysics, The Panum Institute, University of
Copenhagen, DK-2200, Copenhagen N, Denmark.

THE ANION TRANSPORT PROTEIN

The red cell membrane is particularly rich in an integral membrane protein, which mediates a rapid exchange diffusion of bicarbonate and chloride across the cell membrane as an important
partial process in the transport of CO_2 from tissues to lungs
(1,2,3).

The transport protein migrates as band 3 by SDS-polyacrylamide gel electrophoresis (4). To characterize its physiological
function we propose to name it capnophorin. Capnos is the Greek
word for smoke. Carbon dioxide is the fume of oxidative metabolism, and hypercapnemia (accumulation of CO_2 in blood) is the
well known pathophysiological consequence of impaired CO_2 elimi-
nation from the organism.

The transport protein, which constitutes almost 50% of the integral membrane proteins with one million copies per cell (4),
has a molecular weight of 95,000 dalton. About 60% of its molecular mass is closely associated with the lipid membrane, and this
part of the molecule is essential for transport function (5). The
peptide is folded into hairpin-like loops, with at least 7
strands traversing the hydrophobic membrane core (6). This gross
assembly, which seems similar to the membrane arrangement of
bacteriorhodopsin, is fixed in the membrane, so the anion translocation must involve minor conformational changes in specialized regions of the protein. The intramembrane peptide segments
are surprisingly rich in hydrophilic amino acid residues, and
evidence is accumulating that the permeation pathway is composed of residues from several peptide strands (7,8,9).

The knowledge of the structure of capnophorin does not permit
any firm conclusion about the molecular mechanism of transport.
Most of the ideas have been derived from studies of the kinetic
properties of the obligatorily linked 1:1 exchange diffusion of

anions. We feel there is a tendency in the literature to over-
simplify the concept of an anion binding transport site. It is
often assumed that the transport process is facilitated by the recognition and binding of an anion by a single functional group in
the transport protein. Anion binding is believed to trigger a
conformational change transposing the binding site through a narrow permeability barrier, and the exchange cycle is completed,
when the transported anion dissociates and another anion is bound
and translocated in the opposite direction. We are sceptical to
this concept, which is undoubtedly rooted in the similarities of
the kinetics with those of a mobile lipophilic, positively charged carrier, which permeates the membrane only when complexed
with a suitable anion (10,11). Firstly, we do not use the term
'site' in the sense of a single anion binding group. Our concept
of an anion binding site is based on the general model for a
substrate-binding enzyme active center, where several residues
act in concert in substrate binding and in the conformational
changes accompanying the catalytic event. Several side chains
contribute to form the binding site, and the modification of any
of these groups can interfere with the properties of the site.
Secondly, we are also critical to the idea that a transporting
binding site is alternately exposed to the two sides of a narrow
permeability barrier. The transport system is highly asymmetric
in structure and with regard to anion- and inhibitor affinities.
This can well be accounted for by a model, which assumes that
topographically separate anion binding regions at the two sides
of the membrane are interconnected by a well-sealed path of maybe 30-50 Å, which is opened only during the passage of anions in
and out of the membrane (12,13).

KINETIC EVIDENCE OF TRANSPORT AND MODIFIER SITES
   The concentration dependence of chloride self-exchange in experiments where chloride concentrations are varied symmetrically
at the two sides of the membrane is shown in Fig. 1a and analyzed in Fig. 1b. The chloride exchange flux increases with chloride concentration, reaches a maximum at Cl$\sim$150 mM and decreases
by so-called self-inhibition at higher concentrations. A simple
interpretation of this complex concentration dependence was given

by Dalmark (14), who described the relation by means of three parameters: a theoretical maximum flux, J_{max}^o, and two apparent affinities, K_{Cl}, the affinity of chloride for a transport site, and
K_{ClCl}, the affinity for a modifier site with the peculiar function that it arrests transportation in a non-competitive fashion,
when an anion has been bound. The modified maximum flux, J_{max}^{Cl}
(curve 2 in Fig. 1b), is therefore a function of chloride concentration, and the saturation kinetics of the transport process
cannot follow the simple hyperbolic function (curve 1 in Fig. 1b),
as it would in the absence of the inhibitory effect of the modifier site. The experimentally observed flux (Fig. 1a) is thus
described by the equation:

$$J_{Cl} = J_{max}^o \left\{ (1 + K_{Cl}/Cl)\ (1 + Cl/K_{ClCl}) \right\}^{-1}$$

The K_{Cl}-containing term is illustrated by curve 1 in Fig. 1b.
Note that the theoretical maximum flux (J_{max}^o) of 585 pmol cm^{-2}
s^{-1} is more than twice the experimentally observed maximum, which
is found at the chloride concentration where Cl = $(K_{Cl}\ K_{ClCl})^{\frac{1}{2}}$.
The curve numbered 2 in Fig. 1b shows the maximum flux, J_{max}^{Cl}, as
a function of chloride concentration. The value of K_{ClCl} (335 mM)
is the chloride concentration which half-saturates the modifier
site and halves the J_{max}^{Cl} from its theoretical maximum, J_{max}^o.

TABLE 1

APPARENT DISSOCIATION CONSTANTS OF ANION-BINDING TO TRANSPORT
AND MODIFIER SITES (mM)

	Transport site	Modifier site	M/T[b]
Bicarbonate	20	600	30
Chloride	67	335	5
Fluoride	88	337	3.8
Bromide	32	160	5
Iodide	10	60	6
NAP-taurine[a]	0.4	0.02	0.05

Data from (14,15,16).
[a] NAP-taurine: N-(4-Azido-2-nitrophenyl)-2-aminoethylsulfonate
[b] The ratio M/T indicates anion affinity to transport site relative to the affinity for the modifier site.

Fig. 1. Concentration dependence of chloride self-exchange in ghosts at $0^{\circ}C$, pH 7.2.
a. Experimentally observed values
b. 1. Hypothetical transport saturation in absence of the inhibitory modifier effect
   2. Concentration dependence of the modified maximum flux (K_{ClCl}=335 mM).

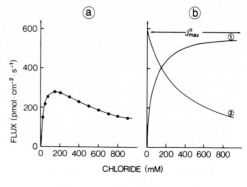

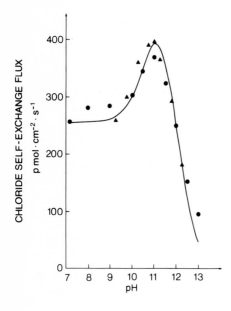

Fig. 2. Chloride self-exchange flux in ghosts at $0^{\circ}C$. $Cl_i = Cl_o$ = 330 mM.
Flux stimulation between pH 9.5 and 11.5 is believed to be related to the deprotonation of a modifier site. From (17).

Fig. 3. Chloride self-exchange flux in ghosts at $0^{\circ}C$ in the absence (o) and presence (x) of DNDS. Cl_i = 165 mM, Cl_o = 16.5 mM. Note absence of stimulation of flux at low Cl_o and reappearance after inhibition with DNDS.

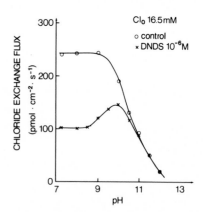
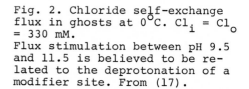

Table 1 shows the values of the two apparent affinity constants for halides, bicarbonate and for NAP-taurine, an organic sulfonate, which exerts a particularly pronounced inhibitor effect by reversible binding to a site, which according to the studies of Knauf (16) appears to be identical with the halide binding modifier site.

TITRATION OF THE MODIFIER SITE

Evidence for the actual existence of an anion binding modifier site exposed to the extracellular phase has been provided by acid-base studies of the transport function (17). It was found that the inhibitory modifier effect decreases at alkaline pH values. Fig. 2 shows the pH dependence of chloride exchange at 330 mM KCl under conditions where the pH was varied in the extracellular phase only. A pronounced stimulation of the self-exchange flux is observed between pH 9.5 and 11.5. A similar stimulation is not observed at lower chloride concentrations, where the anion binding to the protonated modifier site is minimal (cf. Fig. 3). The complex flux curve of Fig. 2 could be resolved into two titratable functions: a) titration of a modifier site with a pK of 10.7 and b) of a transport site with a chloride dependent pK, which reaches a constant value of 12 at chloride concentrations above 150 mM (17). The interpretation assumes that the modifier site loses its inhibitory effect on transport when it is deprotonated. Studies performed at low chloride concentrations showed that flux stimulation between pH 9.5 and 11 in the presence of 60 mM iodide and 0.1 mM NAP-taurine, were of the magnitudes predicted by the reported modifier affinities of the two anions (17).

FUNCTIONAL INTERACTIONS AT TRANSPORT AND MODIFIER SITES

Till this point we have described the modifier and transport sites as distinct separable entities. The reality may be more complicated. In our titration study it became clear that the apparent affinity of chloride for the alleged transport site decreases in the pH-range where the modifier site is deprotonated. Gunn & Fröhlich (18) have shown that the apparent affinities of the transport site(s) differ at the two sides of the membrane.

In the neutral pH range K_{Cl}^{out} is about 2 mM. We found that it increases by a factor of 5-10 to about 16 mM at pH 10.5 (17). Since a decrease of affinity clearly accompanies the titration of the modifier site, it seems likely that the charge(s) of the modifier site contribute(s) to the high apparent affinity as indicated by the low value of K_{Cl}^{out} at neutral pH values.

We have already mentioned that NAP-taurine inhibition showed the anticipated abolition of the inhibitory modifier effect at pH 9.5 - 11. While NAP-taurine binds preferentially to a modifier site (16), the kinetic evidence suggests that DNDS (4,4'-dinitro-stilbene-2,2'-disulfonate) competes with chloride for binding exclusively at the transport site (19). Therefore, it was surprising to find that DNDS inhibition of chloride transport at an extracellular chloride concentration of 16 mM has a pH profile which, as shown in Fig. 3, clearly shows that the disulfonate anion also exerts a modifier effect on transport function. 10^{-6}M DNDS caused 50% inhibition between pH 7.5 and 9 (Fig. 3). However, a flux stimulation was observed at higher pH values, and transport inhibition by 10^{-6} M DNDS was in fact not demonstrable at pH values above 10.5. The result shows that the affinity of DNDS decreases relative to that of chloride (which decreases in itself), suggesting that the titratable group(s) at the modifier site contribute(s) more to binding of DNDS than to binding of chloride to the transport site.

Self-inhibition, or so-called inhibition by substrate, is well known from enzyme kinetics (20). In the enzyme carboxypeptidase, substrate inhibition is ascribed to accumulation of substrate at the entrance of the active site, where positively charged guanidinium groups form a slide guiding the negatively charged substrate to the catalytic center. In analogy with this interpretation it is possible that the physiological role of the modifier site is to feed anions into the transport mechanism and that it in addition increases the anion affinities of the external transport site. Both in the enzyme and in the transport system self-inhibition may reflect a situation where the dissociation step (unloading of product or of the transported ion) is impeded, because there are no vacancies in a region, the modifier site, which is interposed between the functional binding site and the

surrounding aqueous solution. At the end of our contribution we
discuss the possibility that modifier and transport site characteristics reflect different functional states of an external
gating region.

TRANSPORT SITES

Results of experiments like that of Fig. 2 indicated that transport function is critically dependent on the protonation of groups
with an apparent pK of 12 (17). Arginyl residues were the likely
candidates, and this was confirmed by selective modification of
arginyl residues exposed to the extracellular phase (9,21,).
This work is dealt with by Bjerrum (22), it suffices to state
that anion transport can be completely inactivated by covalent
reaction with phenylglyoxal (PG), and that complete inhibition
under the most specific conditions is accompanied by the binding
of sufficient PG to modify only a single arginine per capnophorin molecule.

It is conceivable that the fragile tertiary structure of a protein molecule may change secondary to modification of any of its
many amino acid residues. If that is the case, the effect of arginyl modification is 'inspecific', but it does not seem to be
so. In fact, as shown in Fig. 4, we have found that one third
of the arginines of the cell membrane can be modified without se-
rious effects on membrane transport. Therefore, we believe that
the selectively modified group is directly involved in anion
translocation, maybe forming a part of an external transport
site.

In the experiments of Fig. 4 treatment with PG was carried out
at a pH of 8.3 on both sides of the membrane. Under these conditions anion transport is slowly inactivated (23) as confirmed by
the upper curve in Fig. 4b. The transport inactivation is very
efficiently protected by DNDS (lower curve of Fig. 4b). However,
the binding of PG to the membrane protein was similar, in spite
of the considerable difference of transport inhibition. Fig. 4a
shows that $100 - 110 \times 10^6$ PG-molecules were bound per membrane
after one hour both in the presence and absence of DNDS. This
amount of PG suffices to modify 50×10^6 of the 150 million arginyl residues in the membrane proteins (9). Transport inhibi-

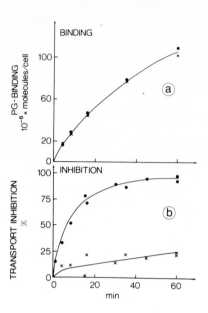

Fig. 4. Phenylglyoxal-binding and
transport inhibition in ghosts
40 mM K_2SO_4, 5 mM KCl, pH 8.3, inside = outside.
a. PG-binding in the absence (●)
and presence (x) of DNDS, 1 mM
b. Inhibition of chloride self-
exchange (per cent of untreated
control sample). PG-treatment with
20 mM PG at 38°C in the absence
(●) and presence (x) of DNDS, 1 mM.

THE EXCHANGE DIFFUSION ZIPPER

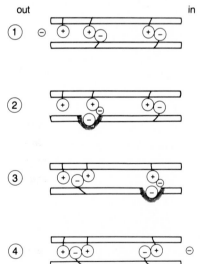

Fig. 5. The exchange diffusion
zipper. Four conformational
states of a two site exchange
gate, 1 to 4 illustrates a half
cycle of influx. States 1 and
4 are stable "in" and "out"
configurations, which intercon-
vert extremely slowly in the
absence of a transportable anion. From (13).

tion was 92-96% after one hour's PG-treatment in the absence of DNDS, whereas inhibition was maximally 20%, when the transport sites had been protected by DNDS during the exposure to PG. The small difference between the two curves in Fig. 4 indicates, in fact, that both the modification which causes inhibition and the protection by DNDS are very specific. If we assume that the differing degrees of inhibition are reflected in the small differences in PG-binding we find that complete inhibition is accompanied by the binding of an excess of 2×10^6 PG-molecules to the capnophorin of a cell membrane, corresponding to the modification of approximately one arginine per transport protein molecule.

We have not been able to show that the modifiable arginine group(s) involved in transport can be recruited to the inside or the outside of the membrane barrier. Modification of internal arginines (by treating alkaline ghosts with PG in a neutral medium) has been performed successfully, but only a partial inhibition of anion transport was obtained (22). Preliminary investigations indicate that groups modified at the outside are located in peptide segments of band 3 different from those labelled by PG-lation from the intracellular side. It has not been possible to demonstrate that the physical properties or the chemical reactivity of the arginyl residues at the external transport site can be manipulated in the presence or absence of anion gradients which are supposed to recruit a mobile transport site to one of the membrane surfaces. In the absence of evidence for recruitment and with the added piece of information that negatively charged carboxylates seem to play an essential role for the function of the transport system (3), we have proposed as a working model that the sequential "ping-pong" kinetics of monovalent anion transport may be accounted for by a zipper model of anion exchange where topographically separate anion binding sites at the two sides of the membrane are alternately accessible to the transported anions (13).

THE ANION EXCHANGE ZIPPER

The paradoxical presence of negatively charged amino acid residues in the anion transport system can be rationalized by as-

suming that salt-bridges within the protein provide the structural basis for gating mechanisms which operate by the attrac-
tive and repulsive forces created by an intruding anion (13,24,
25). A zipper is "a fastener consisting of an interlocking device set along two edges, which can be separated and reunited,
when an attached mobile piece slides between them". In Fig. 5
the interlocking device is provided by intramolecular salt-
bridges, joining the edges of the intramembraneous polypeptide
strands. Oscillations between two stable "empty" configurations
of the transporter are induced by the interaction of the transported anion with the fixed charges in the pathway (13). The
lock of each salt-bridge can be opened only by an anion approach-
ing the side, which exposes the positive charge. The configura-
tional change alters the energy profile of each permeation barrier, which determines the direction of the next translocation.
The affinity for anion binding at each side of the membrane
therefore changes in a cyclic fashion, and the shift between
high and low affinity is responsible for the resemblance of
transport kinetics to the ping-pong, sequential mechanism of a
cyclic, mobile carrier (18). The existence of two separate sites
explains why chemical selectivity and affinity to transported
anions and inhibitor molecules differ at the two sides of the
membrane.

The stabilization energy of an intramolecular salt-bridge in
a relatively low dielectric medium is high (26), and the breaking and making of salt-bridges contribute significantly to the
high activation energy of anion transport. Finally the zipper
model explains why red cell membranes of certain animals may remain as tight to urea and water as a lipid bilayer in spite of
the large amount of hydrophilic groups in integral membrane proteins (27,28). The transport pathway can be opened only by the
transported anion. Hence, capnophorin does not present aqueous
channels that permit nonspecific leakage of ions and polar so-
lutes through the membrane.

TRANSPORT AND MODIFIER SITES - CONCLUSIONS
The 'transport' and 'modifier' sites are unified by the exchange zipper model, if it is assumed that the sites do not re-

present topographically separate entities, but reflect the cyclic variation of anion affinity at the external binding region. In the model of Fig. 5 the high and low affinity states of the external and internal anion binding sites alternate with frequencies between 450 and 85,000 s^{-1} between 0 and 38 ^{O}C (1), i.e. with the frequency of the anion exchange half-cycle. At low temperatures there is a good chance that transport is rate-limited by the conformational change, which is responsible for the affinity shift. According to our interpretation, the inhibitory modifier effect is caused by anion-binding to the non-transporting, 'low affinity' state, arresting the transport system in a 'trans'-configuration. The anion affinities for the external modifier site are shown in Table 1. The high affinity of NAP-taurine means that the organic sulfonate anion binds to the non-transporting configuration of the external binding site with an affinity which is 20-fold higher than the affinity for the inwardly transporting configuration of the site. This interpretation offers a simple explanation both for steric (16) and functional interactions between 'transport' and 'modifier' sites.

REFERENCES

1. Wieth, J.O. and Brahm, J. (1980) in: Membrane Transport in Erythrocytes. Alfred Benzon Sympos. 14. Munksgaard, Copenhagen, pp. 467-487.

2. Crandall, E.D. and Bidani, A. (1981) J. Appl. Physiol. Respirat. Environ. Exercise Physiol. 50, 265-271.

3. Wieth, J.O., Andersen, O.S., Brahm, J., Bjerrum, P.J. and Borders, C.L. Jr. (1982). Phil. Trans. Roy. Soc. Lond. B 299, 383-399.

4. Fairbanks, G., Steck, T.L. and Wallach, D.F.H. (1971) Biochem. 10, 2606-2616.

5. Lepke, S. and Passow, H. (1976) Biochim. Biophys. Acta. 455, 353-370.

6. Tanner, M.J.A., Williams, D.G. and Jenkins, R.E. (1980) Ann. N.Y. Acad. Sci. 341, 455-464.

7. Jennings, M.L. and Passow, H. (1979) Biochim. Biophys. Acta 554, 498-519.

8. Kempf, C., Brock, C., Sigrist, H., Tanner, M.J.A. and Zahler, H. (1981) Biochim. Biophys. Acta 641, 88-98.

9. Bjerrum, P.J., Wieth, J.O. and Borders, C.L. Jr. (1983) J. Gen. Physiol. 81, 453-484.

10. Wieth, J.O. (1972) in: Oxygen Affinity of Hemoglobin and Red Cell Acid Base Status. Alfred Benzon Sympos. 4. Munksgaard, Copenhagen, pp. 265-278.

11. Wieth, J.O. and Tostenson, M.T. (1979) J. Gen. Physiol. 73, 765-788.

12. Passow, H., Kampmann, L., Fasold, H., Jennings, M. and Lepke, S. (1980) in: Membrane Transport in Erythrocytes. Alfred Benzon Sympos. 14. Munksgaard, Copenhagen, pp. 345-367.

13. Wieth, J.O., Bjerrum, P.J., Brahm, J. and Andersen, O.S. (1982) Tokai J. Clin. Exp. Med. 7, Suppl. 91-101.

14. Dalmark, M. (1976) J. Gen. Physiol. 67, 223-234.

15. Wieth, J.O. (1979) J. Physiol. 294, 521-539.

16. Knauf, P.A., Ship, S., Breuer, W., McCulloch, L. and Rothstein, A. (1978) J. Gen. Physiol. 72, 607-630.

17. Wieth, J.O. and Bjerrum, P.J. (1982) J. Gen. Physiol. 79, 253-282.

18. Gunn, R.B. and Fröhlich, O. (1979) J. Gen. Physiol. 74, 351-374.

19. Fröhlich, O. (1982) J. Membrane Biol. 65, 111-123.

20. Laidler, K.J. (1958) The Chemical Kinetics of Enzyme Action. Oxford, At The Clarendon Press, pp. 71-74 and 252-256.

21. Wieth, J.O., Bjerrum, P.J. and Borders, C.L., Jr. (1982) J. Gen. Physiol. 79, 283-312.

22. Bjerrum, P.J. (1983) This Volume.

23. Zaki, L. (1982) in: Protides of the Biological Fluids. Pergamon, N.Y., pp. 279-282.

24. Macara, I.G. and Cantley, L.C. (1981) Biochem. 20, 5695-5701.

25. Wieth, J.O. (1981) in: Sixth School on Biophysics of Membrane Transport, School Proceedings, Vol. 2, Publ. Co. Agricultural Univ. Wrocɫaw, Poland, pp. 59-88.

26. Fersht, A.R. (1971) Cold Spring Harbor Symp. Quant. Biol. 36, 71-73.

27. Brahm, J. and Wieth, J.O. (1977) J. Physiol. 266, 727-749.

28. Wieth, J.O. and Brahm, J. (1977) in: Proceedings of the International Union of Physiological Sciences, XII, p. 126.

IDENTIFICATION AND LOCATION OF AMINO ACID RESIDUES ESSENTIAL FOR ANION TRANSPORT IN RED CELL MEMBRANES

P.J. BJERRUM, Department of Biophysics, University of Copenhagen, The Panum Institute, 2200 Copenhagen N, Denmark

As reviewed in the preceding article (1), functional studies of anion exchange in the red cell membrane indicate that monovalent anion transport is critically dependent on the state of protonation of two classes of titratable groups in the transport protein. Normal function of capnophorin depends on the deprotonation of one set of groups with an apparent pK $\sim$5 (2), and on the protonation of another set with pK $\sim$ 12 (3). The effects of anionic ligands on the apparent pK values (3,4) suggest that the groups form essential parts of a chloride binding site located at the external face of the membrane. It is obviously significant to identify the functionally critical amino acid residues and their location in the transport protein in order to test whether transport models agree or disagree with structural information.

Group specific amino acid reagents have been succesfully used to identify functionally essential amino acid residues in many enzyme active centers (5,6,7). We have adapted such techniques in order to locate functionally essential side chains in capnophorin by selective amino acid modification (9,10).

Possible candidates were chosen from the apparent pK values of the transport function. It was likely that the pK 12 groups were arginyl residues, while the groups with a pK around 5 could be carboxylates of glutamic or aspartic acid residues. To test these conjectures we designed methods suitable for the modification of guanidino- and carboxyl groups exposed to the external side of the erythrocyte membrane.

EXTRACELLULAR ARGINYL MODIFICATION WITH PHENYLGLYOXAL

Modification of guanidino groups was carried out with the arginine specific reagent phenylglyoxal (PG). PG reacts typically with arginyl residues (arg) in proteins with a 2:1 stochiometry (8). The binding is a pseudo first order reaction, rate limited by the first reaction step. PG reacts with the deprotonated form of arginyl residues, and the reaction rate is therefore markedly pH-dependent, increasing with pH.

The inhibition of anion transport with PG was a second order process. By varying the pH and the anionic composition of the reaction medium, the kinetics of transport inactivation were fully compatible with the assumption that phenylglyoxal reacts with functionally essential arginyl residues (9). In labelling experiments, using ^{14}C-PG, indiscriminate modification of 150×10^6

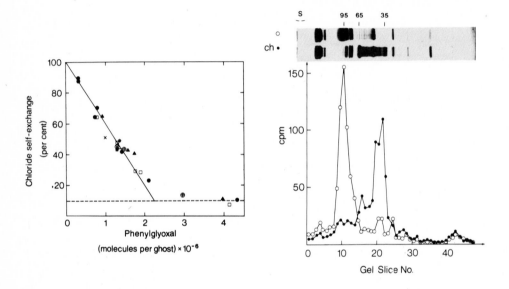

Fig. 1. Anion transport inactivation as a function of phenylglyoxal binding to resealed ghosts treated with 15 mM PG in a sucrose-citrat medium containing 8 mM KCl (pH 10.3, 25°C). The pseudo first order rate constant for transport inactivation was 0.15 s^{-1} (SD 0.02).

Fig. 2. Distribution of ^{14}C-PG in the red cell membrane proteins before (O) and after (●) chymotryptic cleavage of band 3. Labelling conditions, 15 mM PG in 8 mM KCl, pH 10.3, 25°C. The membrane proteins (70 µg) were separated on a 12-25% SDS-polyacrylamide gel. The inhibition of chloride self-exchange in the ghost preparation was 51%. The Figure shows that the major part of the radioactive phenylglyoxal is located in the 35,000 dalton chymotryptic fragment of band 3.

arg per ghost occurred rapidly when extra- and intracellular pH
were both above 10. Selective modification of extracellularly
exposed arginines was achieved when ghosts with a neutral or
acidic intracellular pH were reacted with PG in an alkaline
medium (10). When PG inactivation was performed for short periods of time from the extracellular side of the membrane pH
10.3, with 15 mM PG at 25°C in the absence of chloride, maximal inhibition of anion transport in ghosts was obtained by
binding of about 2 million PG molecules per ghost. (Fig. 1).
The major fraction of these molecules was located in band 3 (as
shown in Fig. 2) indicating that maximal inhibition of anion
transport can be obtained by binding of sufficient PG to modify
only one arginyl residue out of the 44 arginines in the band
3 protein (10).

Location of the Functional Arginine Residues

After treatment of PG-inactivated ghosts with extracellular
chymotrypsin, capnophorin is cleaved to produce a 35,000 dalton
fragment and a 65,000 dalton fragment. ^{14}C-PG is found almost
exclusively in the 35,000 dalton fragment (Fig. 2). The same
labelling pattern was observed when membranes were phenylglyox-
alated under less specific reaction conditions (in the presence
of 165 mM chloride) where 4 million PG molecules are bound per
ghost. This shows that the 35,000 dalton segment contains at
least two highly reactive arginyl residues, and that at least
one of these is involved in transport function. The primary covalent attachment site for DIDS, a lysyl residue of the 65,000
dalton fragment (11,12) resides in a different primary segment
of capnophorin. This indicates that the transporting region of
capnophorin is composed of peptide segments derived from different parts of the primary sequence of band 3.

To further localize the modified arginyl residue, the membranes
were digested extensively with chymotrypsin from both sides of
the membrane (1 mg chymotrypsin per ml, pH 7.6, 38°C). About
half of the radioactivity was found in a membrane-bound fragment
with a molecular weight of about 9,000 dalton (Bjerrum, Brock
and Wieth, unpublished results. The released radioactivity was
independent of the duration of digestion (15-60 min). The radio-

active material in the supernatant was found to have a very low molecular weight as judged from SDS-gel-electrophoresis of the supernatant, it was not retained in the gel, and a small fraction only was found just behind the front of the gel. The result suggests that selective inhibition by PG is due to alternative labelling of two essential arginine groups, one of which is located in the 9,000 dalton chymotryptic fragment, and that modification of one group prevents the labelling of the other, which is lost in the supernatant during the enzymatic treatment. The alternative labelling of one out of two arginyl residues may reflect the localization of these critical groups in a mixed hydrophobic/hydrophilic environment if the groups are so close that they can interact electrostatically with each other (10). In the absence of chloride, the pK of the two groups is lowered (9,10). When PG-lation is performed in the absence of chloride, only one group will be labelled during a brief exposure to phenylglyoxal, if the pK of the non-reacted group jumps to its higher intrinsic pK, when the electrostatic interaction between the two groups is abolished by modification of the first.

Binding of an anion to the external binding site is assumed to reduce the electrostatic interaction between the two arginyl groups, thus increasing their pK values towards 12-13, the intrinsic pK value of non-interacting arginyl sidechains. When the chloride concentration is sufficiently high, electrostatic interaction between the two positive groups is thus shielded, increasing their pK and making them less, but equally, reactive. They are now independent of each other so both react with PG during inactivation of the anion transport protein. This explains why the PG-reaction rate is reduced at high extracellular chloride concentrations (9) and why inactivation of anion transport under such conditions is accompanied by the binding of four PG-molecules, sufficient to modify the two arginyl residues in the 35,000 dalton segment of capnophorin (10).

EXTRACELLULAR CARBOXYL MODIFICATION WITH CARBODIIMIDES

To identify the amino groups responsible for the titratable transport function with a pK of about 5, membranes were chemically modified with carbodiimides (2). Water soluble carbodi-

imides react readily with carboxyl groups of aspartic and glutamic acid, which typically have pK values of 3-5. The experiments
were performed using a membrane impermeable carbodiimide 1-
ethyl-3-(4-azonia-4.4-dimethylpentyl)-carbodiimide (EAC). This
carbodiimide does not permeate the red cell membrane, and was
accordingly ideal for selective modification of extracellularly
exposed carboxyl groups (2).

EAC inactivates the anion transport irreversibly, 50% inactivation of the transport was obtained readily, while inhibition
of the remaining transport function was much slower (Fig. 3)
(2). After 50% transport inactivation with EAC, complete inhibition was attained with sufficient DIDS to label only half of
the capnophorin molecules, showing that DIDS binds preferentially to the non-inhibited transporters (2). The reaction is accelerated in the presence of the aromatic nucleophile tyrosin
ethyl ester (TEE). The first step in the EAC-reaction is the
formation of an unstable O-acyl-isourea compound between the reacted group and EAC. This complex can then either form an internal crosslink with a nearby amino-group or react with a nucleophile which has access to the reaction region.

Location of the Functional Carboxyl Group

By SDS-gel-electrophoresis it was shown that radioactively
labelled TEE is covalently bound to the membrane during the
chemical reaction with EAC (Bjerrum, Wieth and Borders, unpublished results). $1-2 \times 10^6$ TEE molecules were incorporated
in the capnophorin molecules per cell, and a larger fraction
was bound in the phospholipids (3×10^7 molecules per cell).

After chymotrypsin treatment from the extracellular side of
the membrane, transport was still inhibited relative to the control (see Table 1) and the major fraction of the capnophorin-
bound TEE was located in the 35,000 dalton fragment.

The location of the modified carboxyl group(s) in band 3 was
further explored by cleaving the protein with papain, which
splits off a 5-10,000 dalton segment (13). The enzymatic treatment is accompanied by a reduction of chloride exchange by 85-
90% (14). This result was confirmed by the experiments shown in
Table 1, which in addition shows that the magnitude of chloride

TRANSPORT INACTIVATION WITH EAC
(EAC 30 mM, pH 6, 38°C)

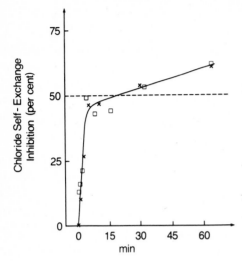

Chloride self-exchange of resealed ghosts
(per cent)

extracellular digestion	control	EAC+TEE treated
none	100.0	69.5
chymotrypsin	98.7	64.0
chymotrypsin + papain	13.2	13.3

Fig. 3. Time cause of irreversible EAC-inactivation of chloride self exchange in resealed ghosts. Resealed ghosts were reacted in a medium (mM): EAC 30, KCl 165, MES 5, pH 5.75, 38°C. The modification was stopped by diluting with an ice cold alkaline medium, final pH between 9 and 10, 0°C. The cells were washed repeatedly with flux medium, 165 mM KCl, 2 mM, phosphat, pH 7.3, before measurement of the chloride transport.

Table 1. Inactivation of the chloride self-exchange of EAC, TEE-reacted resealed ghosts by papain. The ghosts were reacted in a medium (mM): EAC 30, TEE 60, KCl 165, MES 10, pH 5.75 38°C. Extracellular chymotrypsin treatment was performed with 0.5 mg/ml, pH 7.3, 38°C for 45 min. The ghosts were washed with flux medium, 165 mM KCl, 2 mM phosphat, pH 7.3 before the reaction with extracellular papain, 0.1 mg/ml, 5 mM DTT for 1 hour at 38°C. The reaction was stopped by addition of iodoacetic acid (10 mM, 30 min at 20°C). The cells were then washed repeatedly with flux medium before measurement of the chloride transport at pH 7.3, 0°C.

exchange (13% of control) is the same in unmodified and in EAC-treated papain-digested cells.

The identical flux values might indicate that the modified carboxyl group(s) are removed by the papain treatment. This suspicion was confirmed by SDS-gel-electrophoresis of membranes pretreated with EAC + TEE. After successive treatment with chymotrypsin and papain most of the radioactive TEE located in the 35,000 dalton fragment was removed. Because the activity was not recovered at the localization of the 30,000 dalton cleavage product, it must have been released to the supernatant with the peptide(s) removed by the papain digestion (Bjerrum, Wieth and Borders, unpublished results).

The present results suggest that two arginyl residues together with at least one carboxyl group form essential parts of the exterior anion-binding site. Anions which have access to the binding region are assumed to bind closely to the two arginyl groups, whereas the existence of negative functional carboxyl group(s) suggests a role of salt-bridges in molecular configuration and ion translocation. The results thus agree with our proposed model of an external gate in a zipper mechanism (15) regulating the accessibility of the transport pathway from the outside of the membrane.

INTRACELLULAR ARGINYL MODIFICATION WITH PHENYLGLYOXAL

Positively charged residues contributing to form an internal binding site are likely to belong to arginyl or lysine residues, because the transport function is independent of intracellular pH at least to 10.5 (16). To demonstrate the existence of such groups on the intracellular side of the membrane, we have attempted to modify essential arginyl residues exposed to the intracellular phase. The preliminary results show that selective labelling of intracellularly exposed arginyl residues can be obtained by treating ghosts having an alkaline intracellular pH with PG in a neutral medium. PG permeates the membrane rapidly ($t_{\frac{1}{2}}$ 40 msec) (15) and reacts rapidly only with those arginyl residues which are exposed to the alkaline environment. When compared to the extracellular labelling, PG modification on the inside was less specific. Maximal transport inhibition

114

(40-50%) was accompanied by the binding af 15 x 10^6 molecules per ghost. The activity was distributed in all intracellularly exposed membrane proteins with about 25% localized in band 3. After pepsin digestion of labelled membranes, PG activity was found in a 10,000 dalton membrane-bound peptide derived from capnophorin. This fragment cannot be labelled with PG from the outside of the membrane. The peptide seems to be the same as the one which contains a hydrophobic lysine residue essential for anion transport (17). More work will be needed to establish whether the arginyl residues, which can be modified from the inside, form part of an intracellular anion-binding site. It is not likely that the inhibition of transport function is caused by unspecific modification of arginyl residues in capnophorin. The results presented in Fig. 4 of the preceding article (1) clearly indicate that a significant fraction of the arginyls of band 3 can be modified, without consequences for transport function, when the essential residue is "protected" by DNDS during the treatment (1). It is thus clear that chemical modification of the transport protein is a powerful tool for the identification of functionally essential parts of the transport system.

REFERENCES
1. Wieth, J.O., Bjerrum, P.J. (1983) This Volume.
2. Wieth, J.O., Andersen, O.S., Brahm, J., Bjerrum, P.J. and Borders, C.L. Jr. (1982) Phil. Trans. Roy. Soc. Lond. B 299, 383-399.
3. Wieth, J.O. and P.J. Bjerrum (1982) J. Gen. Physiol. 79, 253-282.
4. Milanick, M.A. and Gunn, R.B. (1982) J. Gen. Physiol. 79, 87-113.
5. Riordan, J.F., McElvany, K.D. and Borders, C.L., Jr. (1977) Science 195, 884-886.
6. Borders, C.L., Jr. and Riordan, J.F. (1975) Biochemistry 14, 4699-4704.
7. Riordan, J.F. (1979). Molec. Cell. Biochem. 26, 71-92.
8. Takahashi, K. (1968). J. Biol. Chem. 243, 6171-6179.
9. Wieth, J.O., Bjerrum, P.J. and Borders, C.L., Jr. (1982) J. Gen. Physiol. 79, 283-312.
10. Bjerrum, P.J., Wieth, J.O. and Borders, C.L. Jr. (1983) J. Gen. Physiol. 81, 453-484.

11. Ramjeesingh, M., A. Gaarn, and A. Rothstein (1980) Biochim. Biophys. Acta 599, 127-139.

12. Passow, H., Fasold, H., Gartner, E.M., Legrum, B., Ruffing, W. and Zaki, L. (1980) Ann. N.Y. Acad. Sci. 341, 361-383.

13. Jennings, M.L. and Passow, H. (1979) Biochim. Biophys. Acta 554, 498-519.

14. Jennings, M.L. and Adams M.F. (1981) Biochemistry 20, 7118-7123.

15. Wieth, J.O., Bjerrum, P.J., Brahm, J. and Andersen, O.S. (1982) Tokai J. Clin. Exp. Med. 7, Suppl. 91-101.

16. Funder, J. and Wieth, J.O. (1976) J. Physiol. Lons. 262, 679-698.

17. Kempf, C., Brock, C., Sigrist, H. Tanner, M.J.A. and Zahler, H. (1981) Biochim. Biophys. Acta 641, 88-98.

ANCHORING AND BIOSYNTHESIS OF THE MAJOR INTRINSIC PROTEIN OF THE SMALL-
INTESTINAL BRUSH BORDER MEMBRANE

GIORGIO SEMENZA, HANS WACKER, PAOLA GHERSA AND JOSEF BRUNNER
Laboratorium für Biochemie der Eidgenössischen Technischen Hochschule,
Universitätstrasse 16, CH 8092 Zürich, Switzerland.

The sucrase-isomaltase complex (SI) is the major (or one of the major) intrinsic protein of the small-intestinal brush border membrane, accounting for
approximately 10% of the total protein[1]. The two subunits (a sucrase-maltase
of 120 kDa and an isomaltase-maltase of 140 kDa each consisting of a single
polypeptide chain) are arranged in the membrane as shown in Fig. 1a, i.e.,
sucrase does not interact with the membrane directly, but via isomaltase; the
latter is anchored to the membrane via a highly hydrophobic segment located
at the N-terminal region and spanning the membrane twice. The anchoring segment has a prevailing helical conformation with a β turn around Pro-35; its
N-terminus is located at the extracellular, luminal side; the β turn is (in
all likelyhood) at the cytosolic side. The N-terminus of sucrase also, as
well as the C-termini of both subunits are located at the luminal, extracellular side[2-8].

Any mechanism of biosynthesis and membrane insertion of SI must explain its
positioning, in particular the peripheral location of S and the mode of anchoring of I. In addition, it should accomodate the analogy - indeed partial
homology - of the two subunits and their common or related biological control
mechanism (for reviews, see refs. 5,6).

In 1978 one of us proposed the "one chain-two active sites precursor hypothesis"[9], which can be formulated as follows:

1. Sucrase and isomaltase have arisen phylogenetically by (partial) duplication of an original isomaltase-maltase gene.
2. This would have led first to a gene coding for a single polypeptide chain
carrying two identical domains, each endowed with enzymatic activity: a "double
isomaltase".

118

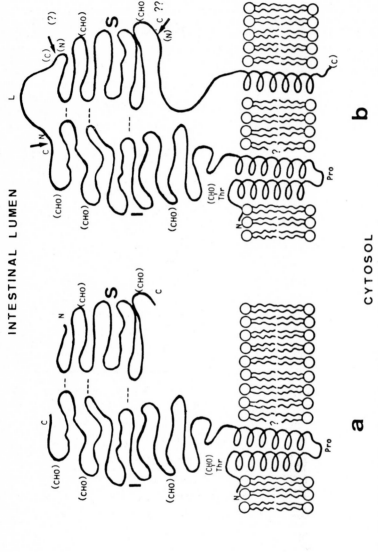

Fig. 1: Positioning of the SI-complex (a) and of pro-SI (b) in the small intestinal brush border membrane. CHO: carbohydrate chains. N and C terminals, respectively of pro-SI and/or of final SI. ↓:site of proteolytic attack. From ref. 6.

3. Subsequent mutation would have transformed one of these domains from an isomaltase-maltase into a sucrase-maltase with the appearance of an (active ?) single-chain-two-active sites precursor ("pro-sucrase-isomaltase").

4. This single-chain-pro-sucrase-isomaltase would be synthesized, glycosylated and inserted in the membrane of the endoplasmic reticulum and transferred, along with other plasma membrane proteins, to the brush border membrane.

5. Post-translational proteolytic modification of this single chain (perhaps by one or more pancreatic proteases), would lead to the two subunits of the "final" SI complex; they would still remain associated via interactions formed during the folding of single-chain pro-sucrase-isomaltase (Fig. 1).

The existence of pro-SI as the immediate, fully enzymatically active precursor of "final" SI is now well established. In fact [10-12] pulse chase experiments show a high molecular weight band appearing in the Golgi membranes first and then in the brush borders; elastase treatment splits it into bands of mobilities close to those of the "final" subunits S and I. In small intestines not exposed to pancreatic juice (because of disconnection of the pancreas from the duodenum in hogs [13], or in rat fetal small-intestinal transplants developing under the skin of adult animals,[14] sucrase and isomaltase activities are associated with a very large polypeptide chain of approximately 260-270 kDa carrying two active sites. Limited proteolysis of this polypeptide chain by pancreatic elastase leads to two subunits similar in size to those of "final" SI.

In the polypeptide chain of pro-SI the N-terminal portion corresponds to I and the C-terminal portion to S (Fig. 1b) as shown by the identity (not mere homology !) of the N-terminal sequences of pro-SI and the corresponding "final" I within each species examined (Fig. 2) [14,15]. No hydrophobic sequence seems to occur in the (hypothetical) loop connecting the I and S portions of pro-SI . Also, it is possible that the original pro-SI has a C-terminus at the cytosolic side which is split from the main protein mass by extracellular proteolysis; except for this, however, the identical N-terminal sequences in pro-SI and I, plus the absence of detectable changes in the enzymatic and in immunological properties in the pro-SI→SI processing strongly indicate that the mode of association of pro-SI with the membrane and its secondary and tertiary structures are similar (at least in part) to those of "final" SI.

```
HOG ProSI   Ala-Arg-Lys-Ser-Phe-Ser-Gly-Leu-Glu-Ile-X-Leu-Ile-Val-Leu-Phe-Ala-Ile-Val-
                        Thr                    CHO

HOG I       Ala-Arg-Lys-Phe-Ser-Gly-Leu-Glu-Ile-X-Leu-Ile-Val-Leu-Phe-Ala-Ile-Val-Leu-Ser-Ile-Ala-Ile-Ala-Leu-Val-Val-Val-X-Ala-Ser-Lys-X-Pro-Ala-Val·
                                           CHO

RAT ProSI   Ala-Lys-Lys-Lys-Phe-Arg-Ala-Leu-Glu-Ile-X-Leu-Ile-Val-Leu-Phe-Ile-Ile-
                                              CHO

RAT I       Ala-Lys-Lys-Lys-Phe-Ser-Ala-Leu-Glu-Ile-X-Leu-Ile-Val-Leu-Phe-Ile-Ile-Val-Leu-Ala-Ile-Ala-Leu-Val-Leu-Val-
                                           CHO

RABBIT I    Ala-Lys-Arg-Lys-Phe-Ser-Gly-Leu-Glu-Ile-Thr-Leu-Ile-Val-Leu-Phe-Val-Ile-Val-Phe-Ile-Ala-Ile-Ala-Leu-Ile-Ala-Leu-Ala-X-X-X-Pro-Ala-Val·
                                           CHO

HOG S       Ile-Lys-Leu-Pro-Ser-Asp-Pro-Ile-Pro-Thr-Leu-Arg-Val-Glu-Val-Lys-Tyr-His-Lys-Asp-Tyr-Val-Leu-Glu-Phe-X-Arg-Tyr-Asp-Pro-Glu-Arg-
                                                      Met      Met Thr      Thr           Met

RAT S       Ile-Lys-Leu-Pro-Ser-Asn-Pro-Ile-Arg-Thr-Leu-Arg-Val-Glu-Val-Thr-Tyr-X-Thr-Asn-Arg-Val-Leu-Glu-Phe-Arg-Ile-Tyr-Arg-Ala-Glu-X-X-Gly-
                                                                                       Gln

RABBIT S    Ile-Asn-Leu-Pro-Ser-Glu-Pro-Glu-
                                  Thr    Thr

            1         5         10        15        20        25        30        35
```

Fig. 2. Partial N-terminal sequences of pig pro-sucrase-isomaltase (pro-SI) and of the isomaltase (I) and sucrase (S) polypeptide chains from "final" SI. Hydrophobic segment beginning at residue 12 in the isomaltase and in the pro-SI. From ref. 15 and references quoted therein.

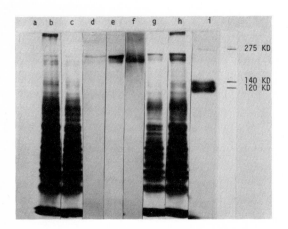

Fig. 3. In vitro synthesis of pro-SI (or a precursor thereof). Fluorograph of ^{35}S-labelled polypeptides synthetized in a reticulocyte lysate pretreated with nuclease in the presence or absence of dog pancreas microsomes, in response to total RNA extracted from rabbit intestinal mucosa. After translation, the synthetized polypeptides were precipitated with anti-SI-antiserum (for details, see text).The SDS-polyacrylamide gels (6%) were exposed for fluorography for 4 days (a-d,g,h) or 2 days (e,f):(a)control (no RNA, no microsomes);b) translation mixture without microsomes; (c)as in (b)centrifugation after incubation (100 000 x g supernatant); (d)immunoprecipitate from (c); (e) mixture of the following 3 immunoprecipitates:(d)+immunoprecipitate from the 100 000 x g pellet (see (c))+immunoprecipitate from (g)(see below); (f) translation mixture with microsomes, spun at 100 000 x g after the incubation (immunoprecipitate of the pellet); (g) translation mixture with microsomes, spun at 100 000 x g after the incubation (supernatant, no immunoprecipitation); (h) translation mixture with microsomes; ^{125}I-SI (120-kDa, sucrase subunit; 140-kDa, isomaltase subunit).

In vitro cell-free translation experiments have provided the final, direct evidence for SI being synthesized as a high molecular weight pro-SI: the translation products which precipitate with polyclonal antibodies directed against "final" SI have apparent molecular weights in the 220-240 kDa range (Fig. 3) and yield similar peptide mappings to those yielded by ^{125}I-labeled "final" SI (Ref. 15).

The minimum sequence of events in the biosynthesis of pro-SI is thus that indicated in Fig. 4. The synthesis of a pre-piece and the segregation at the trans side of the e.r. membrane, which we have indicated in the figure, are suggested by still unpublished experiments from our laboratory and are in keeping with current ideas on biosynthesis and insertion of intrinsic membrane proteins.[14,15]

By pulse-chase experiments of explants in organ cultures Danielsen[12] could follow the sites of biosynthesis and glycosylation of pro-sucrase isomaltase. First of all, pro-SI was always found to be associated with membranes, and at no time to occur in the cytosol, thus confirming Hauri et al.'s previous observations[10]. In all likelyhood, therefore, pro-SI is synthesized on membrane-bound polyribosomes. Glycosylation must take place during or immediately after completion of the polypeptide chain, as shown by the early labeled form of pro-SI being sensitive to endo-β-N-acetylglucosaminidase H. The next labeled form, which has a somewhat larger apparent molecular size, is not susceptible to this enzyme which indicates reglycosylation, in agreement with the current views on protein glycosylation (see e.g., ref. 17). Shortly thereafter, i.e., 60-90 min after the beginning of biosynthesis, pro-SI reaches the microvillus membrane.

The main events in the biosynthesis, insertion, glycosylation and processing of pro-SI→SI seem, therefore, to be well established. They provide a logical explanation of the positioning of "final" SI (Fig. 1), and also of the possible genetic mechanisms underlying human sucrose-isomaltose malabsorption (for a review, see ref. 6). Finally, the partial gene duplication, which is an important element in the "one-chain, two active site precursor" hypothesis, as outlined above, provides an indication pertinent to the phylogeny of this dimeric enzyme. Recently, a high molecular weight "double isomaltase" has been

122

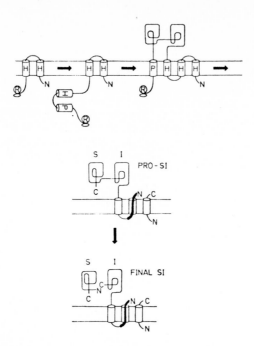

Fig. 4. Suggested minimum mechanism of folding, membrane insertion
and processing (by signalase and pancreatic protease(s) of an
as yet hypothetical pre-pro-sucrase-isomaltase. Note: the proposed scheme puts emphasis on the folding and insertion only,
without indicating any possible interaction with recognition
systems. R, ribosome; H, hydrophobic; P, polar; N, N-terminus;
C, C-terminus; S, sucrase; I, isomaltase. Segments indicated
by the bold line are those the sequence of which is known in
(fig. 2).From ref. 15.

isolated and characterized from the small intestine of puppy Californian
sea lion (Zalophus californianus) (Wacker et al., in preparation). Although
this species is not on the direct phylogenetic linkage, its "double isomaltase" mimicks the properties expected in the ancestral enzyme.

ACKNOWLEDGEMENTS
    The financial support of the SNSF, Berne, is greatfully acknowledged.

123

REFERENCES

1. Kessler, M., Acuto, O., Storelli, C., Murer, H., Müller, M. and Semenza, G. (1978) Biochim. Biophys. Acta 506, 136

2. Brunner, J., Hauser, H., Braun, H., Wilson, K.J., Wacker, H., O'Neill, B., and Semenza, G. (1979) J. Biol. Chem. 254, 1821

3. Spiess, M., Brunner, J., and Semenza, G. (1982) J. Biol. Chem. 257, 2370

4. Bürgi, R., Brunner, J. and Semenza, G. (1983) J. Biol. Chem. in press

5. Semenza, G. (1981) in: Randle, P.J., Steiner, D.F. and Whelan, W.J., Carbohydrate Metabolism and Its Disorders, Academic Press, London, pp. 425-479

6. Semenza, G. (1981) Clin. Gastroent. 10, 691

7. Brunner, J., Wacker, H., and Semenza, G. (1983) Methods in Enzymology, in press.

8. Hauser, H. and Semenza, G. (1983) CRC Critical Rev. in Biochem. in press

9. Semenza, G. (1978) In: Rapoport, S. and Schewe, T. (eds.) Proceedings of the 12th FEBS Meeting, Dresden, 1978, Pergamon Press, Oxford, pp. 21-28

10. Hauri, H.P., Quaroni, A. and Isselbacher, K. (1979) Proc. Natl. Acad. Sci. USA 76, 5183

11. Hauri, H.P., Quaroni, A. and Isselbacher, K. (1980) Proc. Natl. Acad. Sci. USA 77, 6629

12. Danielsen, E.M. (1982) Biochem. J. 204, 639

13. Sjöström, H.,Norén, O., Christiansen, L., Wacker, H. and Semenza, G. (1980) J. Biol. Chem. 255, 11332

14. Hauri, H.P., Wacker, H., Rickli, E.E., Bigler-Meier, B., Quaroni, A. and Semenza, G. (1982) J. Biol. Chem. 257, 4522

15. Sjöström, H., Norén, O., Christiansen, L., Wacker, H., Spiess, M., Bigler-Meier, B., Rickli, E.E. and Semenza, G. (1982) FEBS Letters 148, 321

16. Wacker, H., Jaussi, R., Sonderegger, P., Dokow, M., Ghersa, P., Hauri, H.P., Christen, Ph. and Semenza, G. (1981) FEBS Letters 136, 329

17. Hanover, J.A. and Lennarz, W.J. (1981) Arch. Biochem. Biophys. 211:1

© 1983 Elsevier Science Publishers B.V.
Structure and Function of Membrane Proteins,
E. Quagliariello and F. Palmieri editors.

THE MEMBRANE LOCATION OF THE B800-850 LIGHT-HARVESTING PIGMENT-PROTEIN
COMPLEX FROM *RHODOPSEUDOMONAS SPHAEROIDES*

RICHARD J. COGDELL[1], JANE VALENTINE[1] and J. GORDON LINDSAY[2]
Departments of Botany[1] and Biochemistry[2], The University, Glasgow G12 8QQ, U.K.

INTRODUCTION

The major light-absorbing pigments in bacterial photosynthesis are the
bacteriochlorophylls (Bchl) and the carotenoids (Car).   These pigments are
not 'free' within the photosynthetic membrane but are rather bound to
hydrophobic membrane proteins, forming two major classes of pigment-protein
complexes (the photochemical reaction centres and the light-harvesting or
antenna complexes) (1).

The B800-850 antenna complex from *Rps. sphaeroides* is one of the best
characterised of the bacterial antenna complexes.   This complex is a hydrophobic, integral membrane protein and probably exists *in vivo* as aggregates of
a well-defined 'minimal unit'.   This 'minimal unit' consists of three
molecules of Bchl*a* and one molecule of Car non-covalently bound to two low
molecular weight (5-7 KD) polypeptides (1).   Two of these Bchl*a* molecules
give rise to the 850 nm absorption band, while the third is responsible for
the 800 nm absorption (1).

The aim of this paper was to investigate the membrane location of this
antenna complex using antibodies prepared to the isolated, purified B800-850.

MATERIALS AND METHODS

Cells of *Rps. sphaeroides* strain 2.4.1 were grown anaerobically in the light
with succinate as the sole carbon source.   Following passage of the cells
through a French pressure cell at 10 tons/sq. inch, chromatophores were
isolated by differential centrifugation (2).   The B800-850 antenna complex
was prepared from the chromatophores as described in reference 3.

The purity of the B800-850 antenna complex was checked spectrophotometrically
and by electrophoresis on SDS polyacrylamide gradient gels.

Rabbit antibodies were raised to the isolated B800-850 complex by standard
procedures.

Spheroplasts were prepared by a modification of the procedure described by
Karunairaknam *et al.* (4).   Early log phase cells were treated with lysozyme
and EDTA in Tris buffer.   The spheroplasts were then collected by a low speed
centrifugation, and their integrity checked under the light microscope and by

the release of cytoplasmic malate dehydrogenase. Compared to the release of malate dehydrogenase when the cells are completely disrupted by passage through the French Press the spheroplasts are usually 95% intact.

RESULTS AND DISCUSSION

As a check on our antiserum we initially tested it against purified samples of the B800-850 antenna complex and photochemical reaction centres (also prepared from *Rps. sphaeroides* 2.4.1). Only the antenna sample showed agglutination with the anti-B800-850.

Spheroplasts have 'right-side-out' membrane polarity, while chromatophores are uniformly 'inside-out' (5). In principle, therefore, it is possible to ask whether the B800-850 antenna complex is a transmembrane protein, by challenging both spheroplasts and chromatophores with the antisera prepared against the B800-850 complex. Fig. 1 shows that both spheroplasts and chromatophores are agglutinated by the anti-B800-850 serum. Neither are affected by the control serum.

This result shows that there are antigenic sites recognised by the antiserum to the B800-850 antenna complex present on both sides of the photosynthetic membrane of *Rps. sphaeroides*. In other words the B800-850 complex is present on both sides of the membrane. However, this data do not distinguish between a transmembrane arrangement for the B800-850 complex and the case where the complex is present on both membrane surfaces but does not actually cross the membrane. With monospecific rather than monoclonal antibodies this is a difficult problem to solve. But we are able to obtain some additional data which suggests that the transmembrane option should be preferred (Fig. 2).

We have prepared some spheroplasts and added them in a ten fold excess to a portion of anti-B800-850. The spheroplasts were then removed, after 4 hours incubation, by centrifugation. We then challenged chromatophores with anti-B800-850, which either had or had not been treated with spheroplasts. We reasoned that if the B800-850 complex was a transmembrane protein, then the antigenic sites on each side would be different and that the titre for agglutination of the chromatophores should not be affected by the pre-treatment with spheroplasts. On the other hand, if the antenna complex was not transmembrane but rather only present on both sides then the pre-treatment of the antiserum with spheroplasts would cause a large inhibition of the agglutination reaction with the chromatophores, since similar antigenic sites would be available at both membrane surfaces. It is clear from Fig. 2 that the pretreatment only reduced the titre to chromatophores a little.

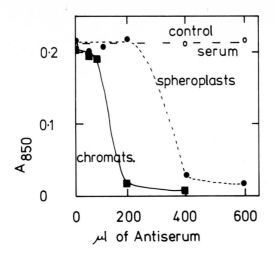

Fig. 1: Titration of chromatophores and spheroplasts from Rps. sphaeroides with antiserum to the isolated B800-850 antenna complex and control serum.

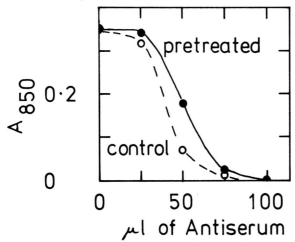

Fig. 2: The effect of spheroplast pre-treatment on the ability of the anti-B800-850 to agglutinate chromatophores.

(Figs. 1 and 2)   The membranes were incubated with the antiserum. Then the precipitated material was removed by centrifugation and the amounts of residual membranes left in solution was monitored spectro-photometrically.

128

Since about 5% of the spheroplasts were ruptured (as judged by the release of malate dehydrogenase) this small shift is easily accounted for.   We therefore favour a transmembrane location for this antenna complex.

Recently the amino acid sequences of a number of the bacterial light-harvesting polypeptides have been determined (6).   These sequences show one striking similarity.   They are polar at each end and very hydrophobic in the middle.   Since they are so small (of the order of 50-60 AA) it has also been suggested from these studies that the light-harvesting polypeptides lie across the membrane, anchored by charged residues at either terminus (6).

ACKNOWLEDGEMENTS

This research was supported by a grant from the SERC.

REFERENCES

1.   Cogdell, R.J. and Thornber, J.P. (1980) FEBS Lett., 122, 1-8.

2.   Jackson, J.B., Crofts, A.R. and Von Stedingk, L.V. (1968) Eur. J. Biochem. 6, 41-54.

3.   Cogdell, R.J. and Crofts, A.R. (1978) Biochim. Biophys. Acta, 502, 409-416.

4.   Karunairatnam, M.C., Spizizen, J. and Gest, H. (1958) Biochim. Biophys. Acta, 29, 649-650.

5.   Takemoto, J. and Bachmann, R.C. (1979) Arch. Biochem. Biophys., 195, 526-534.

6.   Theiler, R., Brunisholz, R., Frank, G., Suter, F. and Zuber, H. (1982) Abs. of the IV Inter. Symposium on Photosynthetic Prokaryotes, Bombannes, France, C40.

PRIMARY STRUCTURE OF TRANSPORT PROTEINS

CYTOCHROME OXIDASE: THE PRIMARY STRUCTURE OF ELECTRON AND
PROTON TRANSLOCATING SUBUNITS AND THEIR HINTS AT MECHANISMS

GERHARD BUSE, GUY C.M. STEFFENS AND LOTHAR MEINECKE
Abteilung Physiologische Chemie, RWTH Aachen, Forckenbeckstraße,
D-5100 Aachen/Germany

INTRODUCTION

The cumulating amount of data on the primary structure of
cytochrome c oxidase [EC 1.9.3.1] obtained either from protein
or DNA sequences of phylogenatically distant organisms allows
first comparative analyses, which lead to conclusions concerning
origin, structural arrangement and functional sites of this
enzyme. The analyses have mainly concerned oxidases from mammals
(man (1) and bovine (2,3,4)), higher plants (maize (5)) and
fungi (yeast (6,7), Neurospora (8)) and have recently been
extended to bacteria such as Paracoccus denitrificans (9).
Fortunately thus they include data from the extreme ends of the
evolutionary scale, which very much substantiates comparative
conclusions.

RESULTS

The mammalian type oxidase. The most complex cytochrome c
oxidase has so far been found in mammals. From the bovine enzyme
there exist rough data on the three dimensional structure (10)
and nearly complete data on the primary structure.*) The protein
chemical analysis of this enzyme revealed the existence of 12
different polypeptides, 3 originating from mitochondrial and 9
from nuclear genes (Table 1). If this enzyme is purified to
10 nMol heme a/ mg protein these components are present in 1:1
stoichiometric amounts with the one exception of VIIIb(Ile),
which has the stoichiometry 2. This stoichiometry has recently
been established by direct Edman degradation of the entire purified oxidase, quantitating the liberated PTH-amino acids
through several cycles. Additionally hydrazinolysis was used in
those cases were the N-terminus is blocked, especially with
VII(Ac-Ala).

---

*For all the details of the protein chemical analysis see the se-
ries of papers in Hoppe Seyler's Z. Physiol. Chem. Last issue:
Biewald, R. and Buse, G., in Vol. 363, 1141-1153 (1982).

From the complete amino acid sequences of all the cytoplasmic chains it is seen that these contribute 6 hydrophobic, probably helical, stretches of about 20 residues length (IV(Ala), IVb(Ala), VIIIa(Ser), 2 VIIIb(Ile), VIIIc(Phe)) to the two membrane domains of the complex while the three mitochrondrial subunits each have two hydrophobic segments close to their N-termini. The rest of the subunit II primary structure is more hydrophilic. Subunits I and III are generally hydrophobic, which prevents a direct recognition of distinct membrane penetrating sections, but characterizes them as 2 hydrophobic core regions of the enzyme complex, which may be brought together with the membrane domains I and II.

TABLE 1

POLYPEPTIDE COMPOSITION OF BEEF HEART CYTOCHROME C OXIDASE

POLYPEPTIDE	M_R	RESIDUES	SYNTHESIS	STOICHIOMETRY	HYDROPHOBIC-DOMAINS	N-TERMINAL SEQUENCES
I	56993^A	514	MIT.	1	> 2, CORE	FORMYL-MET-PHE-ILE-ASN-
II	26049	227	MIT.	1	2	FORMYL-MET-ALA-TYR-PRO-
III	29918^A	261	MIT.	1	> 2, CORE	(MET→)THR-HIS-GLN-
IV	17153	147	CYT.	1	1	ALA-HIS-GLY-SER-
V	12436	109	CYT.	1	NONE	SER-HIS-GLY-SER-
VIA	10670	98	CYT.	1	NONE	ALA-SER-GLY-GLY-
VIB	9419	84	CYT.	1	NONE	ALA-SER-ALA-ALA-
VIC	8480	73	CYT.	1	1	SER-THR-ALA-LEU-
VII	10068	85	CYT.	1	NONE	ACETYL-ALA-GLU-ASP-ILE-
VIIIA	5441	47	CYT.	1	1	SER-HIS-TYR-GLU-
VIIIB	4962	46	CYT.	2	1	ILE-THR-ALA-LYS-
VIIIC	6244	56	CYT.	1	1	PHE-GLU-ASN-ARG-

[A]FROM MtDNA SEQUENCE (ANDERSON ET AL. 1981)

In this connection it may be useful to consider, that purification of the enzyme beyond 10 nMol heme a/ mg protein which appears to be possible either by strong detergent action (11) or by mild alkaline treatment (12) leads to partial or complete loss of some polypeptides, among them subunit III and those cytoplasmic components which sometimes have been considered impurities (VIc(Ser), VII(Ac-Ala)). The question then arises whether this structurally can be related to the loss of one membrane domain and functionally plays a role in arbitrary results concerning the proton pumping activity of the enzyme (13).

As a final result the complete data on the primary structures of all 12 protein components and their stoichiometry allows to sum up amino acid compositions of the individual components and finally of the entire enzyme (functional monomer). Table 2 gives these data. Altogether then the complex consists of 1793 amino acids with Mr close to 200.000.

TABLE 2

AMINO ACID COMPOSITION OF BEEF HEART CYTOCHROME C OXIDASE (COMPLEX IV)

	I	II	III	IV	V	VIa	VIb	VIc	VII	VIIIa	VIIIb	VIIIc		%
Asp	16	10	4	8	9	4	2	3	6	0	1	2	66	3.68
Asn	19	5	6	4	4	4	4	1	5	2	1	3	59	3.29
Thr	37	17	24	5	5	7	7	2	5	1	4	4	122	6.80
Ser	30	22	19	12	4	5	4	4	5	3	5	2	120	6.69
Glu	9	15	8	13	1	8	2	4	4	3	1	4	73	4.07
Gln	6	6	7	3	11	4	0	1	5	1	1	1	47	2.62
Pro	28	13	12	6	7	7	7	1	4	3	4	1	97	5.41
Gly	47	8	21	5	6	9	9	4	4	4	2	6	127	7.08
Ala	40	8	14	13	8	8	8	11	6[++]	3	6	3	134	7.47
Val	38	12	16	11	8	6	2	3	4	2	2	3	109	6.08
Met	34[+]	16[+]	11	4	1	1	0	4	1	2	0	1	75	4.18
Ile	38	18	14	5	7	5	2	2	4	2	2	1	102	5.69
Leu	59	34	31	11	11	7	8	5	2	5	6	9	194	10.82
Tyr	19	11	11	7	4	2	2	4	4	1	2	3	72	4.01
Phe	42	6	24	7	3	1	6	8	4	6	2	2	113	6.30
Lys	9	6	3	18	7	6	3	8	6	4	5	5	85	4.74
His	17	7	17	4	3	4	7	1	1	2	1	2	67	3.74
Arg	8	6	5	5	7	4	7	7	7	2	0	2	60	3.34
Cys	1	2	2	0	1	4	1	0	4	0	0	2	17	0.95
Trp	17	5	12	6	2	2	3	0	4	1	1	0	54	3.01
	514	227	261	147	109	98	84	73	85	47	46	56	1793	99.97
STOICHIOMETRY	1	1	1	1	1	1	1	1	1	1	2	1		

[+] INCLUDING 1 FORMYL-MET
[++] INCLUDING 1 ACETYL-ALA

MR (PROTEIN): 202900 DA

The complete amino acid composition allows for a comparison with amino acid analysis data published previously by different authors (14-16) and in some case gives excellent agreement showing that their preparations are in this respect identical to ours. The

134

TABLE 3

PARTIAL SEQUENCES OF CNBR-FRAGMENTS OF SUBUNITS I AND II OF <u>PARACOCCUS</u> CYTOCHROME OXIDASE.
ALIGNMENT WITH CORRESPONDING SEQUENCES OF MAIZE; YEAST AND BEEF HEART OXIDASE

Subunit I

Beef Heart (4)   277 -Met-Met-Ser-Ile-Gly-Phe-Leu-Gly-Phe-Ile-Val-Trp-Ala-His-

Paracoccus   -Met-Ala-Ala-Ile-Gly-Ile-Leu-Gly-Phe-Val-Val-Trp-Ala-His-

Yeast (7)   -Met-Ala-Ser-Ile-Gly-Leu-Leu-Gly-Phe-Leu-Val-Trp-Ser-His-

Subunit II

Beef Heart (3)   207 -Met-Pro-Ile-Val-Leu-Glu-Leu-Val-Pro-Leu-Lys-Tyr-Phe-Glu-Lys-Trp-

Paracoccus   -Met-Pro-Ile-Val-Lys-Ala-Val-Ser-Glu-Glu-Lys-Tyr-Glu-Ala-Trp-

Yeast (6)   -Met-Pro-Ile-Lys-Ile-Glu-Ala-Val-Ser-Leu-Pro-Lys-Phe-Leu-Trp-

Maize (5)   -Thr-Pro-Ile-Val-Val-Glu-Ala-Val-Thr-Leu-Lys-Asp-Tyr-Ala-Asp-Trp-

tryptophan and tyrosine contents of the individual polypeptides
are in agreement with their relative UV absorbances in gelchro-
matographic separations and again confirms the stoichiometry
given above. The monomer of the oxidase contains 17 cysteine residues, some of them being involved in disulfide formation. The
exact sequence position of these disulfide bridges and the final
proof for the protein sequences of subunits I and III remain however to be completed.

A bacterial cytochrome oxidase. Going from the complex mammalian type enzyme to the 2 subunit, 2 copper, 2 heme a, proton
pumping cytochrome oxidase [EC 1.9.3.1] of Paracoccus denitrificans (17,18) the simple structure of this latter sheds light on
the complex findings with the eucaryotic enzyme. Sequence analyses
with the protein material of both subunits (kindly provided by
B. Ludwig and G. Schatz, Basel) of this bacterial enzyme show
them to be closely related to subunits I and II of the eucaryotic
oxidases. Some 100 amino acid sequences of both subunits have so
far been obtained and give a clear cut evidence for the homology
of this bacterial and the mitochondrial oxidases. Some examples
are given from CNBr fragments of both subunits. These findings
suggest a number of conclusions which will be discussed in the
following sections. (Table 3)

DISCUSSION

General structure of cytochrome c oxidase. From the above data
it is evident, that the eucaryotic cytochrome c oxidase in principal is made up from the two mitochondrial polypeptides I and II
which bind all 4 metal centers and also contain the coupling site
for e^- and H^+ translocation. At least these components must be
considered as true subunits of this enzyme. Possibly they are
anchored together in one of the membrane domains of the complex
and form a big part of the C-domain. Future preparative progress
at least principally may aim at the purification of this two
subunit complex also in the case of eucaryotic oxidases.

Special functions of subunits I and II. The folding pattern of
subunit II (19) shows that the functional core of this protein
is on the cytoplasmic side of the inner membrane. It also has been
identified as the high affinity cytochrome c binding site (20).
In preliminary experiments heme a and copper were found at this

subunit as well as on subunit I (21). The distant relationship
of tis primary structure to blue copper proteins (3), the fact
that only this subunit by sequence comparison is found to possess
invariant cysteine residues, which are involved in formation of
the copper A center (22) point to subunit II as the heme a/copper
A electron accepting site of the oxidase. Correspondingly subunit
I is the candidate for the O_2 activating heme a_3/copper B center.
In complex formation of this bi metal center only N-ligation is
involved (22) in agreement with the absence of invariant cysteine
residues in this subunit. The localization of subunit II on the
C-side of the membrane, as well as the distance between the high
and low potential metal centers (10-25 Å) (23) make on the one
hand an arrangement of both on the C-domain of the membrane likely
with the a_3/Cu_B center "open" to the m side. On the other hand
they ask for protein structures, which might conduct the electrons
between the centers. Indeed the comparative analysis of the
subunit II sequence points to such a possibility: a cluster of
tryptophan and tyrosine residues is found invariant in the structure of this subunit from bacteria to man.

TABLE 4
CYTOCHROME C OXIDASES, SUBUNIT II; TRP,TYR-CLUSTER

				105							
Beef heart	-Gly-His-Gln-Trp-Tyr-Trp-Ser-Tyr-Glu-Tyr-Thr-Asp-
Maize	-Gly-His-Gln-Trp-Tyr-Trp-Ser-Tyr-Glu-Tyr-Ser-Asp-
Yeast	-Gly-Tyr-Gln-Trp-Tyr-Trp-Lys-Tyr-Glu-Tyr-Ser-Asn-
Paracoccus	-Gly-His-Gln-Trp-Tyr-Trp-Ser-Tyr-Glu-Tyr-Ala-Asn-

Model building shows that these aromatic residues can easily be
arranged to form a stacking π-bonding system of 3 to 6 members
spanning distances from 10 to 20 Å. It is tempting to speculate
that the integration of tyrosine (105) in this structure provides
the possibility of donating or accepting H^+ coupled to its
benzoid/quinoid structure change during e^- transfer. The aromatic
cluster probably evolved by fusion of two genetic domains.
    What is the function of subunit III? Though the Paracoccus
oxidase works as a proton pump (17) apparantly without a protein
like the mitochondrial subunit III, from its close relationship
to mitochondria it is hard to accept that the corresponding gene

should be absent in this bacterium. It therefore may be the special
needs of the structural arrangements in narrow mitochondrial
christae which arrange subunit III to the oxidase complex making
proton pumping more efficient when the coupled H^+ -via the coupling device in the C-domain- is channeled from the mitochondrial
matrix to the F_o segment of the ATP synthase.

Origin of mitochondria. The close homology between the oxidase
subunits from Paracoccus denitrificans and the eucaryotic species
leaves little doubt about the origin of mitochondria from early
aerobic bacteria. Calculation of the relative distances between
the bacterium and the eucaryotic groups even suggests that the
rise of the symbiosis was a determinating event at the start of
the phylogenetic radiation of the eucaryotes some 1.5 billion
years ago. It is thus not surprising that the nuclear control
over the symbiosis is different in the eucaryotic phyla and needs
a different set of nuclear genes in each case. Further differentiation occurred along with tissue specification in verte-
brates (24).

ACKNOWLEDGEMENT

This work was supported by the Sonderforschungsbereich 160
of the Deutsche Forschungsgemeinschaft.

REFERENCES

1. Anderson, S., Bankier, A.T., Barrell, B.G., de Bruijn, M.H.L.,
    Coulson, A.R., Drouin, J., Eperon, I.C., Nierlich, D.P., Roe,
    B.A., Sanger, F., Schreier, P.H., Smith, A.J.H., Staden, R.
    and Young, I.G. (1981) Nature 290, 457-465.

2. Buse, G., Steffens, G.J., Steffens, G.C.M., Sacher, R. and
    Erdweg, M. (1982) in: Chien Ho (Ed.), Electron Transport
    and Oxygen Utilization, Elsevier North Holland, Inc.

3. Steffens, G.J. and Buse, G. (1979) Hoppe-Seyler's Z. Physiol.
    Chem. 360, 613-619.

4. Anderson, S., de Bruijn, M.H.L., Coulson, A.R., Eperon, I.C.,
    Sanger, F. and Young, I.G. (1982) J. Mol. Biol. 156, 683-717.

5. Fox, T.D. and Leaver, C.J. (1981) Cell 26, 315-323.

6. Coruzzi, G. and Tzagoloff, A. (1979) J. Biol. Chem. 254,
    9324-9330.

7. Bonitz. S.G., Coruzzi, G., Thalenfeld, B.E., Tzagoloff, A. and
    Macino, G. (1980) J. Biol. Chem. 255, 11927-11941.

8. Burger, G., Scriven, C., Machleidt, W. and Werner, S. (1982)
    The EMBO Journal 1, 1385-1391.

138

9. Steffens, G.C.M., Buse, G., Oppliger, H. and Ludwig, B. Manuscript in preparation.

10. Fuller, S.D., Capaldi, R.A. and Henderson, R. (1979) J. Mol. Biol. _134_, 305-327.

11. Georgvich, G., Darley-Usmar, V.M., Malatesta, F. and Capaldi, R.A. (1983) Biochemistry _22_, 1318-1322.

12. Saraste, M., Pentillä, T. and Wikström, M. (1981) Eur. J. Biochem. _115_, 261-268.

13. Wikström, M. and Krab, K. (1979) Biochim. Biophys. Acta _549_, 177-222.

14. Kuboyama, M., Yong, F.C. and King, T.E. (1972) J. Biol. Chem. _247_, 6375-6383.

15. Verheul, F.E.A.M., Boonman, J.C.P., Draijer, J.W., Muijsers, A.O., Borden, D., Tarr, G.E. and Margoliash, E. (1979) Biochim. Biophys. Acta _548_, 397-416.

16. Matsubara, H., Orii, Y. and Okunuki, K. (1965) Biochim. Biophys. Acta _97_, 61-67.

17. Ludwig, B. and Schatz, G. (1980) Proc. Nat. Acad. Sc. _77_, 196-200.

18. Solioz, M., Carafoli, E. and Ludwig, B. (1982) J. Biol. Chem. _257_, 1579-1582.

19. Bisson, R., Steffens, G.C.M. and Buse, G. (1982) J. Biol. Chem. _257_, 6716-6720.

20. Bisson, R., Azzi, A., Gutweniger, H., Colonna, R., Montecucco, C. and Zanotti, K. (1978) J. Biol. Chem. _253_, 1874-1880.

21. Mason, H.S. in: Biological Oxidations 34. Mosbacher Kolloquium 1983, Springer Verlag, in press.

22. Malmström, B. in: Biological Oxidations 34. Mosbacher Kolloquium 1983, Springer Verlag, in press.

23. Dockter, M.E., Steinemann, A. and Schatz, G. (1978) J. Biol. Chem. _253_, 311-317.

24. Kadenbach, B., Hartmann, R., Glanville, R. and Buse, G. (1982) FEBS Letters _138_, 236-238.

ON THE LOCATION OF PROSTHETIC GROUPS IN CYTOCHROME $\underline{aa_3}$ AND $\underline{bc_1}$

MATTI SARASTE AND MÅRTEN WIKSTRÖM
Department of Medical Chemistry, University of Helsinki,
Siltavuorenpenger 10, SF-00170 Helsinki 17, Finland

INTRODUCTION

Mitochondrial DNA contains the genes for apocytochrome $\underline{b}$ and for cytochrome oxidase subunits I, II and III. For each the nucleotide sequences have
been determined from several species. The predicted amino acid sequences are
analysed here in three ways: (1) they are aligned to find evolutionarily
invariant spots, (2) hydropathic profiles are used to predict the folding of
polypeptides in the membrane, and (3) possible haem-binding sites are considered on the basis of invariancy, homologies to other haemoproteins and
the known spectroscopic properties. These considerations lead to a probable
if not yet unambiguous localisation of the two haems in cytochrome oxidase
and to two alternative models for haem-binding sites in cytochrome $\underline{b}$.

MATERIALS AND METHODS

Five sequences of cytochrome oxidase subunit I (human, bovine, mouse,
Saccharomyces and Neurospora) and six sequences of subunit II (human, bovine,
mouse, rat, Saccharomyces and corn) were used; these are compiled and aligned
in ref. 1. Five apocytochrome $\underline{b}$ sequences of the human, bovine, mouse, Saccharomyces and Aspergillus proteins are taken from refs 2-6, respectively.
The numbering of all sequences is based on their alignments. Hydropathic
profiles were calculated using the index of Kyte and Doolittle (7).

RESULTS AND DISCUSSION

1. Cytochrome $\underline{b}$. Hydropathic profile of apocytochrome $\underline{b}$ suggests that
this polypeptide traverses the membrane bilayer nine times: nine long hydrophobic segments alternate with more hydrophilic short sequences as shown in
Fig. 1A. Our schematic interpretation of the hydroplot (Fig. 1B) illustrates
the membrane-traversing segments as α-helices.

Much of the earlier data proposed that a monomeric $\underline{bc_1}$ complex contains
two cytochrome $\underline{b}$ subunits each binding one haem. However, re-evaluation (8)
using the corrected molecular weight prefers only a single cytochrome $\underline{b}$
subunit which must then bind two protohaems. Physical properties of the

140

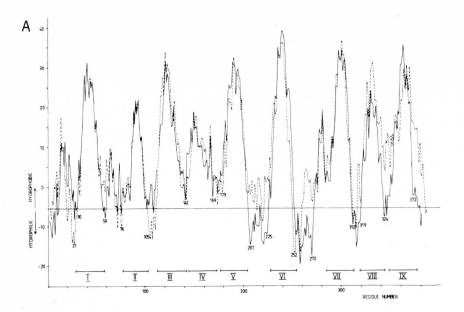

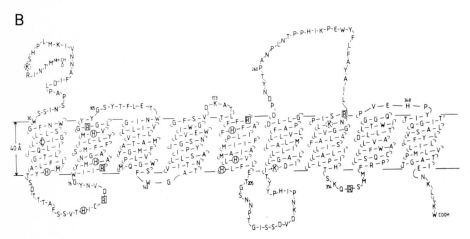

Fig. 1. Hydropathic profile of apocytochrome b from bovine (————) and yeast
(----) mitochondria (A). Relatively more hydrophobic sequences point upwards,
and the solid line gives a mean hydropathy for a peptide of 11 residues used
in the calculations. Nine predicted trans-membrane segments are indicated
with Roman numerals. (B) Schematic model showing the path of the apocytochrome
b polypeptide (bovine) in membrane. Invariant histidines, methionines, lysines,
arginines and a cysteine are indicated.

haems are different in situ but the differences appear to vanish when cytochrome b is isolated from the complex (8). The axial ligands to the haems
are not known. EPR spectrum of the ferric b cytochrome gives unusually high
g_z values 3.4 and 3.8 (9). The model compounds as reference, these values do
not fit to a bis-imidazole haem but suggest, perhaps, that a histidine and
an amino group (lysine) could be the axial ligands (10).

Four invariant histidines in apocytochrome b are predicted to reside
within the bilayer and three others on one membrane surface (Fig. 1B). The
embedded histidines are found in pairs in two trans-membrane segments II and V.
In each helix the histidines are separated by 13 residues and can therefore be
located on the same side of helix. Thus two bis-imidazole haems could be pos-
itioned above each other and sandwiched between two helices. The Fe-Fe distance would be about 20 Å and each haem iron would be about 10 Å from the
nearest membrane/water interphase. Two invariant arginines in segment II are
located so that they could bind propionate side chains of the protohaems
facing out of the membrane. This model could automatically explain the spectroscopic observation (11) that the haem planes are perpendicular to the
membrane.

The high g_z values might be explained by increased electronegativity of
imidazole nitrogen in lipophilic environment rendering it similar to an
"aqueous" amino group as a ligand. If it turned out, however, that bis-imidazole haems are impossible, an alternative model can be easily depicted from
Fig. 1. Two invariant lysines are found at topographically opposite ends of
segments VI and VII. These lysines and some of the invariant histidines could
be the imidazole/amino ligands to the two haems. Here three or four transmembrane segments (e.g. I or V and VI, II or V and VII) would be needed to
sandwich the haems.

2. Cytochrome oxidase. The bis-imidazole nature of the haem of cytochrome a and the proximal histidine ligand to the oxygen-binding haem a_3 are
well established (ref. 1). Again, both haems of cytochrome oxidase appear
to have their planes perpendicular to the membrane (11).

Subunit I has 11 invariant histidines. The hydropathic profile of the
protein suggests up to 12 transmembrane segments as shown in Fig. 2. The
invariant histidines are clustered to one side of the membrane, five of them
being slightly buried in it, two at the water/membrane interphase and four in
more aqueous domains. A weak homology is found between the sequences around
two invariant histidines (his-66 and his-156, Fig. 2) and the distal and

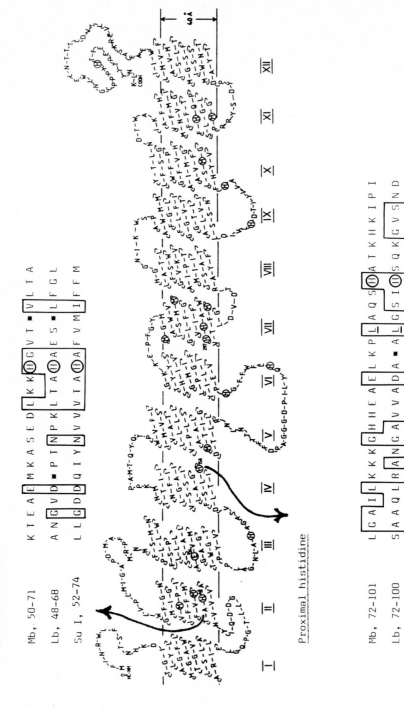

Distal histidine

Mb, 50-71   K T E A E M K A S E D L K K H G V T · V L T A
Lb, 48-68   A N G V D · P T N P K L T A H A E S · L F G L
Su I, 52-74   L L G D D Q I Y N V V T A H A F V M I F F M

Proximal histidine

Mb, 72-101   L G A I L K K K G H H E A E L K P L A Q S H A T K H K I P I
Lb, 72-100   S A A Q L R A N G A V V A D A · A L G S I H A L G K G V S N D
Su I, 137-163 L A G N L · A H A G A S V D L · T I F S L H A · G V S S I

proximal histidines in myoglobin and leghaemoglobin. This is not so surprising, because spectroscopic studies suggest striking similarities between the oxygen-binding sites in cytochrome oxidase and the globins (see (1)). These histidines in subunit I are both predicted to reside about 10 Å from the same membrane surface in the transmembrane segments II and IV, and haem a_3 would be consequently sandwiched between them. "Distal" his-66 and two nearby invariant methionines are potential ligands to the copper of oxygen-binding site (Cu_b).

The C-terminal part of subunit II shows homology to blue copper proteins and contains a site for Cu_A. Two cysteines and at least one histidine are the ligands to Cu_A (1, 12). The haem of cytochrome a has been proposed to bind to subunit II (13). If so, the sparsity of invariant histidines uniquely places its centre at the membrane interphase on cytoplasmic side of the inner mitochondrial membrane (Fig. 3).

The haem-haem distance in cytochrome oxidase is deduced from spectroscopic data to be about 8 Å as projected to the membrane normal (14). If the haem of cytochrome a is bound to subunit II as shown in Fig. 3, the proposed binding

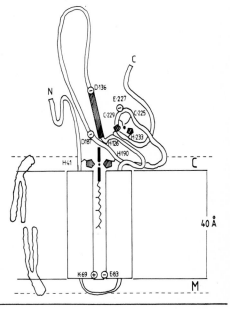

Fig. 3. Subunit II of cytochrome oxidase arranges a hydrophobic hairpin into the membrane. Four invariant histidines are indicated in the figure; his-126 is, however, substituted to tyrosine in the yeast protein. Binding site for Cu_A and a possible site for haem a of cytochrome a are shown, see (1,12).

Fig. 2. A schematic model of cytochrome oxidase subunit I (bovine) fold in the membrane; twelve trans-membrane segments are suggested by a hydroplot (1). Invariant histidines and methionines and a membrane-embedded glutamic acid are circled. Residue numbers refer to the aligned sequences. Homologies around two invariant histidines (his-66 and his-156) and the distal and proximal histidine of myoglobin (Mb) and leghaemoglobin (Lb) are included. See ref. 1.

site of haem $\underline{a}_3$ should also be close to the cytoplasmic side of the membrane. However, cytochrome $\underline{a}$ centre could alternatively reside in subunit I. For instance, two invariant histidines (his-239 and his-381) are found at the membrane/water interphase on the same side as the haem $\underline{a}_3$ centre (Fig. 2). Welinger and Mikkelsen (15) have recently proposed his-239 to be a good cancidate for a haem ligand, because amino acid sequence surrounding it shows apparent similarities to other haemoproteins.

ACKNOWLEDGEMENTS

This work has been supported by grants from Academy of Finland and the Sigrid Jusélius Foundation.

REFERENCES

1. Wikström, M., Saraste, M. and Penttilä, T. (1983) in: Martinosi, A. (Ed.), The Enzymes of Biological Membranes, in press.

2. Andersson, S., Bankier, A.T., Barrell, B.G. et al. (1981) Nature 290, 457.

3. Andersson, S., de Bruijn, M.H.L., Coulson, A.R. et al. (1982) J. Mol. Biol. 156, 683.

4. Bibb, M.J., Van Etten, R.A., Wright, C.T. et al. (1981) Cell 26, 167.

5. Nobrega, F.G. and Tzagolof, A. (1980) J. Biol. Chem. 255, 9828.

6. Waring, R.B., Davies, R.W., Lee, S. et al. (1981) Cell 27, 4.

7. Kyte, J. and Doolittle, R.F. (1982) J. Mol. Biol. 157, 105.

8. Von Jagow, G., Engel, W.D., Schägger, H. et al. (1981) in: Palmieri, F. et al. (Eds.) Vectorial Reactions in Electron and Ion Transport, Elsevier/North Holland, Amsterdam, pp. 149-161.

9. DeVries, S., Albrecht, S.P.J. and Leeuwerik, F.J. (1979) Biochim. Biophys. Acta 546, 316.

10. Peisach, J. (1978) in: Dutton, P.L. et al. (Eds.) Frontiers of Biological Energetics, Vol II, Academic Press, New York, pp. 873-881.

11. Erecińska, M., Wilson, D.F. and Blasie, J.K. (1979) Biochim. Biophys. Acta 501, 63.

12. Millett, F., de Jong, C., Paulson, L. and Capaldi, R.A. (1983) Biochemistry 22, 546.

13. Winter, D.B., Bruyninckx, W.J., Foulke, F.G. et al. (1980) J. Biol. Chem. 255, 11408.

14. Onishi, T., LoBrutto, R., Salerno, J.C. et al. (1982) J. Biol. Chem. 257, 14821.

15. Welinger, K.G. and Mikkelsen, L. (1983) FEBS Lett., in press.

PROBING THE STRUCTURE OF THE ADP/ATP CARRIER WITH PYRIDOXAL PHOSPHATE

WERNER BOGNER, HEINRICH AQUILA AND MARTIN KLINGENBERG
Institut für Physikalische Biochemie der Universität München,
Goethestrasse 33, 8000 München 2, FRG

INTRODUCTION

The ADP/ATP carrier from beef heart mitochondria has a relatively high content of lysine groups, i.e. 22 lysines out of a total of 297 amino acids. The primary structure (1) revealed that these lysine groups are distributed all over the amino acids sequence. Therefore, favorable conditions appear to exist to us the lysine reactivity with covalently binding reagents to probe not only functional but also structural parameters of the ADP/ATP carrier. Whereas lysine can be assumed to participate in the cationic binding site for the highly negatively charged substrates ADP and ATP, a more general use of the lysine reactivity in terms of probing various structural parameters of the carrier will be stressed in the work reported here.

For this purpose the most useful lysine reagent appeared to be pyridoxal phosphate (PLP), since it fulfills relatively well two prerequisites; it is relatively innocuous and also hydrophilic, thus not penetrating into hydrophobic areas or through the membrane (2). On the other hand, because of the relatively small size it encounters little steric hinderance.

So far PLP has been mainly used in specific site directed lysine modifications. In particular PLP has a tendency to react at sites which bind phosphate or phosphate containing substrates (3). In the present study we shall use PLP as a more general lysine reagent thus producing information on a number of structural parameters of the ADP/ATP carrier. The principle was to expose the ADP/ATP carrier either in the original membrane or in the isolated state to ^{3}H-PLP and then after isolation of the carrier protein from the membranes, to determine the localization of the incorporated PLP in the primary sequence.

Abbreviations: ATR, atractylate; BKA, bongkrekate; CAT, carboxy-atractylate; HTS, hydroxylapatite, PLP pyridoxal phosphate; SMP, submitochondrial particles; TX-100, Triton X-100.

METHODS

PLP was tritiated according to (4). Bovine heart mitochondria were prepared and loaded with CAT, ATR or BKA (5,6).Submitochondrial particles were prepared from CAT or BKA loaded mitochondria (7). PLP (^{3}H-PLP) incubations of mitochondria or submitochondrial particles were performed in 50 mM triethanolamine-HCl, 250 mM sucrose (8). The PLP modified carrier as the CAT protein complex was isolated conventionally (9). Modified ATR and BKA protein complex was obtained by a revised method. Mainly following the isolation procedure of BKA protein complex, the second Brij 58 extraction step was omitted. The insoluble material after the TX-100 solubilization was removed by stirring the solution 5 min at 0°C with 1 ml hydroxylapatite (HTS) per 20 mg protein. After short centrifugation (about 3 min) the supernatant was subjected to HTS chromatography as the main purification step.

For ^{3}H-PLP incubation of routinely prepared CAT or ATR protein, the pH was raised to 8.0 by 50 mM triethanolamin-HCl final concentration. ^{3}H-PLP was added to 6 mM final concentration and reaction was performed at 20°C for 10 min in the dark. The Schiff's base was reduced by an equimolar amount of NaBH$_4$.

The ^{3}H-PLP modified protein was denatured and, after extraction of phospholipids, carboxymethylated and citraconylated. Cleavage by thermolysin, decitraconylation and separation of peptides on "reversed phase" was done according to (1).Fractions were pooled in respect to the ^{3}H-radioactivity of the elution profile and subjected to preparative fingerprinting on cellulose plates.The pyridoxyl-lysine containing peptides were identified by their blue fluorescence under UV lamp (350 nm), scratched out and eluted. Aliquots of the samples were subjected to manual sequencing (10), amino acid analysis and determination of the radioactive yield.

Separation of ^{3}H-PLP modified peptides from protein in the ATR bound state required an additional purification step. The large peak of radioactive peptides ($\approx$70%) appearing early on "reversed phase", was further subjected to ion exchange chromatography on 0.5 x 30 cm SP-Sephadex C-25 column (H$^+$-state). After elution with gradients of 50-250 mM pyridin-acetate, pH 2.8-4.0 and 250 mM - 2.0 M pyridin-acetate, pH 4.0-5.0, fractions were pooled and further subjected to fingerprinting as described

For the localization of the positions of ^{3}H-PLP incorporation the peptides were sequenced manually and the thiohydantoin derivatives identified on polyamide sheets. Radioactivity of aliquots of the butylacetate extract showed exactly the modified lysine position. However, because of the poor extractability of the thiazolinon derivative of N-ε-pyridoxyl-lysine (11), some tailing is seen frequently in the following degradation steps (8).

RESULTS AND DISCUSSION

General remarks

In a number of experiments it was first determined to what degree PLP incorporation may influence functional parameters of the carrier. It was shown that PLP inhibits to a considerable extent (about 60%) the ADP/ATP exchange, and at about the same range of concentration (2-5 mM PLP), it also inhibits binding of these specific inhibitor ligands such as ^{3}H-CAT, ^{3}H-ATR and ^{3}H-BKA in mitochondria. These findings indicate that PLP interferes with one or more lysine groups at the binding center of the carrier, which we believe is competent both for binding the substrates and transport, as well as for binding the anionic inhibitors. If the ADP/ATP carrier in the mitochondria was first loaded by these ligands, the same exposure to PLP did not remove ^{3}H-CAT or ^{3}H-BKA to a marked extent. However, ATR could be displaced by higher concentrations of PLP (6 mM), ATR is known to have weaker affinity to the ADP/ATP carrier as CAT.

When the ADP/ATP carrier was isolated as the ^{3}H-CAT protein complex, PLP up to 8 mM was unable to remove this ligand. Under the same conditions ^{3}H-ATR and ^{3}H-BKA were released to about 80-90% from the corresponding isolated proteins. Thus there is a clear difference concerning the influence of PLP on the BKA binding between mitochondria and the isolated protein. This is not surprising since the isolated BKA protein complex is known to be relatively labile (6), whereas in the mitochondria it is obviously more stabilized. The removal of BKA reflects the loss of the native m-state structure. The removal of ATR from the isolated protein by PLP, however, could be shown not to disturb the basic c-conformation. For this purpose the ATR protein was prepared in a UV transparent detergent (3-lauramido-N,N-dimethylpropylamine-oxide) where well defined CD-signals originating from tryptophane

are a quantitative criterion for the c-conformation.Even after 80%
removal of ATR on incubation with PLP, the isolated protein did
not show a marked change of the characteristic tryptophane CD-
bands. This indicates that in this instance the c-conformation is
not essentially altered by the PLP incorporation.

By incorporating the PLP into the ADP/ATP carrier in a variety
of well defined states and by comparing these different states,
conclusions can be made concerning the localization of the PLP
modified lysine groups with respect to the membrane and in the
active center. Also the influence of the conformation change
between the c- and m-state upon this distribution can be demonstrated. The following four structural and functional features are
thus explored by the PLP incorporation.

1. Sidedness, direction of lysine either to the c- or m-face of
the membrane; based on the comparison of the PLP incorporation to
mitochondria and to submitochondrial particles (SMP).

2. The conformation difference between the c- and m-state as
revealed by different lysine reactivities; lysine reactivities
differ when the carrier in mitochondria is loaded either with CAT
or BKA.

3. Phospholipid headgroup contact region with the mitochondria
(neckline annulus); difference of lysine accessibility of the
carrier in situ and after isolation in the protein detergent
micelle.

4. Binding center; specific protection of lysine by ligands.

Inside-outside distribution

The distribution of lysine groups towards the c- or m-face of
the inner mitochondrial membrane was examined by incorporation of
PLP into mitochondria and sonic particles. For this purpose CAT
loaded mitochondria were reacted with PLP and the incorporation
compared. The inverted membranes were derived by sonication from
mitochondria loaded prior with CAT. The data summarized in Figure
1 where the incorporation of mitochondria is the mean value of
three and those on SMP's from two different preparations.Generally
more lysine groups are accessible in mitochondria. Thus lysine
groups 42, 198, 205, 259, 262 and 267 are only labeled in mitochondria. Groups 48 and 106 are labeled both in mitochondria and

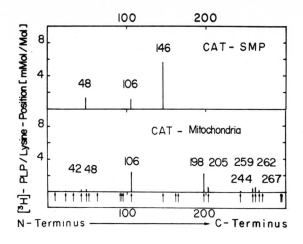

N-Terminus ⟶ C-Terminus

Fig.1. Incorporation of
[3]H-PLP in lysine posi-
tions of the ADP/ATP
carrier in the c-state.
The protein is labeled
from the inner side of
the inner mitochondrial
membrane in sonic par-
ticles and from the
cytosolic side in mito-
chondria.The abscissa
represents the primary
structure (1-297), the
positions of lysines
are marked by arrows.
The ordinate shows the
quantity of labeling
expressed as mmoles
peptide per moles
carrier protein before
encymic cleavage.

SMP's. The only group which is very strong and distinctly labeled
in SMP's is lysine 146. Thus in the CAT-SMP's only three and in
CAT-mitochondria only eight lysine groups out of the total of 22
are accessible to PLP incorporation.

With these data the following folding pattern has been proposed
which closely follows the model published by us recently (8). The
folding incorporates the existence of three, possible four largely
hydrophobic amino acid stretches of 18 to 23 residues and which
are a possible candidate for transmembrane α-helices. The sidedness specificity of lysine incorporation and the transmembrane α-
helices can be well fitted together, as shown in Figure 2. There
are four helices, helix I from 72 to 90, helix II from 111 to 130,
helix III from 171 to 190 and helix IV from 209 to 230. At the
cytosolic face lysine 106 fits well in the bridge section between
helix I and II by placing the terminal of helix I and the start of
helix II on the c-side. Lysine 198 and 209 are both part of the
link between helices III and IV in the c-side by placing the terminal on 190 and start 209 to this face. A much less easily accessible cluster of lysines exists on 259, 262 to 267 near the c-
terminal, also accessible from the c-side. This arrangement of the
helices can also be reconciled with the accessibility of lysine
from the m-side. Thus lysine 48 and 106 are barely labeled whereas
lysine 146 is the most easily accessible and has been situated
between the terminal of helix II and the start of helix III.

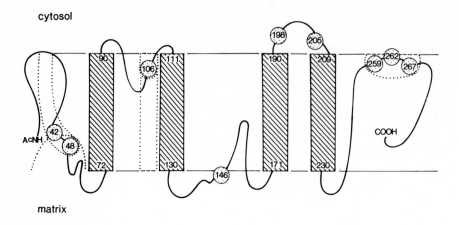

Fig. 2. Two-dimensional transmembrane folding pattern of the ADP/ATP carrier in the c-state. The four hydrophobic stretches are shown, representing possible α-helices. This inside-outside distribution is based on the surface labeling experiments in CAT-bound submitochondrial particles and mitochondria. Reactive lysines are emphasized by circles.

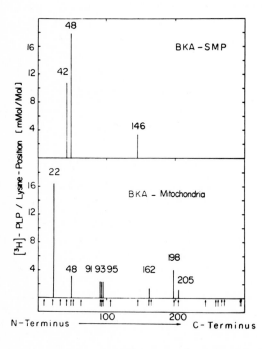

Fig. 3. Incorporation of H-PLP in the BKA bound m-state into the ADP/ATP carrier in sonic particles and mitochondria. The abscissa again represents the primary structure, the ordinate shows the quantity of labeling per position (see Fig. 1).

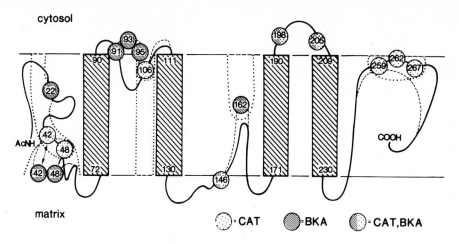

Fig. 4. Possible spanning of the ADP/ATP carrier across the membrane as derived from lysine labeling in the c- and m-state in sonic particles and mitochondria. The dotted circles represent lysines, labeled in c-state (see Fig.2), the hatched circles indicate the newly or more exposed positions of the m-state.Dotted lines show possible pore forming regions, implicated by reactivity from both sides of the membrane. Dashed lines link exclusively m-state reactive lysines to the sequence.

Conformational change between c- and m-state

By loading with BKA the ADP/ATP carrier can be brought into the m-conformation. The PLP incorporation has been studied to BKA loaded mitochondria and SMP derived therefrom. The localization of the PLP in the sequence of the ADP/ATP carrier under these conditions is shown in Fig. 3 for a single preparation of each BKA loaded mitochondria and SMP's. In this case the labeling pattern shows considerable differences to the CAT carrier complex. There are new PLP incorporations into lysine 22, 91, 93, 95 and 162. On the other hand position 106 and 259, 262 and 267 are missing. In SMP's the strong incorporation to position 42 and 48 is most striking. Again labeling of 146, similar as in the CAT, is singularly outstanding.

Combining both the sidedness and differences in the two conformational states in a single folding model, two additional features are introduced in Fig.4. Lysines which incorporate PLP only weakly and lysines which change their incorporation between the c- and m-state are visualized to be localized within the delimitations of the membrane surface, however, within hydrophobic areas path or

channels of the protein. By assuming these hydrophilic invagina-
tions we have in mind that these are part of the hydrophilic translocation path across the membrane which must exist in the carrier.
Conformational changes are depicted by moving the lysines between
regions of different accessbilities.

With these novel features added to the presently fashionable
game of transmembrane folding, we make the following assignments.
Lysine 22 is a special case and will be discussed below. Lysine
42 and 48 are more accessible in the m-state by moving more to the
surface. The cluster 91, 93 and 95 has moved to a more accessible
position at the c-side in the m-state. As a result of the same
change in this region lysine 106 becomes inaccessible. Lysine 162
causes more problems. It is separated from the start of helix III
on the m-side by 15 amino acids but is accessible only from the
c-side. Therefore, we place the lysine 162 in the membrane
although this will be revised somewhat by the following data.Again
this amino acid is available to the c-side only in the m-state. In
contrast the cluster 259, 262 and 267 becomes inaccessible in the
m-state.

Phospholipid headgroup/neckline region

A number of new lysine groups became accessible by comparing the
carrier in the detergent micelle to the membrane. On the other
hand there is good evidence that the carrier retains its original
conformation of the c-state as defined by the CAT protein complex.
Therefore, the appearance of new lysines is interpreted as a
removal of the phospholipid headgroups, which form necklines protecting the protein in the membrane. Moreover, negatively charged
phospholipid headgroups, in particular on the m-side cardiolipin, may form protective salt bridges to a number of lysines
located in this area. These headgroups may be only partially replaced by the hydrophilic loose sponge-like polyoxyethylene TX-100
(12). By introducing this new idea into the folding game, the
data can be quite well fitted into the folding model.

In Figure 5 the lysine incorporation to the isolated CAT protein
complex in TX-100 is shown and compared to that in the original
membrane in mitochondria and SMP's. There are seven lysine groups
which become more accessible to PLP incorporation in the isolated
protein. Reactivity of position 42, 48, 106 and 262 appears to be

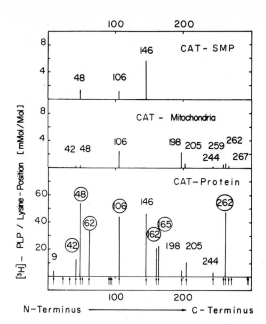

Fig. 5. Comparison of ^{3}H-PLP incorporation into the ADP/ATP carrier in the CAT bound c-state in the membrane and isolated in detergent. The abscissa represents the primary structure, the 23 lysine positions are marked by arrows. The ordinate shows the quantity of labeling (see Fig. 1).The position numbers in circles call special attention to the possible "neckline" location of these lysines.

strongly facilitated. In position 62, 162 and 165 entirely newly labeled lysines are found. For that reason we place lysine 42 and 48 near the"neckline" region where at the same time they are supposed to be close to a conformational sensitive area,as shown in Figure 6.

Lysine 62 is clearly blocked by phospholipid headgroups. 106 is similar as the lysine 42 near hydrophilic pore but also largely covered by the "neckline" forces. The lysine 162 and 165 seem to resist this simple classification,since 162 is only accessible from the c-side in the m-state and on the other hand 165 is too short a distance to the 171 start of the helix III,in order to place both these amino acids into the "neckline" at the c-region. Here is a clear contradiction in the interpretation which must await further studies. 262 can without difficulties be placed into the C-side "neckline." However, it is also part of the strongly conformation sensitive cluster.

It has not been possible to study lysine PLP incorporation into the isolated BKA protein complexes. Because of the high lability of this complex, conditions required to incorporate PLP would denature this very labile complex and as a result of the

cytosol

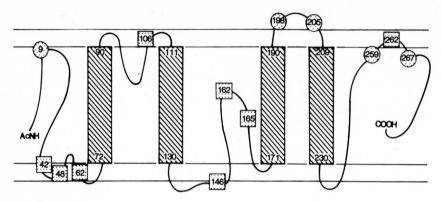

Fig. 6. Possible spanning of the ADP/ATP carrier across the inner membrane in the c-state. Lysine numbers in circles show no significant protecting effect of the membrane comparing modification of carrier in the membrane and isolated protein. The squared positions, located between the two narrow lines ("neckline annulus") become much more accessible in the isolated carrier.

unfolding give artefacts which cannot be interpreted in terms of the native folding structure.

Binding center lysine assignments

By comparing the PLP incorporation into the CAT and ATR loaded mitochondria, we arrive at results which might be interpreted in terms of lysine involvements to ligand binding. Whereas CAT is retained on PLP incorporation, ATR is removed from the complex. The difference in lysine accessibility is tentatively interpreted in terms of lysine group being unmasked by the ligand removal. As shown in Figure 7, differently from CAT loaded mitochondria, in ATR loading appears lysine 22 and, to a smaller extent, lysine 146. Similar differences have been obtained by comparing the incorporation of PLP to the isolated CAT and ATR carrier complexes. In the latter case it has been shown by CD-studies that PLP incorporation does not markedly alter the native conformational state of the carrier although ATR has been removed. Thus the appearance of the additional lysine 22 and 146 is not to represent a conformational change by PLP induced ATR removal but rather in an

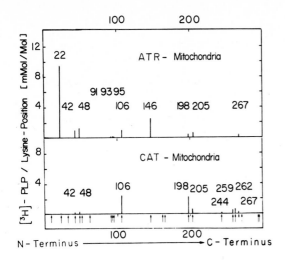

Fig. 7. Comparison of ATR and CAT bound mitochondria (both in c-state). The abscissa shows the primary structure of the carrier, lysines are represented by arrows. The ordinate indicates the amount of labeling (see Figure 1).

unmasking and displacement of the ATR ligand. Therefore, we like to conclude that these groups are involved or close to the binding center of the nucleotide carrier.

ACKNOWLEDGEMENTS

This work was supported by grants from the Deutsche Forschungsgemeinschaft (Aqu 1/3 and Kl 134/22). The technical assistance of Heidi Gross, Christl Knott and Christine Kraus is gratefully acknowledged.

REFERENCES

1. Aquila, H., Misra, D., Eulitz, M. and Klingenberg, M. (1982) Hoppe Seyler's Z. Physiol. Chem. 363, 345-349.

2. Canbantchik, Z.I., Balshin, M., Breuer, W. and Rothstein, A. (1975) J. Biol. Chem. 250, 5130-5136.

3. Glazer, A. N. (1976) in: Neurath H. and Hill, R.L. (Eds.) The Proteins, Vol. 2, Academic Press, New York, p. 74.

4. Stock, A., Ortanderl, F. and Pfleiderer, G. (1966) Biochem. Zeitschrift 344, 353-360.

5. Klingenberg, M., Grebe, K. and Scherer, B. (1975) Eur. J. Biochem. 52, 351-363.

6. Aquila, H., Eiermann, W., Babel, W. and Klingenberg, M. (1978) Eur. J. Biochem. 85, 549-560.

7. Klingenberg, M. (1977) Eur. J. Biochem. 76, 553-565.

8. Bogner, W., Aquila, H. and Klingenberg, M. (1982) FEBS Lett. 146, 259-261.

9. Riccio, P., Aquila, H. and Klingenberg, M. (1975) FEBS Lett. 56, 129-138.

10. Chang, J. Y., Brauer, D. and Wittmann-Liebold, Б. (1978) FEBS Lett. 93, 205-214.

11. Tanase, S., Kojima, H. and Morino, Y. (1979) Biochemistry 18, 3002-3007.

12. Hackenberg, H. and Klingenberg, M. (1980) Biochemistry 19, 548-555.

© 1983 Elsevier Science Publishers B.V.
Structure and Function of Membrane Proteins,
E. Quagliariello and F. Palmieri editors.

SEQUENCING OF LARGE MEMBRANE PROTEINS: THE Ca^{2+}-ATPase OF
SARCOPLASMIC RETICULUM

GEOFFREY ALLEN
Department of Molecular Biology, The Wellcome Research
Laboratories, Langley Court, Beckenham, Kent BR3 3BS, U.K.

INTRODUCTION

Membrane proteins can be considered to fall into three
structural classes: extrinsic or non-integral proteins, bound to
the surfaces of membranes through polar interactions with the
head groups of phospholipids or through an intermediary protein,
integral membrane proteins that consist of globular portions
attached to the membrane through a relatively small hydrophobic
tail, and integral membrane proteins in which a substantial portion
of the structure is embedded within and traverses the membrane
bilayer. Fig. 1 shows these three classes schematically.

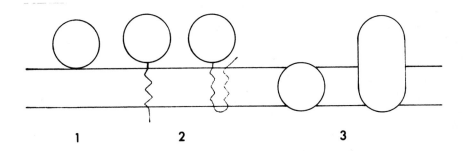

Class 1 included erythrocyte glyceraldehyde phosphate
dehydrogenase, and a variant surface glycoprotein of Trypanosoma
brucei of which we have recently determined the sequence (1).
Numerous examples of the second class are known, including the
haemagglutinin and the neuraminidase of influenza virus, erythrocyte glycophorin, cytochrome b_5, major histocompatibility
antigens and the small intestinal sucrase-isomaltase. While many
of these proteins, including influenza haemagglutinin (2) and
glycophorin (3) have a hydrophobic segment spanning the bilayer
close to the carboxy terminus, others, such as influenza neuramin-

idase (4) and small intestinal sucrase-isomaltase (5) have a
hydrophobic segment close to the amino terminus, and in the latter
the polypeptide chain probably crosses the membrane twice (5).
Class 3 is exemplified by many proteins concerned with the transport
of ions and molecules across membranes, such as bovine and bacterial rhodopsins, the erythrocyte anion-transport protein and
the calcium ion transporting ATPase of sarcoplasmic reticulum.

The determination of protein sequences of the first class is
generally straightforward, using methods developed for the study
of soluble globular proteins. The second class is typified by the
presence of an uncharged and often highly hydrophobic sequence of
about 20 residues, lacking proline residues and likely to form a
single alpha helix spanning the bilayer. For example, rabbit small
intestinal isomaltase has a sequence of 20 residues that consists
solely of leucine, isoleucine, valine, phenylalanine and alanine (5).
In other proteins of this class an occasional hydrophilic residue
is present within the hydrophobic segment. Sequence determinations
of proteins of class 2 may often be accomplished following cleavage
of the globular portion away from the hydrophobic tail by perform-
ing automated sequence analyses through the hydrophobic peptides
and treating the soluble portion as for class 1.

The third class of protein also has extensive hydrophobic
regions, but from the limited data presently available these seem
in general rather less hydrophobic than those in the second class,
with charged residues and proline residues not infrequent. These
observations are not surprising, given the potential for satisfying
hydrogen bonding requirements between adjacent helices and the
likelihood of the presence of hydrophilic pores through which ions
and molecules are transported. However, the properties of the memb-
rane-buried segments are generally such that considerable difficulties are encountered in the determination of their sequences,
mainly because of aggregation and insolubility in solvents used
for most protein sequence work. To bring out the contrast between
sequence determinations of class 1 and class 3 proteins, the
sequencing of a variant surface glycoprotein, VSG 117, from
Trypanosoma brucei will be described in outline.

SEQUENCE DETERMINATION OF VSG 117

African salivarian trypanosomes, including T. brucei, possess
a surface coat consisting essentially of a single glycoprotein, or
VSG, that is one of a large number of antigenically distinct glyco-
proteins that can be sequentially expressed, allowing the organisms
to evade the immune response of the mammalian host. The glycoprotein
can be released from the membrane surface in the absence of
detergent and is a non-integral membrane protein.

The entire sequence of 470 residues of VSG 117 was determined
by manual sequencing of peptides isolated from a number of digests
by standard methods, apart from overlaps at seven positions that
were determined from the nucleotide sequence of cDNA (1,6). The
cDNA sequence also revealed that a precursor had an amino-terminal
'signal' sequence and a carboxy-terminal hydrophobic extension (6).
The hydrophobic tail presumably acts as a membrane anchor during
processing and assembly of the coat. It is removed prior to
maturation, since the alpha carboxyl group of Asp-470 bears a
unique carbohydrate structure linked through an ethanolamine
moiety (7).

The arrangement of disulphide bonds in the glycoprotein has
been determined (8). Four disulphide bonds in the carboxy-terminal
region inhibit protease digestion, and the determination of these
disulphide bonds presented the only real technical challenge in
this work.

PARTIAL SEQUENCE DETERMINATION OF THE Ca^{2+}-ATPase OF RABBIT
SARCOPLASMIC RETICULUM

In contrast, the relatively large proportion of the polypeptide chain of the Ca^{2+}-ATPase embedded within the bilayer has
caused much difficulty in chain cleavage, peptide purification and
sequencing. The sequences within the extramembranous portions of
the molecule have mostly been determined (9-12), but the determination of the sequences of intramembranous portions has advanced
little beyond that reported earlier (13).

Preparation of protein for sequence analysis

Rabbit skeletal muscle sarcoplasmic reticulum was isolated as
vesicles by fractional precipitation and centrifugation and the
Ca^{2+}-ATPase lipoprotein was purified by salt precipitation of

deoxycholate solubilized vesicles (14). The protein was reduced
and carboxymethylated in 6M guanidinium chloride containing
taurodeoxycholate, immediately after isolation. Following dialysis,
the lipoprotein was treated with a tenfold weight of sodium dodecyl
sulphate (SDS) and the protein was further purified and delipidated
on a column of Sepharose 6B in the presence of 0.1% SDS, 1mM
mercaptoethanol and 1mM EDTA. It was found that unless reducing
conditions were maintained during the isolation of the lipoprotein
irreversible loss of cysteine residues occurred, possibly through
reaction with products of autoxidation of lipids.

Sequence determination
      Most of the sequence work was performed on the succinylated
protein. Tryptic digestion at arginine, other protease digestions,
solubility of large peptides, purification of these on DEAE-cellulose columns and sequence analysis were all aided by succinylation.
Despite these observations, in all digests studied between 30% and
50% of the peptide material was aggregated in aqueous solution and
emerged from Sephadex G-100 columns at the void volume.
      Non-aggregated peptides were isolated by standard methods from
digests by trypsin, pepsin, chymotrypsin, staphylococcal protease,
thermolysin and cyanogen bromide, and a total of 835 residues were
placed in unique sequences, of which 575 were present in a set of
five sequences comprising the amino- and carboxy-termini and three
long sequences within the protein (12).
      The precise molecular weight is not known, but on the assumpt-
ion that it is 110,000, around 180 residues remained to be identified, and a large number of overlaps remained to be determined.
Within these 180 residues are 18 of the 19 tryptophans, but 30%
of the residues are polar (15).
      Aggregated peptides from tryptic and cyanogen bromide digests
were partially resolved by gel filtration on a column of Biogel
P-100 in phenol-acetic acid-water (1:1:1, by vol.) and further
resolution was obtained by ion-exchange chromatography on DEAE-
cellulose DE52 in phenol-water-formamide (2:2:1, by vol.) in a
gradient of triethylamine-acetic acid buffer. Final purification
of some components was achieved by SDS-polyacrylamide gel electrophoresis. Partial sequence analysis was performed on some of these
fragments, of size estimated at 30 - 60 residues, but was rather

unsuccessful, owing to apparently blocked amino-termini (probably resulting from the use of formamide in phenol solution), inefficient subfragmentation with pepsin and other proteases and low recoveries. A number of other solvent systems have been used for the separation of membrane peptides by other workers, and some examples are given in Table 1.

TABLE 1

EXAMPLES OF CHROMATOGRAPHIC SEPARATION OF MEMBRANE PEPTIDES

Solvent	Stationary phase	Reference
methanol/chloroform/NH$_3$/ acetic acid/formic acid gradient system	silica gel	16
80% formic acid	Biogel P-30	16, 17
0.1% trifluoroacetic acid, acetonitrile gradient	reverse-phase HPLC beads	17
88% formic acid/ethanol (30/70)	Sephadex LH20	18
5% formic acid, gradient of 40% - 95% ethanol	reverse-phase HPLC beads	18
3% Ammonyx LO detergent	CM-cellulose	19

Increased sensitivity in methods of sequence analysis have allowed SDS-polyacrylamide gel electrophoresis and HPLC to become suitable methods for peptide purification. Some recent results on membrane-associated peptides of the Ca^{2+}-ATPase derived by cleavage at tryptophan residues and sequenced using a gas-phase sequencer have extended the sequence information on the protein (N.M. Green, personal communication).

However, it has become clear that nucleotide sequencing of cloned cDNA is the method of choice for the determination of the primary structures of membrane proteins. Oligonucleotide probes have been synthesized on the basis of amino acid sequences determined within ATPase peptides and are being used to probe cDNA banks derived from rabbit muscle mRNA. The first indications of success have been obtained and it is expected that the sequence of the protein will be completed by this route.

This approach requires supplementing with direct studies on the protein to identify any post-translational modifications (e.g. cleavage of pre- and pro- pieces, glycosylation and phosphorylation) active site residues and other aspects such as organisation of the peptide chain with respect to the membrane, and a considerable body of such information has been accumulated for the Ca^{2+}-ATPase.

Active site and nucleotide binding site

The active site aspartic acid residue that is phosphorylated during the enzymic reaction was identified by using $[\gamma^{32}P]$-ATP (20). The mixed anhydride was unstable, particularly under alkaline conditions, and it was necessary to devise rapid systems for the purification of labelled tryptic and peptic peptides.

Fluorescein isothiocyanate reacts rapidly with a single lysine residue in the Ca^{2+}-ATPase. The reaction leads to inhibition of ATPase activity and is prevented by bound nucleotides: the lysine is presumably within a high-affinity nucleotide binding site. The lysine has been identified within a long sequence in the protein, 164 residues from the active site aspartic acid and close to a trypsin-sensitive site in the enzyme (21).

A model for the Ca^{2+}-ATPase

A schematic model for the Ca^{2+}-ATPase is presented in Fig. 2. The native ATPase has two trypsin-sensitive sites (22) and the positioning of three long extramembranous fragments of known sequence (12) was aided by amino-terminal sequencing of the large tryptic fragments (23). The transmembrane segments are represented as helices by analogy with those in bacteriorhodopsin, and because circular dichroic measurements indicate a high content of alpha helices in membranes divested of the head portion by extensive proteolysis (15). No particular structure is adduced to the head, but predictions are that both β-sheet and α-helix are present (12).

Reithmeier and MacLennan (24) have located the amino terminus to the cytoplasmic face of the membrane. The biosynthesis of the protein involves translation on membrane-bound ribosomes (25) and there is no signal peptide cleavage (24,26), the initiator methion-

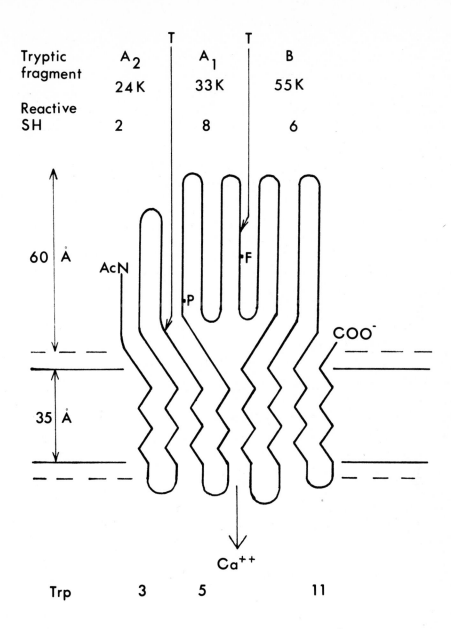

Fig. 2. A schematic model of the Ca^{2+}-ATPase of rabbit sarcoplasmic reticulum. For details see the text. The cytoplasmic face of the membrane is at the top. T, trypsin-sensitive sites.

ine being acetylated (27). The location of the carboxy-terminus is not known; if it lies on the lumen side there might be seven transmembrane helices, a motif demonstrated in bacteriorhodopsin. The number of transmembrane helices is not yet known.

All thiol groups reactive with N-ethyl maleimide are on the cytoplasmic face (24) and peptides covering the bulk of the sequences shown as lying on the cytoplasmic face were released from vesicles by extended tryptic digestion (15). The two or three disulphide bonds are within large tryptic fragment B (28) but their precise location is not known; possibly they are within the membrane.

The active site aspartic acid residue (P) and the reactive lysine (F) are shown; although separated by 164 residues in the linear sequence, they are presumably juxtaposed in the native structure. Secondary structural predictions indicate that a nucleotide binding fold may be present in the sequence following the reactive lysine (21). All but one of the 19 tryptophan residues appear to be within membrane-bound sequences, and fluorescence quenching studies indicate that a large proportion of these are accessible to lipid-soluble agents (N.M. Green, personal communication).

A number of other observations are consistent with the proposed model. The products of limited tryptic digestion are only separable in the presence of denaturing detergent (29). All three of the large tryptic fragments are labelled by a hydrophobic nitrene reagent (15) indicating penetration of the membrane. Antibodies to each of the large tryptic fragments bind to the cytoplasmic face (30). Electron microscopic studies show projections from the cytoplasmic face of the dimensions shown, as well as intramembranous particles.

The Ca^{2+}-ATPase is shown as a monomer. Although ATPase activity is retained in certain detergent solutions by monomers (31) there are differences in properties between monomers and aggregates that indicate the importance of protein-protein interactions. Calcium transport is not, of course, demonstrable following solubilization. The recent description of two-dimensional arrays of the ATPase in vanadate-treated vesicles (32) confirms that protein-protein interactions occur. It is to be hoped that these arrays will serve for high resolution electron microscopic analysis.

ACKNOWLEDGEMENTS

The work on the Ca^{2+}-ATPase was performed while the author was at the National Institute for Medical Research, Mill Hill, London. I thank Dr. N.M. Green for recent unpublished results and the proposed model of the ATPase, and Dr. D.H. MacLennan for communicating results prior to publication.

REFERENCES

1. Allen, G., Gurnett, L.P. and Cross, G.A.M. (1982) J.Mol.Biol. 157, 527-546.

2. Waterfield, M.D., Espelie, K., Elder, K. and Skehel, J.J. (1979) Brit. Med. Bulletin, 35, 57-63.

3. Tomita, M., Furthmayr, H. and Marchesi, V.T. (1978) Biochemistry, 17, 4756-4770.

4. Fields, S., Winter, G. and Brownlee, G.G. (1981) Nature, 290, 213-217.

5. Sjöström, H., Norén, O., Christiansen, L.A., Wacker, H., Spiess, M., Bigler-Meier, B., Rickli, E.E. and Semenza, G. (1982) FEBS Lett. 148, 321-325.

6. Boothroyd, J.C., Paynter, C.A., Coleman, S.L. and Cross, G.A.M. (1982) J.Mol. Biol. 157, 547-556.

7. Holder, A.A. (1983) Biochem.J. 209, 261-262.

8. Allen, G. And Gurnett, L.P. (1983) Biochem.J. 209, 481-487.

9. Allen, G. (1980) Biochem.J. 187, 545-563.

10. Allen, G. (1980) Biochem.J., 187, 565-575.

11. Allen, G., Bottomley, R.C. and Trinnaman, B.J. (1980) Biochem. J. 187, 577-589.

12. Allen, G., Trinnaman, B.J. and Green, N.M. (1980) Biochem.J. 187, 591-616.

13. Allen, G. (1978) FEBS Symp. 45, 159-168.

14. MacLennan, D.H. (1970) J.Biol.Chem. 245, 4508-4518.

15. Green, N.M., Allen, G. and Hebdon, G.M. (1980) Ann.N.Y.Acad. Sci. 358, 149-158.

16. Aquila, H., Misra, D., Eulitz, M. and Klingenberg, M. (1982) Hoppe-Seyler's Z.Physiol.Chem. 363, 345-349.

17. Ovchinnikov, Yu.A. (1982) FEBS Lett. 148, 179-191.

18. Gerber, G.E., Anderegg, R.J., Herlihy, W.C., Gray, C.P., Biemann, K. and Khorana, H.G. (1979) Proc.Natl.Acad.Sci.U.S.A. 76, 227-231.

19.Mullen, E. and Akhtar, M. (1983) Biochem.J. 211, 45-54.

20. Allen, G. and Green, N.M. (1976) FEBS Lett. 63, 188-192.

21. Mitchinson, C., Wilderspin, A.F., Trinnaman, B.J. and Green, N.M. (1982) FEBS Lett. 146, 87-92.

22. Thorley-Lawson, D.A. and Green, N.M. (1973) Eur.J.Biochem. <u>40</u>, 403-413.

23. Klip, A. and MacLennan, D.H. (1978) in Frontiers of Biological Energetics, vol. 2 (Dutton, L., Leigh, J. and Scarpa, A., eds.) pp 1137-1147, Academic Press, New York.

24. Reithmeier, R.A.F. and MacLennan, D.H. (1981) J.Biol.Chem. <u>256</u>, 5957-5960.

25. Greenway, D.C. and MacLennan, D.H. (1978) Can.J.Biochem. <u>56</u>, 452-456.

26. Mostov, K.E., DeFoor, P., Fleischer, S. and Blobel, G. (1981) Nature, <u>292</u>, 87-88.

27. Reithmeier, R.A.F., deLeon, S. and MacLennan, D.H. (1980) J.Biol.Chem. <u>255</u>, 11839-11846.

28. Thorley-Lawson, D.A. and Green, N.M. (1975) Biochem.J. <u>167</u>, 739-748.

29. Rizzolo, L.J. and Tanford, C. (1978) Biochemistry, <u>17</u>, 4049-4054.

30. Stewart, P.S., MacLennan, D.H. and Shamoo, A.E. (1976) J.Biol.Chem. <u>251</u>, 712-729.

31. Møller, J.V., Lind, K.E. and Andersen, J.P. (1980) J.Biol.Chem. <u>255</u>, 1912-1920.

32. Dux, L. and Martonosi, A. (1983) J.Biol.Chem. <u>258</u>, 2599-2603.

STRUCTURE AND ASSEMBLY OF F_1F_o ATPASE COMPLEX

JOHN E. WALKER, NICHOLAS J. GAY, MICHAEL J. RUNSWICK, VICTOR L.J. TYBULEWICZ AND
GUNNAR FALK
Laboratory of Molecular Biology, The Medical Research Council Centre,
Hills Road, Cambridge CB2 2QH, England

Bacterial proton translocating ATP-synthase couples the electrochemical
potential difference for H^+ across the cytoplasmic membrane to synthesis of ATP
from ADP and Pi [1,2]. The enzyme is bound to the inner membrane of Escherichia
coli. As depicted in Fig. 1 it is made of two sectors, commonly called F_1 and
F_o. F_1 is an ATPase. It is on the cytoplasmic surface of the membrane, from
which it can be easily dislodged. F_o, the membrane sector, contains a transmembrane proton channel. F_1 is composed of five different polypeptides α, β, γ,
δ and ε assembled with stoichiometries of $3\alpha:3\beta:1\gamma:1\delta:1\epsilon$ [3,4]. F_o comprises
three proteins a, b and c. A range of stoichiometries has been proposed [see
ref. 4], a recent estimate being 1a:2b:10c: [5].

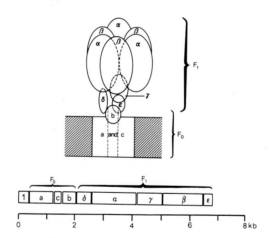

Fig. 1. Possible arrangement of subunits in E. coli H^+-ATPase complex and order
of genes in unc operon. a and c are components of proton channel (dashed)
through the lipid bilayer (hatched).

Mitochondria and chloroplasts also contain proton translocating F_1F_o ATP-
synthases with many structural features in common with the bacterial enzyme [for
reviews see refs. 6-8]. However, the mitochondrial enzyme is more complex than
the bacterial one. A number of extra subunits have been characterised; a
protein inhibitor binds to the F_1 portion and at least two extra subunits are

associated with F_O, the oligomycin sensitivity conferral protein (oscp) [9-12] and factor 6 (F_6) [13]. The E. coli complex is encoded in the unc operon which we have sequenced at the nucleotide level and identified the genes for the eight subunits of the enzyme [summarised in 14]. The order of the genes is shown in Fig. 1.

Structure of F_O subunits

Information about the structure of the F_O subunits can be deduced from their protein sequences. For instance, hydrophobic segments that might form trans-membrane helices of about 25-30 largely non-polar amino acids can be detected with the computer program HYDROPLOT [15].

Subunit a

Application of this method to the a subunit has shown that although the sequences of a subunit are poorly conserved in bacteria, mitochondria and chloroplasts, the hydrophobicity is conserved, consistent with the a subunits having closely related structures. The E. coli protein has seven distinct hydropbobic regions I-VII; all except II have counterparts in the mitochondrial proteins. In the latter proteins, region II appears to be less hydrophobic than in the bacterial protein (see Fig. 2). Regions I and III-VI are of lengths (27-28 residues) consistent with their being able to form transmembranous α-helices and the mitochondrial proteins would be expected to be similarly folded. Region II is less clear. It is conceivable that it also could traverse the membrane, although this would place a number of charged residues within the bilayer. Alternatively it could form an external loop near the surface of the membrane. An important consequence of this uncertainty is that it is not possible to decide if the amino and carboxyl terminals are on the same or opposite sides of the bilayer.

Subunit c

Comparison of the sequence of E. coli subunit c with counterparts from mito-chondrial and chloroplasts demonstrates a wide sequence diversity similar to that described in subunit a. Although the proteins are clearly homologous, sequence is strictly conserved in only six places. However, residue 66 (Asp or Glu) is functionally most important; in the native complex this acidic residue reacts with DCCD [16], thereby blocking proton translocation. A number of authors have suggested that subunit c folds into a hairpin structure in the membrane (see 16], an idea supported by the hydropathy profile (Fig. 3).

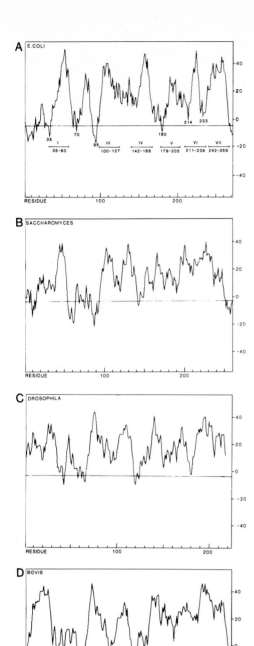

Fig. 2. Hydrophobic profiles
of H$^+$-ATPase subunit a from
E. coli and three mitochondria.
The sequences is plotted along
the horizontal axis. Vertical
axis is a measure of hydrophobicity calculated with span 11.
I and III-VII represent regions
proposed to form transmembrane
helices.

170

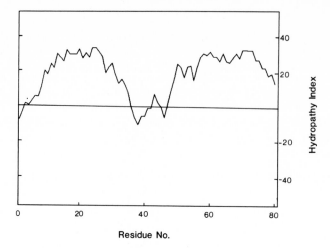

Fig. 3. Hydrophobic profile of subunit c.

This illustrates that the sequence has two hydrophobic segments linked by a conserved hydrophilic sequence containing proline. The other two conserved sequences are found in the middle of the two arms.

Subunit b

The sequence of the E. coli protein shows it to be an amphiphilic protein. It is hydrophobic around the N-terminus and highly charged in the remainder (Fig. 4). We have proposed a model in which the N-terminal segment lies within the membrane and the rest outside [17]. In this model the N-terminal region

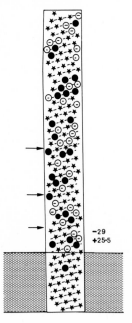

Fig. 4. Proposed secondary structure of membrane subunit b. Sequence is plotted on a helix surface and reads upwards, left to right. ⊖ , Asp or Glu; ●, Lys, Arg or His; *, other. Arrows denote possible breaks in α-helix.

forms a single transmembrane α-helix making contacts with a and c. The extra-
membrane section contains two adjacent stretches of 31 amino acids of related
sequence (the long repeat); within the second of these a further internal
homology has been detected (the short repeat). It has been proposed that these
duplicated stretches of polypeptide fold into two α-helices with many common
features able to make contacts with F_1 subunits. It would be expected to make
important contacts with the δ and ϵ subunits in particular since it is known
that they are required for binding of F_1 to F_o. The amino acid compositions of
these interacting segments in b support the view that their interactions with F_1
will be α-helix rather than α-helix to β-sheet. This is consistent with the
high α-helicity of δ proposed from optical measurements of the subunits from
thermophile PS3 F_1-ATPase and also predicted from its primary sequence (see
below). Hoppe et al. [18] reported that E. coli protein b in F_1 depleted
membranes is very sensitive to proteolytic degradation and F_1 protects b in
these membranes from proteolysis. Additionally, a photoreactive hydrophobic
phospholipid derivative reacts with Lys-21. These experiments support the model
for b proposed above.

Thus, protein b appears to occupy a central position in the ATP-synthase
complex in providing a structural link between F_1 and F_o. In this position it
may also be important in H^+ translocation. It may also be important in biosyn-
thetic assembly.

F_1 subunits

The functional properties of F_1 subunits have been discussed by Fillingame [4].
α and β both contain nucleotide binding sites. It has been suggested that β
contains the catalytic sites of the enzyme (evidence summarised by Futai and
Kanazawa [19]). However, point mutations in α or β result in loss of ATPase
activity and so α also plays an important role. This is emphasised by the
finding that the sequences of α and β are homologous and therefore will have
very similar protein folds. In the following section we discuss the significance of this homology and describe attempts to delineate sequences important
for binding nucleotides.

α and β subunits

The importance of the β subunit in the F_1-ATPase complex is emphasised by the
high degree of sequence conservation between E. coli bovine mitochondria and
maize and spinach chloroplasts (Table 1). These include conserved residues
identified in chemical labelling studies as being important in catalysis. Thus,
labelling of Tyr-385 of the bovine enzyme with p-fluorosulphonyl [^{14}C]-benzoyl-
5'-adenosine inactivated F_1-ATPase [20]. The F_1-ATPase is also inactivated by

TABLE 1

PERCENTAGE SIMILARITY MATRIX FOR β SUBUNITS OF ATP-SYNTHASE
The values were calculated from the sequences shown in ref.
[14].

	1	2	3	4
1. E. coli				
2. Beef mitochondria	70			
3. Maize chloroplasts	64	66		
4. Spinach chloroplasts	65	66	88	

DCCD. In the bovine enzyme the label was located at Glu-215 [21]; the label in
TF_1 inactivated with DCCD was localised to a glutamic acid residue in a peptide
identical to residues 200-205 of the E. coli protein [22].

Sequence studies of the bovine α subunit show that it also is very similar to
its bacterial counterpart (V.L.J. Tybulewicz and J.E. Walker, unpublished work).

In contrast, the homology detected between E. coli α and β subunits [23] is
weak. It persists throughout most of their peptide chains but is strongest
from residues 101 to 208.

Since α and β have related sequences, and therefore probably a common fold, it
is highly likely that they will have common functions. The best defined
biochemically is that they both bind adenine nucleotides [for reviews see refs.
24-26]. We have identified sequences in α and β that may contribute to nucleotide binding by comparison of their sequences with other proteins that employ
ATP in catalysis [23]. In this search we used the computer program DIAGON to
look for the most similar regions between pairs of proteins. In this way we
detected related sequences in α, β and adenylate kinase. Importantly, the
sequences in adenylate kinase form part of the AMP binding pocket [27]. Related
sequences were also detected in a number of other proteins including myosin in
the 25 kDal peptide. This peptide forms part of the S_1 head [28]; chemical
labelling studies and proteolysis studies have implicated sequences in it as
being important for binding ATP [29; R. Yount, personal communication]. In a
third example phosphofructokinase the related sequences also form the part of
the catalytic site believed to bind ATP [30].

The homologous sequences in α and β detected in this study clearly are not
the only residues involved in binding adenine nucleotides. This is illustrated
by a recent photoaffinity labelling study of F_1-ATPase with 8-azido ATP and 8-
azido ADP [31]. Under particular conditions the label can be localised predomi-

nantly (about 85%) on the β subunit and causes total inactivation of ATP hydrolysis [32]. Excess of ATP protects against reaction. By chemical degradation the label has been localised in two specific regions of the polypeptide chain. The label in one region is localised on three amino acids, lysine-301, isoleucine-304 and tyrosine-311. The second region is the amino terminal tryptic arginine peptide. The region corresponding to N-terminal region of the bovine β protein is largely absent from the bacterial enzyme. It can be removed from the bovine protein by limited proteolysis of BF_1 (J.E.W., unpublished work) and so appears to be an exposed tail presumably able to be near the nucleotide binding site of the enzyme.

Assembly

(i) Gene order and assembly of ATP-synthase

The order of genes in the unc operon is striking. It is related to the structure of ATP-synthase in so far as genes for F_o components a, c and b are clustered at the promoter proximal end and are followed by genes for F_1 subunits in the order δ:α:γ:β:ε. It is noteworthy that the order β:ε is also found in maize and spinach chloroplasts [33,34] and in the photosynthetic bacterium Rhodopseudomonas blastica (unpublished work). Clustering of genes with related functions is a feature of the genetic maps of coliphages (T4, lambda and other phages [35]); in lambda [36], morphogenetic genes for head proteins form a cluster adjacent to a cluster for tail genes; in T4, genes for baseplate head and tail also form clusters. It has been suggested that these gene clusters might reflect the evolutionary origin of present day assembly genes by tandem duplication and divergence of ancestral assembly genes [37]. In addition, or alternatively, clustering could reflect the selective advantage of minimising recombination between genes for proteins that interact structurally so as to decrease the probability of non-viable hybrids in interstrain matings [37–39]. In lambda the similarity between the order of action of gene products during assembly (J,I,L,K,H,G,M,V,U,Z) and the arrangement of genes in the lambda genome (J,I,K,L,M,H,G,V,U,Z) is a particularly good example in support of this idea. A third possibility is that proteins that interact with each other are synthesised near to each other in the cell [36]. All of these possibilities also apply to the unc operon. For example, evolutionary relationships between genes are apparent, α being related to β and related sequences being present in γ and ε, and ε and b, and evidence has been presented for interaction between subunits from all adjacent pairs of genes in the unc operon. However, the gene order is not in accord with the order of action of gene products proposed by Cox et al. [40], discussed below. A fourth explanation for the gene order in the lambda morphogenetic locus was advanced by Parkinson [41], namely that the genes might be arranged in such a way that their temporal sequence of expression would parallel their requirement during morphogenesis. Subsequently in lambda this explanation was discounted [36]. The sequence of assembly proposed by Cox et al. [40] suggests that the first events are association of β and then α with the membrane already containing a. Only after association of one α subunit and one β subunit can b enter the membrane. This is based upon the finding that a mutant in the β subunit did not contain b in its membranes. Mutants in the α subunit exerted a similar effect. A further β subunit is then added. γ, δ and ε associate with the 2β:1α structure before addition of further α and β subunits. The terminal stages of the pathway are described as tentative [40]. An alternative interpretation to these results is that in the absence of α or β, F_1 cannot assemble and that incorporation of b is dependent upon an assembled F_1. F_o assembles subsequent to this event.

(ii) Does gene 1 product influence assembly?

Although the region of the unc operon called gene 1 or uncI [14] has the characteristics of a gene, convincing demonstration of the gene product is still lacking. Schneider and Attendorf [42] have demonstrated the presence of small amounts of a 14 kD protein in preparations of F_o and have suggested that it might correspond to gene 1 product. However, protein sequence information to confirm this suggestion is lacking at present.

These difficulties notwithstanding, it is possible from the putative gene sequence to make predictions about the structure of its gene product. The protein would have a net positive charge of +11 although the protein overall is rather hydrophobic. Analysis of its sequence with HYDROPLOT (Fig. 5) suggests

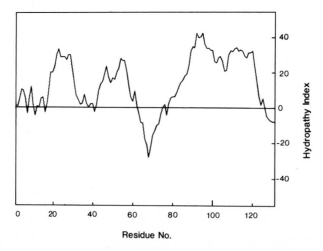

Fig. 5. Hydrophobic profile of uncI gene product

the presence of transmembrane α-helical segments and a folding pattern based upon this has been proposed [14]. This speculative model is of a membrane protein in which the positive charges cluster to one side of the membrane.

On the basis of these criteria it was suggested that the protein might fulfil a role in assembly by providing a site to anchor the unc transcript near to its 5' end, such that the F_o polypeptides might be produced near to each other and to the membrane. It is envisaged that this would aid efficient assembly of the enzyme; this might be important for the organism under difficult growth conditions. This model has little experimental basis. A deletion mutant (SD42) of gene 1 was constructed [43] for which a clear phenotype has not been found. However, its reduced growth yield in comparison with wild type is consistent with a role in efficient assembly of the enzyme complex.

REFERENCES

1. Mitchell, P. (1981) in Mitochondria and Microsomes (Lee, C.P., Schatz, G. and Dallner, G., eds.), pp. 427-457, Addison-Wesley Publishing Co., Reading, Mass., USA.

2. Nicholls, D. (1981) in Bioenergetics. Academic Press, London.

3. Bragg, P.D. and Hou, C. (1975) Arch. Biochem. Biophys. 167, 311-321.

4. Fillingame, R.H. (1981) Curr. Topics Bioenerg. 11, 35-106.

5. Foster, D.L. and Fillingame, R.H. (1982) J. Biol. Chem. 257, 2009-2015.

6. Racker, E. (1981) in Mitochondria and Microsomes (Lee, C.P., Schatz., G. and Dallner, G., eds.), pp. 337-356, Addison-Wesley Publishing Co., Reading, Mass. USA.

7. Senior, A.D. (1979) in Membrane Proteins in Energy Transduction (Capaldi, R.A., ed.), pp. 233-276, Dekker Inc., New York and Basel.

8. Nelson, N. (1981) Curr. Topics Bioenerg. 11, 1-34.

9. Kagawa, Y. and Racker, E. (1966) J. Biol. Chem. 241, 2461-2482.

10. MacLennan, D.H. and Tzagoloff, A. (1968) Biochemistry 7, 1603-1610.

11. Tzagoloff, A., MacLennan, D.H. and Byington, K.H. (1968) Biochemistry 7, 1596-1602.

12. Senior, A.E. (1971) Bioenergetics 2, 141-150.

13. Kanner, B.I., Serrano, R., Kandrach, M.A. and Racker, E. (1976) Biochem. Biophys. Res. Commun. 69, 1050-1056.

14. Walker, J.E., Saraste, M. and Gay, N.J. (1983) Biochim. Biophys. Acta Bioenerg. Rev. (in press).

15. Kyte, J. and Doolittle, R.F. (1982) J. Mol. Biol. 157, 105-132.

16. Sebald, W. and Hoppe, J. (1981) Curr. Topics Bioenerg. 12, 2-64.

17. Walker, J.E., Saraste, M. and Gay, N.J. (1982) Nature 298, 867-869.

18. Hoppe, J., Friedl, P., Schairer, H.U., Sebald, W. and Jørgensen, B.B. (1982) EMBO J. 2, 105-110.

19. Futai, M. and Kanazawa, H. (1980) Curr. Topics Bioenerg. 10, 181-215.

20. Esch, F.S. and Allison, W.A. (1978) J. Biol. Chem. 253, 6100-6106.

21. Yoshida, M., Poser, J.W., Allison, W.S. and Esch, F.S. (1981) J. Biol. Chem. 256, 148-153.

22. Yoshida, M., Allison, W.J. and Esch, F.S. (1981) Fed. Proc. 40, 1734.

23. Walker, J.E., Saraste, M., Runswick, M.J. and Gay, N.J. 1982) EMBO J. 1, 945-951.

24. Harris, D.A. (1978) Biochim. Biophys. Acta. 463, 245-273.

25. Slater, E.C., Kemp, A., van der Kraan, I., Muller, J.L.M., Roveri, O.A., Verschoor, G.J., Wagenvoord, R.J. and Wielders, J.P.M. (1979) FEBS. Lett. 103, 7-11.

26. Cross, R.L. (1981) Ann. Rev. Biochem. 50, 681-714.

27. Pai, E.F., Sachsenheimer, W., Schirmer, R.H. and Schultz, G.E. (1977) J. Mol. Biol. 114, 37-45.

28. Bálint, M., Sréter, F.A., Wolf, I., Nagy, B. and Gergely, J. (1975) J. Biol. Chem. 250, 6168-6177.

29. Szilagyi, L., Bálint, M., Sréter, F.A. and Gergely, J. (1979) Biochem. Biophys. Res. Commun. 87, 936-945.

30. Evans, P.R., Farrants, G.W. and Hudson, P.J. (1981) Phil. Trans. R. Soc. Lond. B. 293, 5B-62.

31. Hollemans, M., Runswick, M.J. and Walker, J.E. (1983) J. Biol. Chem. (in press).

32. Wagenvoord, R.J., Kemp, A. and Slater, E.C. (1980) Biochim. Biophys. Acta 593, 204-211.

33. Krebbers, E.T., Larrinua, I.M., McIntosh, L. and Bogorad, L. (1982) Nucleic Acids Res. 10, 4985-5002.

34. Kurawski, G., Bottomley, W. and Whitfield, P.R. (1982) Proc. Nat. Acad. Sci. USA 79, 6260-6264.

35. Wood, W.B. and King, J. (1979) Comprehensive Virology 13, 581-633.

36. Katsura, I., in "Lambda II" (Hendrix, R.W., Roberts, J.W. and Stahl, F.W., eds.), Cold Spring Harbor (in press).

37. Casjens, S. and Hendrix, R. (1974) J. Mol. Biol. 90, 20-23.

38. Kikuchi, Y. and King, J. (1975) J. Mol. Biol. 99, 695-716.

39. King, J. and Laemmli, U.K. (1973) J. Mol. Biol. 75, 315-337.

40. Cox, G.B., Downie, J.A., Langman, L., Senior, A.E., Ask, G., Fayle, D.R.H. and Gibson, F. (1981) J. Bact. 148, 30-42.

41. Parkinson, J.S. (1968) Genetics 59, 311-325.

42. Schneider, E. and Altendorf, K. (1982) Eur. J. Biochem. 126, 149-153.

43. Gay, N.J. (1983) J. Bact. (submitted for publication).

STRUCTURE AND FUNCTION OF ATP SYNTHASE

E.C. SLATER, M. HOLLEMANS AND J.A. BERDEN
Laboratory of Biochemistry, B.C.P. Jansen Institute, University of Amsterdam,
P.O. Box 20151, 1000 HD Amsterdam (The Netherlands)

TERTIARY STRUCTURE

The structure - or, at least, its rough shape - of the ATP synthase has been
known longer than its function. It was first described as a tripartite unit by
Fernandez Moran in electron micrographs of bovine-heart mitochondria after
negative staining [1]. Indeed, this was one of the first great successes of
the negative-staining technique.

The tripartie unit, consisting of a base-piece, a stalk and a knob was
first interpreted by Green to represent the four multi-subunit enzymes required
for the oxidation of NADH and succinate by oxygen. Nearly 20 years ago, however,
at the first ISOX meeting, the results of collaborative research between
E. Racker, B. Chance and D.F. Parsons, the electron microscopist of the
team, were reported [2] in which it was clearly shown that the knob is the
coupling factor F_1, which when separated from the base-piece is able to catalyse either the hydrolysis of ATP, or when a suitable energy source is present,
its synthesis [3]. The base-piece is now identified with the membrane moiety,
F_o [4]. McLennan and Asai [5] proposed that the stalk, which the electron
micrographs clearly show as an attachment between F_1 and F_o, is identical
with a protein OSCP (oligomycin-sensitivity conferring protein) which is
required for the attachment of F_1 to F_o and is also required to make the ATPase
activity of F_1 sensitive to oligomycin which does not bind to F_1 itself, but
to F_o. The 'o' stands, in fact, for oligomycin *not* for zero or nought. Since
no further information has come available concerning the composition of the
stalk, McLennan and Asai's proposal is still a valid one, although it is not
yet been proven (see [6]).

Indeed, we know no more about the tertiary structure of the stalk or F_o
than at the time of Fernandez Moran. I sincerely hope that the new methods
of crystallizing membrane proteins that will be reported at this meeting will
be directed towards F_o.

The knob F_1 is quite hydrophilic and has been crystallized, and X-ray
diffraction of crystals of rat-liver F_1 to 9 Å resolution has been analysed
by Amzel [7]. He concludes that the molecule consists of two equivalent halves,
each formed by three regions of approximately equal size, that is a total of

six regions of equal size, with a central hole. Amzel gives two possible subunit arrangements compatible with the subunit stoicheiometry which is now - after a period of doubt - established to be $\alpha_3\beta_3\gamma\delta\epsilon$. The six blocks correspond to the six regions revealed by the X-ray diffraction pattern. In one possible arrangement, each of the blocks contains one α or β subunit, but one block represents a combination of a β subunit with one or more of the smaller subunits, and a second block an α subunit with the other smaller subunit(s). In an alternative arrangement, the other four blocks contain domains from both of the major subunits. Electron microscopy of bovine-heart F_1 is consistent with this picture. Tiedge and coworkers [8] have proposed that the central hole is occupied by the γ subunit.

The important conclusion to be drawn from the two-fold symmetry in Amzel's model, despite the inherent asymmetry of the subunit composition, is that the major subunits exist in at least two different environments in the enzyme and that different copies are functionally not equivalent. As he aptly remarks "This observation appears to provide a natural explanation of the complicated binding and labelling data of F_1 ATPases" [7].

SUBUNIT STRUCTURE

The operon directing the synthesis of the ATP synthase in *E. coli* - called the unc operon by Gibson [9] - has been sequenced by Walker [10] and Kamazawa [11,12]. It contains 9 open reading frames. Eight of the gene products have been identified as components of the synthase, 5(α, β, γ, δ and ϵ) in F_1 and 3(a, b and c) in F_0. The ninth open reading frame has been called Gene I by Walker [10] (see Table I).

The mitochondrial ATPase complex contains at least three additional small subunits. The ϵ subunit, which is part of F_1, has, according to Walker [13], no counterpart in *E. coli*. OSCP [14] (also called F_5 and Fc_1 [15]), like the homologous [13] δ subunit in *E. coli*, is involved in binding F_1 to F_0. An additional subunit (F_6 or Fc_2) is also required for binding of F_1 to F_0 in mitochondria [15,16], but not in *E. coli*. Mitochondria contain, furthermore, an inhibitory subunit [17], whose function in *E. coli* is fulfilled by the ϵ subunit, but these subunits are not structurally homologous.

Subunits a and c clearly have their counterparts in mitochondria. These subunits, usually called ATPase 6 and 9, respectively, are coded for by the mitochondrial genome in yeast. The subunit ATPase 4, which is poorly characterized, may correspond to subunit b in *E. coli*.

The latest component to join F_0 in mitochondria is a 5.8 KDa subunit that is coded for by the mitochondrial genome in yeast. Since a similar sequence

TABLE 1

SUBUNITS OF ATP SYNTHASE

E. coli			Mitochondria		
Gene	Gene product	Mol. wt. x 10^3	Gene	Gene product	Mol. wt. x 10^3
Established					
unc A	α	55.4		α	55
unc D	β	50.4		β	52.3
unc G	γ	31.5		γ	32
unc C	ε	14.3		δ	12^b - 17^a
				ε	6^a - 10^b
unc H	δ	19.3		OSCP (F_5,Fc_1)	18
				F_6 (Fc_2)	10
				I	10
unc B	a	30.3	oli 2	ATPase 6	24.6^a - 28.3^b
unc F	b	17.2		ATPase 4	*c* 27
unc E	c	8.3	oli 1	ATPase 9	7.6
Proposed					
gene 1	?	14.0	aap 1 (URF A6L)	ATPase 8	5.8^b - 7.9^a
				UBP	30
				Factor B	12
				Unnamed	12

[a]Bovine.
[b]Yeast.

(although considerably longer) is also present in human and bovine mitochondrial DNA - in the URF designated A6L [18] - this protein is very likely expressed in mammalian mitochondrial ATP synthase.

The subunit was first described in 1978 by Guérin and colleagues [19] in
Bordeaux as a mitochondrially synthesized phosphate-binding proteolipid running
at about 10 KDa on SDS gels. Recently, this group has isolated it from both
intact yeast mitochondria and an ATP synthase preparation and determined the
amino acid composition [20]. Linnane and coworkers [21,22] in Melbourne
discovered mutants of a newly identified mitochondrial gene, denoted aap 1
(ATPase-associated protein), in which this

subunit (denoted subunit 8 in earlier work) is absent in both SDS extracts of mitochondria and immunoprecipitates with anti-holo-ATPase antibody. Furthermore, ATPase subunit 6 is absent from the immunoprecipitates although present in total mitochondrial extracts. It seems, then, that subunit 8 is necessary for the assembly of subunit 6 into the ATP synthase.

The molecular weight, calculated from the amino acid sequence read off from the DNA, is only 5.8 KDa but a comparison of the amino acid composition calculated from this sequence and that determined by the French group [20] (recalculated from the same molecular weight) leaves little doubt that it is the same protein. The corresponding sequences coded for by the URF A6L [18] in human and bovine mitochondrial DNA are considerably longer (68 and 66 amino acids, respectively) and show little homology, apart from the starting sequence and positively charged residues near the C-terminal. Indeed, URF A6L is the least conserved of the major reading frames in the human and bovine genomes [18]. The molecular weight of the bovine protein is 7927 compared with 5800 for the yeast protein. The gene product of gene 1 in *E. coli*, which, like aap 1 in yeast and URF A6L in mammalian mitochondria, precedes the gene for ATPase 6, is much larger.

Three additional components have been proposed in mitochondrial F_o, but none can be considered to be firmly established. By sucrose-gradient centrifugation, Berden and Henneke [23] have removed Hanstein's [24] uncoupler-binding protein from ATP synthase without affecting its ability to catalyse an ATP-P_i exchange. However, since the exchange activity in Berden's preparation is very low, the possibility that this subunit is necessary in the intact system remains open, especially as treatment of sub-mitochondrial particles with trypsin, which drastically lowers the exchange activity, seems specifically to affect this subunit. Factor B, defined by Sanadi [25] as a thiol-containing protein required for the ATP-P_i exchange but not for ATPase, is not detectable (with radioactive N-ethylmaleimide) in Berden's preparations. The ATP-P_i exchange is not inhibited by N-ethylmaleimide, nor is it stimulated by adding Factor B prepared by Sanadi's method even after extraction with NH_4OH which removes Factor B, or when the ATP synthase is prepared from particles deficient in Factor B. However, a second 12 KDa subunit, not labelled with N-ethylmaleimide, is always present in ATP synthase preparations.

PRIMARY AND SECONDARY STRUCTURES

The primary structures of all 8 (or 9?) subunits of the *E. coli* ATPase have been established via DNA sequencing [10-12]. The structure of the bovine β subunit has also been published by Runswick and Walker [12] and that of the chloroplast (via the chloroplast DNA) by Bogorad [26] (in maize) and Whitfield [27] (in spinach).

Since the tertiary structure of these subunits is not known, it is important to obtain as much information as possible from the primary structure. Sebald and Hoppe [28] and Senior [29] have modified the Chou-Fasman rules [30] to make them applicable to membrane proteins and applied them to the F_o subunits. This matter is dealt with by others elsewhere in this volume.

FUNCTION OF ATP SYNTHASE

The function of the ATP synthase in aerobic organisms is to utilize energetic protons originating from electron transfer in the respiratory chain for the synthesis of ATP from ADP and phosphate. In some anaerobic organisms, it has the converse function, namely to couple the hydrolysis of ATP to the supply of energetic protons.

There are 6 binding sites for ATP and ADP. The photo-affinity label 8-azido-ATP labels 2 β subunits and 2 α subunits [31]. The label fluorosulfonylbenzoyl-adenosine labels 3 β subunits [32]. Two of the 6 binding sites bind ATP and ADP much more firmly than the other 4 and are not accessible to 8-azido-ATP [31]. One of these tight binding sites is presumably on the third β subunit that can be labelled with the fluorosulfonyl label, the other is probably on the α subunit, but there is no conclusive evidence for this. At least one catalytic site is on a β subunit. A regulatory site (or sites) is (are) also present [33,34], probably on α subunits, but this is not firmly established. Binding of F_1 to F_o alters the catalytic site in such a way that hydrolysis of ATP is compulsorily linked to the production of energetic protons, and becomes sensitive to inhibitors (oligomycin, DCCD) that block the production of energetic protons. Friedl and coworkers [35] have recently shown that in the *E. coli* enzyme both subunits a and b, individually and together, bind F_1 but that subunit c, which does not bind F_1, is necessary for inhibition by DCCD. The delivery of energetic protons from the respiratory chain to the enzyme drastically alters the conformation of F_1, for example by weakening the binding of adenine nucleotide to the tight sites [36]. As predicted earlier by Paul Boyer [37], energy is not necessary for ATP synthesis per se. Addition of phosphate and ADP to chloroplast F_1 results in the synthesis of enzyme-bound ATP, just as is the case with myosin (see below).

A number of groups including those of Richard Guillory [38], Pierre Vignais

[39], Gunter Schäfer [40], Hans-Jochen Schäfer [41], and Rob Wagenvoord and Bertus Kemp [42] in our laboratory have been using photo-affinity azido compounds to locate the adenine nucleotide-binding sites in F_1. Wagenvoord and coworkers have used 8-azido-ATP which has the advantage that the nitreno group which is formed from the azide when the compound is irradiated is only one atom away from the adenine ring. It is very important to check that, under the experimental conditions used, the chosen photo-label is indeed a photo-affinity label, that is that it binds specifically to the site to be investigated, in this case the catalytic site. Wagenvoord [43] showed in different ways that 8-azido-ATP satisfies this criterion. The strongest evidence is that it is hydrolysed by F_1 and it competes with ATP for the binding site. This is shown by the fact that the rates of hydrolysis at different ATP concentrations in the presence of different amounts of 8-azido-ATP are the same as those calculated assuming that the two substrates compete for the same site with the K_m's and V_{max}'s determined for the two substrates separately.

When F_1 is irradiated with radioactive 8-azido-ATP, virtually all the label, except for a small amount in the γ subunit, is found in the α-β region, mostly on the β subunit, although some is on the α [42]. When F_1 was irradiated with different concentrations of azido-ATP for different periods, a linear relation was found between the amount of bound label and the degree of inactivation. Complete inactivation was obtained when 2 molecules azido-ATP - or rather nitreno-ATP - were bound [42].

Having satisfied ourselves that 8-azido-ATP is a good affinity label for the ATP-catalytic site on the β subunit, it was obviously of interest to use it to determine where the adenine ring binds to the enzyme, since only one nitrogen atom bridges the 8-carbon atom of the adenine to F_1 in the nitreno-F_1. To locate the amino acids to which the nitreno-adenine is bound, we joined forces with John Walker at the MRC Laboratory in Cambridge and the results of these studies are reported in a paper now in press [44].

After covalently labelling F_1 with 8-nitreno-[2-^{3}H]ATP, the subunits were fractioned on a Sepharose 6B column in the presence of 8 M urea. 90% of the bound label was found in the peak containing the α and β subunits. Fractionation of this peak on DEAE-cellulose also in the presence of 8 M urea showed that 65% of the label was on the β subunits and 25% on the α subunits. Fractionation of the cyanogen bromide digest of the fraction containing the β subunit showed that the label was confined to three peaks. The first corresponds to partial cleavage products, the second to a region containing 5 previously characterized peptides. By high-pressure liquid chromatography a

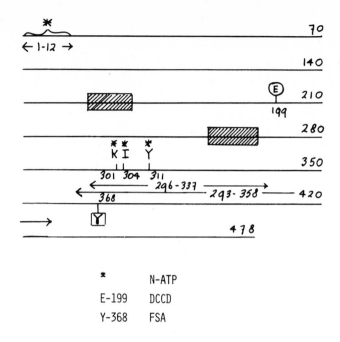

Fig. 1. Labelling of β subunit of F$_1$. The arrows mark the polypeptides labelled with 8-nitreno-ATP, the asterisks indicate the labelled amino acids (K = lysine, I = isoleucine, Y = tyrosine). The binding site of DCCD (Glu-199) [46] and p-fluorosulfonylbenzoyl-5'-adenosine (Tyr-368) [32] are also shown. The cross-hatched regions are weakly homologous to other enzymes reacting with ATP [45].

radioactive peptide 293-358 was isolated (see Fig. 1). The third radioactive peak did not contain amino acids and was not investigated further.

A second digest of the labelled β subunit was made with trypsin after first modifying the amino groups with succinic anhydride. Radioactivity was localized in three peaks. The first was shown by high-pressure liquid chromatography to contain the sequnce 296-337. The second does not contain amino acids. The third contains a mixture of several polypeptides. By further fractionation by high-pressure liquid chromatography a radioactive peptide 1-12 was obtained.

Thus the label was confined to two regions, 1-12 and 296-337 (see Fig. 1). Many other peptides were isolated from both digests and shown to be non-radioactive, notably the peptide 357-372 which contains Tyr - 368, which was identified by Esch and Allison [32] as the binding site of p-fluorosulfonyl-benzoyl-5'-adenosine.

In order to determine the site of attachment of radioactivity, the peptide 297-337 was immobilized on porous glass and degraded in a solid phase micro-sequencer. Radioactivity was released at Lys-301, Ile-304 and Tyr-311. The radioactivity released in the first cycle is probably some persistent non-covalently bound label.

The precise site of attachment of label in peptide 1-12 was not determined. It is unlikely to be important for binding ATP; this region is not homologous to the equivalent regions in the chloroplast enzyme, it is almost entirely absent from the *E. coli* enzyme and residues 1-5 are removed by mild trypsin hydrolysis without affecting the ATPase activity. It is concluded, then, that the specific binding site of the adenine at the catalytic site is close to Lys-301, Ile-304 and Tyr-311. This tyrosine is different from that to which *p*-flu-orosulfonylbenzoyl-5'-adenosine binds, which is not surprising since if the adenine ring of this compound binds in the same region as that in ATP, one must expect that the fluorosulfonyl group would label the ATP-binding site in the region of the γ-phosphate.

The 8-C of adenine does not bind near two regions that, on the basis of comparisons with other ATP-reacting proteins, have been suggested to contri-bute to the nucleotide-binding fold [45]. However, this does not exclude the possibility that these residues contribute to the binding fold. The binding site is also quite far from the glutamic acid to which DCCD binds [46], which has also been suggested as being in the catalytic site. Gromet-Elhanan has shown, however, that this region is specifically involved in binding the β subunit to the rest of the ATP synthase [47].

It would be very interesting to know the binding site of another type of photo-affinity label, 3'-benzoylbenzoic acid ATP, introduced recently by Coleman [48]. In this case, the label is on the ribose ring and is separated from it by a carbonyl group, a benzene ring and a carbon subsituent in the benzene ring. Benzoylbenzoic ATP is a good substrate for the enzyme.

The adenine-binding region is well-conserved in F_1 isolated from spinach and maize chloroplast, bovine mitochondria and *E. coli* (Fig. 2). The iso-leucine and tyrosine residues are conserved, the lysine is not.

The tyrosine to which the fluorosulfonyl compound binds is also in a well-conserved region. It is noticeable that acidic residues are quite close to the tyrosines at both sites.

Kozlov [49] has shown that iodide quenches the fluorescence of etheno-ATP bound to F_1, from which he concludes that the adenine ring is bound near the surface of the molecule, which, as he says, is surprising since the hetero-

ADENOSINE TRIPHOSPHATE-BINDING SITE IN F_1

		*		*		*
Bovine	K	GSITS	V	QA	I	YVPADDLT
E. coli	T	GSITS	V	QA	V	YVPADDLT
Spinach	E	GSITS	I	QA	V	YVPADDLT
Maize	L	GSITS	L	QA	V	YVPADDLT

p-FLUOROSULFONYLBENZOYL-5'-ADENOSINE-BINDING SITE IN F_1

					*		
Bovine	P	NI	VG	S	EHY	DV	A
E. coli	P	LV	VG	Q	EHY	DT	A
Spinach	P	RI	VG	E	EHY	EI	A
Maize	P	RI	VG	N	EHY	ET	A

Fig. 2. Homologous 8-azido-ATP and p-fluorosulfonylbenzoyl-5'-adenosine-binding sites in β subunit of F_1. The p-fluorosulfonylbenzoyl-5'-adenosine-binding site was determined in the bovine enzyme by Esch and Allison [32]. The homologous sequences are taken from [13], [10], [27] and [26] for the bovine, E. coli, spinach and maize β subunits, respectively.

cyclic base is the most hydrophobic moiety of the ATP. In fact, however, the adenine-binding site shown here is quite hydrophilic, which probably explains Kozlov's findings.

MECHANISM OF ACTION

I do not have time to speculate about the mechanism of action of the ATP synthase. In my opinion, the most important development in recent years is the increasing support for the idea, most explicitly propounded by Paul Boyer [37,50], that the energy from electron transfer is not used primarily to make the β-γ phosphoryl bond of ATP but to promote binding of ADP and phosphate, and dissociation of ATP. This hypothesis received its stimulus from three directions: the insensitivity of the esterification reaction, as measured by ^{18}O exchange, to uncouplers of oxidative phosphorylation [51], the existence of firmly bound ADP and ATP in isolated F_1 [52], and most recently, and perhaps most convincingly, by the demonstration by Feldman and Sigman [53] in Paul Boyer's laboratory that at pH 6.0 and high $^{32}P_i$ concentration, in the absence of membranes, ADP bound to isolated chloroplast F_1 is phosphorylated to enzyme-bound radioactive ATP, which is not accessible to hexokinase. Since addition of cold ATP led to a rapid loss of radioactivity from the bound ATP, it is concluded that the enzyme-bound ATP was formed at a catalytic site. How the various catalytic and regulatory binding sites for ADP and ATP are involved in the synthesis of ATP and its release in the complete ATP synthase is as yet by no means clear, but progress is being made [50].

ACKNOWLEDGEMENTS

This work was supported in part by the Netherlands Organization for the Advancement of Pure Research (ZWO) under the auspices of the Netherlands Foundation for Chemical Research (SON). One of us (M.H.) thanks EMBO for a fellowship.

REFERENCES

1. Fernandez-Moran, H., Oda, T., Blair, P.V. and Green, D.E. (1964) J. Cell Biol. 22, 63-100.
2. Racker, E., Tyler, D.D., Estabrook, R.W., Conover, T.E., Parsons, D.F. and Chance, B. (1965) in: Oxidases and Related Redox Systems (King, T.E., Mason, H.S. and Morrison, M., eds.) Vol. 2, pp. 1077-1094, Wiley, New York.
3. Pullman, M.E., Penefsky, H.S., Datta, A. and Racker, E. (1960) J. Biol. Chem. 235, 3322-3329.
4. Kagaya, Y. and Racker, E. (1966) J. Biol. Chem. 241, 2461-2466.
5. MacLennan, D.H. and Asai, J. (1968) Biochem. Biophys. Res. Commun. 33, 441-447.
6. Racker, E. (1974) in: Molecular Oxygen in Biology (Hayaishi, O., ed.) pp. 339-361, North-Holland, Amsterdam.
7. Amzel, L.M., McKinney, M., Narayanan, P. and Pedersen, P.L. (1982) Proc. Natl. Acad. Sci. USA 79, 5852-5856.
8. Tiedge, H., Schäfer, G. and Mayer, F. (1983) Eur. J. Biochem. 132, 37-45.
9. Gibson, F. (1982) Proc. Roy. Soc. B 215, 1-18.
10. Saraste, M., Gay, N.J., Eberle, A., Runswick, M.J. and Walker, J.E. (1981) Nucleic Acids Res. 9, 5287-5296.
11. Kanazawa, H., Mabuchi, K., Kayano, T., Tamura, F. and Futai, M. (1981) Biochem. Biophys. Res. Commun. 100, 219-225.
12. Kanazawa, H., Kayano, T., Mabuchi, K., Tamura, F. and Futai, M. (1981) Biochem. Biophys. Res. Commun. 103, 604-612.
13. Runswick, M.J. and Walker, J.E. (1983) J. Biol. Chem. 258, 3081-3089.
14. Tzagoloff, A., Byington, K.H. and MacLennan, D.H. (1968) J. Biol. Chem. 243, 2405-2412.
15. Knowles, A.F., Guillory, R.J. and Racker, E. (1971) J. Biol. Chem. 246, 2672-2679.
16. Vǎdineanu, A., Berden, J.A. and Slater, E.C. (1976) Biochim. Biophys. Acta 449, 468-479.
17. Pullman, M.E. and Monroy, G.C. (1963) J. Biol. Chem. 238, 3762-3769.

18. Anderson, S., De Bruijn, M.H.L., Coulson, A.R., Eperon, I.C., Sanger, F. and Young, I.G. (1982) J. Mol. Biol. 156, 683-717.

19. Guérin, M. and Napias, C. (1978) Biochemistry 17, 2510-2516.

20. Velours, J., Esparza, M. and Guérin, B. (1982) Biochem. Biophys. Res. Commun. 109, 1192-1199.

21. Orian, J.M., Murphy, M. and Marzuki, S. (1981) Biochim. Biophys. Acta 652, 234-239.

22. Novitski, C.E., Macreadie, I.G., Maxwell, R.J., Lukins, H.B., Linnane, A.W. and Nagley, P. (1983) in: Manipulation and Expression of Genes in Eukaryotes (Nagley, P., Linnane, A.W., Peacock, W.J. and Papeman, J.A., eds.), Academic Press, Sydney, pp. 257-263.

23. Berden, J.A. and Henneke, M.A.C. (1981) FEBS Lett. 126, 211-214.

24. Hanstein, W.G. (1976) Biochim. Biophys. Acta 456, 129-148.

25. Sanadi, D.R. (1982) Biochim. Biophys. Acta 683, 39-56.

26. Krebbers, A.T., Larrinun, I.M., McIntosh, L. and Bogorad, L. (1982) Nucleic Acids Res. 10, 4985-5002.

27. Zurawski, G., Bottomley, W. and Whitfeld, P.R. (1982) Proc. Natl. Acad. Sci. USA 79, 6260-6264.

28. Sebald, W. and Hoppe, J. (1981) Curr. Top. Bioenerg. 12, 2-64.

29. Senior, A.E. (1983) Biochim. Biophys. Acta 726, 81-95.

30. Chou, P.Y. and Fasman, G.D. (1978) Adv. Enzymol. 47, 45-148.

31. Wagenvoord, R.J., Kemp, A. and Slater, E.C. (1980) Biochim. Biophys. Acta 593, 204-211.

32. Esch, F.S. and Allison, W.S. (1978) J. Biol. Chem. 253, 6100-6106.

33. Schuster, S.M., Ebel, R.E. and Lardy, H.A. (1975) J. Biol. Chem. 250, 7848-7853.

34. Recktenwald, D. and Hess, B. (1977) FEBS Lett. 76, 25-28.

35. Friedl, P., Hoppe, J., Gunsalus, J.P., Michelsen, O., Von Meyenburg, K. and Schairer, H.V. (1983) EMBO J. 2, 99-103.

36. Harris, E.J. and Slater, E.C. (1975) Biochim. Biophys. Acta 387, 335-348.

37. Kayalar, C., Rosing, J. and Boyer, P.D. (1977) J. Biol. Chem. 252, 2486-2491.

38. Russell, J., Jeng, S.J. and Guillory, R.J. (1976) Biochem. Biophys. Res. Commun. 70, 1225-1234.

39. Lunardi, J., Lauquin, G.J.M. and Vignais, P.V. (1977) FEBS Lett. 80, 317-323.

40. Weber, J., Lücken, U., Tiedge, H., Onur, G. and Schäfer, G. (1982) Second EBEC, 63-64.

41. Scheurich, P., Schäfer, H.J. and Dose, K. (1978) Eur. J. Biochem. 88, 253-257.

42. Wagenvoord, R.J., Van der Kraan, I. and Kemp, A. (1977) Biochim. Biophys. Acta 460, 17-24.

43. Wagenvoord, R.J., Van der Kraan, I. and Kemp, A. (1979) Biochim. Biophys. Acta 548, 85-95.

44. Hollemans, M., Runswick, M.J., Fearnley, I.M. and Walker, J.E. (1982) J. Biol. Chem., in the press.

45. Walker, J.E., Saraste, M., Runswick, M.J. and Gay, N.Y. (1982) EMBO J. 1, 945-951.

46. Esch, F.S., Böhlen, P., Otsuka, A.S., Yoshida, M. and Allison, W.S. (1981) J. Biol. Chem. 256, 9084-9089.

47. Khananshvili, D. and Gromet-Elhanan, Z. (1983) J. Biol. Chem. 258, 3720-3725.

48. Williams, N. and Coleman, P.S. (1982) J. Biol. Chem. 257, 2834-2841.

49. Kozlov, I.A. and Novikova, I.Yu. (1982) FEBS Lett. 150, 381-384.

50. Boyer, P.D., Kohlbrenner, W.E., Smith, L.T. and Feldman, R.E. (1982) Second EBEC, 23-24.

51. Boyer, P.D., Cross, R.L. and Momsen, W. (1973) Proc. Natl. Acad. Sci. USA 70, 2837-2839.

52. Harris, D.A., Rosing, J., Van de Stadt, R.J. and Slater, E.C. (1973) Biochim. Biophys. Acta 314, 149-153.

53. Feldman, R.I. and Sigman, D.S. (1982) J. Biol. Chem. 257, 1676-1683.

STRUCTURE – FUNCTION RELATIONSHIP IN MEMBRANE PROTEINS

© 1983 Elsevier Science Publishers B.V.
Structure and Function of Membrane Proteins,
E. Quagliariello and F. Palmieri editors.

CRYSTALLIZATION OF MEMBRANE PROTEINS AND ACTUAL STATE OF STRUCTURE
ANALYSIS OF PHOTOSYNTHETIC REACTION CENTRE CRYSTALS

H. Michel, J. Deisenhofer, K. Miki, K. Weyer and F. Lottspeich

Max Planck Institut für Biochemie, D-8033 Martinsried, West
Germany

INTRODUCTION

X-ray crystallography is still the only way to determine the
spatial structure of proteins at high resolution. Crystallography
depends on the availability of large (this means a size of several
microns in each dimension), well-ordered three-dimensional crystals. So far the main area without crystals has been the area of
the membrane proteins. The reason for the difficulties to crystallize membrane proteins resides in the amphipatic nature of the
proteins`surface. Those surface domains exposed to the aqueous
phases on both sides of the membrane are hydrophilic like the surface of globular proteins, whereas the surface domains which are
in contact with the alkane chains of the lipids must be highly
hydrophobic.

Very recently use of the small detergent ß-D-octylglucopyranoside allowed crystallization of the two integral membrane proteins
bacteriorhodopsin (1) and porin (2) by precipitation using salts
or polyethylenglycol. The success with this detergent fits well
into the working hypothesis (1,3) that the detergent is still
present in the crystals and bound to the hydrophobic protein surface domains with its hydrophopic tail. The recent determinations
of the detergent contents of the crystals and the analyses of the
crystal packing (4,5) substantiate this view.

A further improvement in the crystallization procedure was
achieved by the introduction of small amphiphilic molecules like
heptane-1,2,3-triol (3,4). These small amphiphilic molecules are
thought to replace those detergent molecules which are too large
to fit perfectly into cavities of the proteins crystal lattice.
The small amphiphilic molecules may also change shape and size of
the detergent micelles which incorporate the detergent-solubilized
protein.Larger polar surface parts of the membrane protein might
then be available for polar protein-protein interaction in the

crystallization process.

The application of this "small amphiphile concept" has recently allowed the crystallization of photosynthetic reaction centres from Rhodopseudomonas viridis, a bacteriochlorophyll b containing purple photosynthetic bacterium (6). The crystals of this membrane protein complex diffract X-rays to at least 2.5 Å resolution. An X-ray structure analysis is underway. In this communication we summarize the recent progress in our crystallization procedures for membrane proteins, and describe the actual state of X-ray structure analysis and protein chemistry of photosynthetic reaction centres.

RESULTS AND DISCUSSION

Crystallization

Useful combinations of precipitating agents, detergents and small amphiphilic molecules. Table 1 summarizes the combinations of detergents and salts which could be used to crystallize bacteriorhodopsin or photosynthetic reaction centres. Only in the case of bacteriorhodopsin in combination with ß-D-octylglucopyranoside could crystals be obtained in the absence of small amphiphilic molecules. Use of ammonium sulphate as precipitating agent in most cases leads to cube-like crystals, which must be disordered due to their unfavourable crystal packing (4). Crystallization in the presence of small amphiphilic molecules leads to small amphiphilic molecules which are better ordered. When sodium phosphate is used as precipitating agent only the hexagonal columns are observed. However, switching from ß-D-octylglucopyranoside to ß-D-nonylglucopyranoside or decanoyl-N-methylglucamide (7) as detergent in combination with the small amphiphilic molecule heptane-1,2,3-triol and sodium phosphate as precipitating agent leads to even better ordered crystals. With synchroton radiation a resolution of 3.3 Å along the long axis of the crystals was observed (experiments done together with H. Bartunik and D. Oesterhelt). Unfortunately the resolution in the two perpendicular directions is limited to 6 - 7 Å not allowing a high resolution structure determination.

To obtain crystals of photosynthetic reaction centres so far only the conditions given in table 1 could be applied. Quite

Table I: The successful combinations of detergents, precipitating agents and small amphiphilic molecules.

prot.:	det.:	precipitating agent:	comment:
bR	OG	ammonium sulphate or sodium phosphate	addition of small amphiphilic molecules improves crystal quality
bR	NG	sodium phosphate only	heptane-1,2,3-triol (Oxyl, Bobingen) had to be present
bR	DMG	sodium phosphate only	heptane-1,2,3-triol had to be present
RC	LDAO	ammonium sulphate only	heptane-1,2,3-triol had to be present

abbreviations: bR, bacteriorhodopsin; OG, ß-D-octylglucopyranoside (Riedl de Haen, Hanover, or Sigma, Munich); ß-D-nonylglucopyranoside (Calbiochem, Gießen); DMG, decanoyl-N-methylglucamide (Oxyl, Bobingen); LDAO, N,N-dimethyldodecylamine-N-oxide (Fluka, Neu-Ulm); RC, photosynthetic reaction centres.
Crystallization was achieved by vapour diffusion against 3 - 4 molar ammonium sulphate at pH 5.7 in the case of bacteriorhodopsin and against 2.2 - 2.5 molar ammonium sulphate at pH 5.5-7.5 in the case of photosynthetic reaction centres.

unexpectedly the heptane-1,2,3-triol isomer with amelting point of $80^\circ C$, probably corresponding to the threo-form, had to be used to achieve crystallization.

As it is seen in three of the four successful detergent/salt combinations the small amphiphilic molecule heptane-1,2,3-triol had to be present, demonstrating the value of the "small amphiphile concept". By far the best crystals, those of the photosynthetic reaction centres have been obtained using N,N-dimethyldodecylamine-N-oxide as detergent. Among those detergents in the table this detergent has the smallest polar headgroup, which might fit better into cavities of the proteins`crystal lattice. Unfortunately, due to the zwitterionic nature of the headgroup it denatures the more sensitive membrane proteins like bacteriorhodopsin.

Phase separation. Increasing the salt concentration of detergent solutions often leads to a phase separation into a viscous phase, which is enrichedin detergent and protein, and a salt containing, aqueous phase. In the crystallization of porin the crystals were obtained in the detergent phase (2,5). In the crystallization of

bacteriorhodopsin and photosynthetic reaction centres the phase
separation had to be avoided: bacteriorhodopsin loses its retinal
with a halftime of about 20 hours and forms an amorphous precipitate; bacteriochlorophyll b containing reaction centres lose their
bacteriochlorophyll and a green degradation product of bacteriochlorophyll b is observed in the detergent phase. Therefore we
expect that crystallization of membrane proteins in detergent
phases is restricted to the most stable membrane proteins.

Actual state of structure determination of photosynthetic reaction centres

  The crystallized photosynthetic reaction centres contain four
different protein subunits, one being a cytochrome of the c-type
containing four heme groups (8, K. Weyer, unpublished). The other
three subunits correspond to the so called H (heavy), M (medium)
and L (light) subunits of the well known reaction centres from
the bacteriochlorophyll a containing purple photosynthetic bacteria Rhodopseudomonas sphaeroides or Rhodospirillum rubrum. On
sodium dodecylsulphate polyacrylamide gels the subunits from the
Rhodopseudomonas viridis reaction centres show apparent molecular
weights of 38-40,000 (cytochrome), 35,000 (H-subunit), 28,000 (M-
subunit) and 24,000 (L-subunit).

  X-ray crystallography. The crystals belong to the tetragonal
space group $P4_12_12$ or its enantiomorph with the rather large unit
cell dimension of 223.5 x 223.5 x 113.6 Å. The asymmetric unit is
probably occupied by only one reaction centre molecule, since protein determinations using crystals of known volume yield a "molecular weight" of 136,000 22,800 (n =4) per asymmetric unit.

  A set of native X-ray diffraction data has been collected by
45° of rotationaround the c-axis and of 15° degree around the a-
or b-axis. To collect the data the crystals could be rotated only
by 0.5^O per photograph to avoid overlap of the reflections.
Two thirds of the reflections were only partially recorded on each
film and had to be added together. 204,361 reflections were measured with significant intensity. This corresponds to 54,347 crystallographic independent reflections (84.7% of all possible reflect-

ions to 2.9 Å resolution). After scaling and addition of the partially recorded reflections from subsequent films the R_{merge}-value
was 8.37 %.

To calculate an electron density map we must know the phase for
each reflection. In our case the only method to obtain phase information is the method of heavy atom isomorphous replacement.
Possible heavy atom derivatives have been obtained by conventional soaking of the crystals. Half-complete data sets have been
collected from five possible derivatives and are evaluated at
present.

Protein chemistry and partial amino acid sequences: For the
interpretation of the electron density map and refinement of the
structure it is of great advantage to know the amino acid sequences. Nowadays the quickest way seems to be to isolate and to
sequence the gene(s). However, protein chemistry and chemical
sequencing of the protein is still necessary in order to synthe-
size oligonucleotide probes or to prove the identity of an isolated gene.

To isolate the subunits the denatured cytochrome and the H-subunit were separated from the L- and M-subunits by molecular
sieve chromatography. The cytochrome could then be separated from
the H-subunit by chromatography on OH-apatite. The separation of
the L and M subunits was more complicated. These two subunits are
very similar in all respects tested. In addition, they are very
hydrophobic and tend to aggregate. The only way to separate these
subunits was preparative polyacrylamide electrophoresis in sodium
dodecylsulphate containing buffers.

We could sequence three of the four protein subunits, namely
the cytochrome, the M- and the L-subunit, from the N-terminus by
liquid phase sequencing (Edman-degradation). The N-terminus of
the H-subunit seems to be blocked. Table 2 shows the obtained
N-terminal amino acid sequences, and compares the sequences with
those published for the corresponding subunits from Rhodopseudomonas sphaeroides. These are the only published amino acid sequences for any reaction centres. The L-subunits show extreme sequence
homologies. From the first 26 amino acid only four are different.
This similarity may reflect the fact that the L-subunit carries
the bacteriochlorophylls, the bacteriopheophytins and the

quinone-iron complex as was demonstrated in the case of <u>Rhodospi-</u>
<u>rillum</u> <u>rubrum</u> (9). If the same is true for <u>Rhodopseudomonas</u>
<u>sphaeroides</u> and <u>Rhodopseudomonas</u> <u>viridis</u> then there is not much
space for evolutionary changes of the amino acid sequences.
Interestingly, the first 23 amino acids are preferentially hydrophilic. Then however, with the exception of one serine and one
tyrosine only hydrophobic amino acids are found, indicating that
this sequence must be located in the interior of the membrane
or the reaction centre.

Table II: NH_2-terminal amino acid sequences of three of the subunits from <u>Rhodopseudomonas</u> <u>viridis</u> reaction centres compared
with those from <u>Rhodopseudomonas</u> <u>sphaeroides</u> (10).

L-subunits:

Rp. vir. NH_2-ala-leu-leu-ser-phe-^{6}glu-arg-lys-tyr-arg-^{11}val-arg-
Rp. sph. NH_2-ala-leu-leu-ser-phe-^{6}glu-arg-lys-tyr-arg-^{11}val-pro-

Rp. vir.   gly-gly-thr-^{16}leu-ile-gly-gly-asp-^{21}leu-phe-asp-phe-
Rp. sph.   gly-gly-thr-^{16}leu-val-gly-gly-asn-^{21}leu-phe-asp-phe-

Rp. vir.   trp-^{26}val-gly-pro-tyr-phe-^{31}val-gly-phe-phe-gly-^{36}val-
Rp. sph.   (his)-^{26}val-/not published

Rp. vir.   (ser)-ala-ile-phe-^{41}phe-ile-phe-leu-gly-^{46}val-

M-subunits:

Rp. vir. NH_2-ala-asp-tyr-gln-thr-^{6}ile-tyr-thr-gln-ile-^{11}gln-ala-
Rp. sph. NH_2-ala-glu-tyr-gln-asn-^{6}ile-phe-phe-ser-gln-^{11}gln-val-

Rp. vir.   arg-gly-pro-^{16}(his)-ile-thr-val-(ser)-^{21}gly-glu-trp-
Rp. sph.   arg-gly-pro-^{16} ala -asp-leu-gly- met -^{21}thr-glu-asp-

Rp. vir.   gly-asp-^{26}asn-asp-arg-val-gly-
Rp. sph.   val-asn-^{26}leu-ala-asn-/ not published

Cytochrome (no corresponding subunit in <u>Rp</u>. <u>spaeroides</u>):
NH_2-ala-phe-glu-pro-pro-^{6}pro-ala-thr-thr-thr-^{11}gln-thr-gly-phe-
arg-^{16}gly-leu-(x)-met-gly-^{21}glu-val-leu-(x)-pro-^{26}ala-thr-val-

The M-subunits show much less sequence homology. From the first 15 amino acids 9 are identical. Then however no sequence homology at all is found.

Acknowledgements: We thank Prof. R. Huber and Prof. D. Oester-helt for generous help and support.

REFERENCES

1. Michel, H. and Oesterhelt, D.(1980) Proc. Natl. Acad. Sci. U.S.A. 77, 1283-1285

2. Garavito, R. M. and Rosenbusch, J.P.(1980) J. Cell Biol. 86, 327-329

3. Michel,H.(1983) Trends Biochem. Sci. 8, 56-59

4. Michel,H.(1982) EMBO Journal 1, 1267-1271

5. Garavito, R. M., Jenkins, J., Jansonius, J.N., Karlson, R. and Rosenbusch, J.P.(1983) J. Mol. Biol. 164, 313-327

6. Michel, H.(1982) J. Mol. Biol. 158, 567-572

7. Hildreth, J.E.K. (1982) Biochem. J. 207, 363-366

8. Thornber, J.P., Cogdell, R.J.,Seftor, R.E. and Webster, G.D. (1980) Biochim. Biophys. Acta 593, 60-75

9. Gimenez-Gallego, G., Suanzes, P. and Ramirez, J.M. (1982) FEBS Lett. 149, 59-62

10. Sutton, M.R., Rosen, D., Feher, G. and Steiner, L.A. (1982) Biochemistry 21, 3842-3849

PROPERTIES OF ANOMERIC OCTYLGLUCOSIDES USED IN THE CRYSTALLIZATION OF MEMBRANE PROTEINS

D.L. DORSET[1] AND J.P. ROSENBUSCH[2,3]

[1]The Medical Foundation, Buffalo, NY 14243 (U.S.A.), [2]European Molecular Biology Laboratory, 6900 Heidelberg (W. Germany) and [3]Biozentrum, University of Basel (Switzerland).

INTRODUCTION

Integral membrane proteins in their native environment are embedded in a quasi-solid state within the hydrophobic core of lipid bilayers. For membrane proteins other than those arranged regularly in 2-dimensional crystalline lattices *in vivo*, purification, characterization and reconstitution is re-quired for high resolution studies. To this end, 3-dimensional crystals of protein-detergent complexes are optimal, whereas 2-dimensional lattices, con-taining phospholipids as amphiphiles, are useful for correlating structural with functional observations. Since solution biochemistry is not applicable to such proteins, transformation into an isotropic, monodisperse form is a prerogative for all subsequent steps. Detergents forming highly dynamic protein-detergent mixed micelles are best suited to achieve such monodisperse solutions. For an appropriate selection of amphiphiles, knowledge of the properties of the detergents displacing the native lipids is required both in the presence or absence of protein. The surfactants which proved most useful exhibit high critical concentrations of micellization as well as well-defined phase diagrams. The critical (temperature-dependent) phenomena and their effect on the properties of proteins vary widely for different surfactants. The colloidal properties of several detergents in various states in the phase diagram have a profound effect on the crystallization of membrane proteins. Thus, small changes in the head or tail groups of alkyl-oligooxyethylenes strongly affect their temperature of demixing into two unmiscible, isotropic phases (1). This phenomenon is likely to be critical also with respect to membrane protein crystallization (2).

In the present context, we shall concentrate on the large effect that slight differences in stereochemistry can have on the physical properties of detergents, such as the α and β anomers of 1-O-n-octyl-glucopyranoside. Both of these compounds have been used for the crystallization of a membrane spanning, pore-forming protein, porin from *E. coli* membranes (3,4). While the β - anomer is highly soluble , and exhibits neither critical temperatures of crystallization, nor of demixing, α-octylglucoside has a critical temperature of crystallization (T_c) at 42°C (3,5) and is, correspondingly, poorly soluble at room temperature. In an attempt to explain these greatly different solubility properties, electron diffraction structural investigations have been performed (5). They revealed an equally significant difference in crystallization behaviour. Differential thermal analysis, on the other hand, indicated that the melting behaviour of both anomers is similar, i.e. small endotherms occur between 55°-60°C and are followed by a major transition near 70°. Complete melting was observed above 100°C. Construction of heated sample stages for a powder X-ray diffraction apparatus and an electron microscope allowed structural characterizations of these thermal events, as discussed below for both anomers. The methods used in these studies were essentially as described previously (5).

RESULTS

1-O-n-Octyl-α-D-glucopyranoside

Unit cell parameters obtained from electron diffraction measurements on solution-grown, as well as epitaxially-grown microcrystals of the α-anomer revealed that the crystal structure is identical to the n-decyl homolog (6), i.e. a = 5.12Å, b = 7.74Å, c = 19.6Å. The very slight changes in the X-ray powder diffraction pattern at 60°C revealed that the small endotherm does not signify loss of crystallinity, but possibly a slight rearrangement in alkyl chain packing. At the major endotherm (above 72°C), the X-ray long spacing increases from 19.6Å to 23.3Å, and wide angle data indicate melting of the chains. The resultant smectic liquid crystalline state is birefringent. Contrary to the proposal of other workers for related material (7), the bipolar monolayer packing should be retained after this transition because the bilayer suggested by them would require an unusual cross-sectional area for the alkyl chains. Increase of the long unit cell spacing, moreover, can be explained by a transition from inclined to upright long chain axes.

Recrystallization of this material from the liquid crystalline form produced two polymorphic forms - one identical to the initial crystalline form, another with a 43.8Å long spacing which has not been further characterized. Crystallization from the melt (117°), on the other hand, yielded only the first crystal form.

1-O-n-Octyl-β-D-glucopyranoside

It has been more difficult to understand the crystal structure of the highly water-soluble β-anomer. Since electron diffraction patterns from microcrystals grown from ethanol or molten naphthalene (5) represented the same projection with unit cell spacings d_{100} = 4.88Å, d_{010} = 29.4Å, it was thought that crystal growth from solution gave an unusual structure with long chain axes parallel to the best developed crystal face. Later indexing of powder X-ray diffraction data with these zonal dimensions, however, revealed a longer 34.5Å axis to be perpendicular to this crystal face. Yet, since the low angle X-ray data also seemed to indicate a lamellar repeat for the 29.4Å spacing, the chains were still presumed to lie in this axial direction.

An analysis more consistent with other physical data resulted after examination of the crystal structure of 1-O-phenyl-β-D-glucopyranoside (8). Expansion of the longest unit cell long spacing (29.01Å) in this structure, due to replacement of an octyl group for a phenyl (8.89Å-2.76Å = 6.13Å), gave a value close to the 34.5Å spacing for the octyl compound. The 29.4Å spacing in β-octyl glucoside, moreover, is about three times that of the 9.75Å axis in the phenyl analog and corresponds to a hydrogen-bonded network of hydrated sugar rings. Density measurements gave a result similar to the α-anomer ($\rho \simeq 1.24g/cm^3$ vs $1.22g/cm^3$) and, with the measured unit cell constants above, would account for twelve monohydrated octyl glucoside molecules in this volume which is three times the contents of the phenyl analog structure. This, again, is consistent with the above argument. The projected alkyl chain cross-sectional area therefore could be deduced to be (4.88Å x 29.4Å)/6 or 23.9Å. The 4.88Å axis of the octyl structure is also nearly 4.89Å axis of the phenyl structure.

At room temperature, the octyl chains in the β-anomer will thus closely parallel the longest unit cell axis. Since the zonal cross-sectional area denotes chains which are nearly melted (or at least slightly tilted to the surface normal), hydrogen bonding interactions in the sugar moieties account for most of the crystalline packing forces. This is indicated by the enthalpy of the major transition near 68°C which is only a third of that formed for the α-anomer (5). Since we found that electron diffraction patterns on oriented microcrystals heated from -140°C to +85°C reveal neither major changes in crystallinity nor diffraction intensity, the loss of birefringence above 68°C possibly denotes a change in chain tilt and/or further randomization of the chain packing. The resultant increase of respective unit cell constants to 4.96Å and 32.2Å would give a chain cross-sectional area of 26.6Å, slightly larger than expected for liquid alkane chain packing (9).

If the structural analogy between n-octyl and phenyl analogs holds, the crystal structure of the β-octyl glucoside proposed would also incorporate a bipolar monolayer packing of detergent molecules which exhibits lesser order than that of the α-anomer. The liquid crystalline state of the β-anomer may therefore be smectic and thus explain the viscosity of the bulk material at this transition (10). Recrystallization of the detergent from this state restores the original crystalline form.

CONCLUSIONS

We conclude that crystalline behaviour for the anomeric 1-0-n-octyl-D-gluco-pyranosides is consistent with the different solubilities of the two anomers, α and β, and is itself an expression of the often discussed anomeric effect. Due to the steric requirements for derivates of D-glucose formed on the glyco-sidic carbon (11), more sugar hydroxyl-groups are involved in hydrogen-bonding in β-anomers than in α-anomers, as can be appreciated readily by a review of representative crystal structures. The hydrogen bonding scheme of the β-anomer can be further supplemented by hydration (12), as demonstrated by the crystal structures of methyl (13) and phenyl (8) β-D-glucopyranosides. For the octyl glucosides considered here, the difference in alkyl chain crystallinity also contributes to the solubility difference. Thus the α-anomer, with its fewer hydrogen bonds in the polar regions but with highly crystallized octyl chains behaves as a typical amphiphile, whereas the nearly melted chains in the β-anomer, in conjunction with an extensive hydrogen bonding network accounts for the good water solubility of the β-anomer.

ACKNOWLEDGEMENT

Research supported by Grant GM-21047 from the National Institute of General Medical Sciences.

REFERENCES

1. Zulauf, M. and Rosenbusch, J.P. (1983) J. Phys. Chem. 87, 856-882.
2. Garavito, R.M. and Jenkins, J.A. (1983), These Proceedings.
3. Garavito, R.M. and Rosenbusch, J.P. (1980) J. Cell Biol. 86, 327-329.
4. Garavito, R.M., Jenkins, J.A., Jansonius, J.N., Karlsson, R. and Rosenbusch, J.P. (1983) J. Mol. Biol. 164, 313-327.
5. Dorset, D.L. and Rosenbusch, J.P. (1981) Chem. Phys. Lipids 29, 299-307.
6. Moews, P.C. and Knox, J.R. (1976) J. Amer. Chem. Soc. 98, 6628-6633.

7. Carter, D.C., Ruble, J.R. and Jeffrey, G.A. (1982) Carbohydrate Res. $\underline{102}$, 59-67.

8. Jones, P.G., Sheldrick, G.M., Clegg, W., Kirby, A.J. and Glenn, R. (1982) Z. Krist. $\underline{160}$, 269-274.

9. Stewart, G.W. (1928) Phys. Rev. $\underline{31}$, 174-179.

10. Grabo, M. (1982) Ph.D. Thesis, University of Basel.

11. Jeffrey, G.A., Pople, J.A., Binkley, J.S. and Vishveshwara, S. (1978) J. Amer. Chem. Soc. $\underline{100}$, 373-379.

12. Jeffrey, G.A. and Takagi, S. (1977) Acta Cryst. $\underline{B33}$, 738-742.

13. Jeffrey, G.A., McMullan, R.K. and Takagi, S. (1977) Acta Cryst. $\underline{B33}$, 728-737.

© 1983 Elsevier Science Publishers B.V.
Structure and Function of Membrane Proteins,
E. Quagliariello and F. Palmieri editors.

THE PREPARATION OF MEMBRANE PROTEIN CRYSTALS FOR X-RAY
DIFFRACTION.

R.M. GARAVITO AND J.A. JENKINS
Biozentrum der Universität Basel, CH-4056 Basel, Switzerland

INTRODUCTION

The lack of suitable specimens for X-ray crystallography or high
resolution electron microscopy has hindered the study of integral
membrane proteins. However, several recent reports (5,7,9,12)
demonstrate that, in principle, single crystals of membrane
proteins can be obtained. The methods use standard crystallization
techniques for soluble proteins but adapted for use with
detergent-containing solutions. The proper choice of detergent is
necessary to obtain a protein preparation that is monodisperse,
isotropic, and stable (15). These are necessary conditions for
crystallization. However, the characterization of bacteriorhodopsin (10) and matrix porin (5) crystals suggest that the
detergent may play a direct role in crystal formation. This may
stem from the fact that it is the protein-detergent mixed micelle
which crystallizes and is thus affected by the micellar behavior
of the detergent (11,15).

We have continued to explore methods to crystallize integral
membrane proteins. To date, the outer membrane proteins from *E.
coli*, matrix porin (OmpF), maltoporin (LamB), and conjugin (OmpA)
have been obtained in Basel as crystalline preparations (a
collaboration with J.P. Rosenbusch, J.-M. Neuhaus and U. Hinz in
Basel and EMBL, Heidelberg; 4,5,6). Microcrystal formation has
also been observed with the photosynthetic reaction center from *R.
spheroides* (a collaboration with G. Feher, University of
California, San Diego). In this report we discuss the general
methods for crystallization, detergent behavior, and the effects
of polar organic additives.

GENERAL METHODS

Vapor diffusion has been the method of choice for matrix porin
(5), *R. viridis* photosynthetic reaction center (9) and
bacteriorhodopsin (10). The "hanging drop" form of vapor diffusion
has not been used primarily since detergent solutions at or above

the *critical micellar concentration* (CMC) have little surface
tension. The CMC of the detergent also determines, in part, the
starting conditions for vapour diffusion. Ammonium sulfate or
polyethylene glycol (PEG) have been used as the precipitants, the
latter often with added NaCl. The presence of detergent in the
crystallizing medium does cause phase separation (5,7,10,12) at
high precipitant concentrations, perhaps more so with PEG than
with ammonium sulfate. Nevertheless, crystallization with minimal
phase separation can be obtained with both precipitants. PEG does
have the advantage that the ionic strength of the system can be
kept below 1 molar.

DETERGENT ASPECTS

The choice of detergent(s) is a critical factor for good
crystallization. Nonionic and zwitterionic detergents which are
pure and monodisperse have proven most successful. Experiments on
bacteriorhodopsin and matix porin grown from β-octyl glucoside
reveal that a substantial amount of detergent remains bound to the
protein in the crystal (5,10) suggesting that the presence of the
detergent layer about the protein does not interfere with
crystallization. This detergent region of the protein-detergent
mixed micelle should form two distinct zones equivalent to the
hydrophobic core and the hydrated head-group, or mantle, of a pure
detergent micelle. We have done several experiments with
detergents having a n-octyl tail (thus, minimally affecting the
hydrophobic core) but different head group moieties (5,6; R.M.
Garavito, unpublished results). The results clearly show that
changes in the detergent head-group can radically alter the
crystallization potential, and even space group obtained, when all
other variables are held constant (Table I, compare b-c, d-f).
However, changes in the hydrophobic tail can affect crystallization as well. Bacteriorhodopsin which crystallizes well in β-
octyl glucoside does not in β-nonyl glucoside (11). Maltoporin
will not crystallize if β-octyl glucoside in experiment (f) in
Table I is replaced by β-hexyl glucoside.

TABLE 1

DETERGENT EFFECTS ON CRYSTALLIZATION: INITIAL VAPOUR DIFFUSION CONDITIONS[1]

Crystals	pH	Detergent(s)[2,3]
Matrix porin		
a. $P4_2$	7.0	1% β-OG, 0.2% C_8-POE
b. $P6_322$	9.0	1% β-OG, 0.2% C_8-POE
c. P1	9.0	2% C_8-HESO
Maltoporin[4]		
d. P1	7.0	0.25% C_8-POE, 0.25% β-OG
e. C222	7.0	0.3% C_8E_4
f. C222 (improved)	7.0	0.3% C_8E_4, 0.2% β-OG

1. Initially 7.2% (w/v) PEG, 0.2M NaCl; Reservoir 18% PEG, 0.5M NaCl.
2. Weight percent.
3. Detergent abbreviations: β-OG, β-octyl glucoside; C_8E_4, octyl tetraoxyethylene; C_8-POE, octyl polyoxyethylene (see ref. 15); and C_8-HESO, octyl hydroxyethylsulfoxide.
4. See ref. 6.

TABLE 2

SMALL AMPHIPHILE EFFECTS

Amphiphile	Amount	Precipitant[1]	Protein[2]	Result[3]
heptane-1,2,3-triol	5%(w/v)	AS	RC	C
piperidine-2-carbonic acid	6%(w/v)	AS	bR	IC
ethylglycol butyl-ether	5%(v/v)	PEG	LamB	sc
glycolate butylester	5%(v/v)	PEG	LamB	sc
heptane-1,2,3-triol	5%(w/v)	PEG	LamB	sc

1. AS, ammonium sulfate; PEG, polyethylene glycol.
2. bR, bacteriorhodopsin; RC, phtosynthetic reaction center; LamB, maltoporin.
3. C, induces crystallization; IC, improves crystal quality; sc, causes space group changes.

In the PEG system, the crystallization often occurs along with
phase separation into a detergent-protein rich phase and a PEG-
rich phase (5,7). However, a two phase system is not obligatory
for the crystallization of either maltoporin (6) or, as recently
discovered, matrix porin (R.M. Garavito and J.A. Jenkins,
unpublished results). What is interesting is that crystallization
often occurs around conditions where this demixing phenomenon does
occur. The starting conditions for (e) in Table I would demix at
∿ 28°C (M. Zulauf, personal communication). As vapour diffusion
progresses, the temperature for phase separation thus decreases
until it is below room temperature. We (this paper, 15) suggest
that the process of crystal formation may rely on the phenomenon
of micelle aggregation that is the basis for the phase separation
(2,8,16, see below).

POLAR ORGANIC ADDITIVES

Michel (9,10,11) has reported that the addition of compounds
which could interact with detergent micelles affected the
crystallization of bacteriorhodopsin and *R. viridis* photosynthetic
reaction center. The most effective compound, or small amphiphile,
he found was 1,2,3-heptane-triol (11, Table 2).

We have studied the effect of polar organic compounds on the
crystallization of maltoporin (6). Their efficacy in producing
distinct changes in crystal formation was dependant on their
hydrophobic moiety: $R-(CH_2)_3CH_3 \gg R(CH_2)_2CH_3 > R-CH_2(CH_3)_2 > R-CH_2-CH_3$. The most effective compounds we have yet found are
ethylene glycol butylether and glycolate butylester. The addition
of these compounds to maltoporin preparations (Table 2) can
produce substantial changes in crystal habit and space group. The
high efficacy of compounds containing an n-butyl moiety is
probably due to their ability to completely insert into detergent
micelles and thus affect global changes in micelle structure and
behavior. This speculation is supported by the work of Benjamin
(1) on dodecyldimethylamine oxide.

GENERAL CONCLUSIONS

The most important conclusion we can draw is that the detergent
plays an unprecedented role in membrane protein crystallization:
given the same ionic conditions and pH, changing to a new

detergent or mixing two detergents can cause the appearance of crystals or produce   new crystal forms. The explanation may lie in the hybrid nature of the detergent-protein mixed micelle; it is part soluble protein and part detergent micelle. Changes in detergent structure will radically alter micellar structure (e.g. micelle size or aggregation number) and behavior (13), perhaps enhancing the accessibility for interaction of the protein surfaces. These *intra*micellar aspects are partly embodied in the "small amphiphile concept" as described by Michel (11).

Factors which affect micelle structure also affect micelle-micelle interactions through changes in micellar hydration (8,15). Micelle clustering is induced by salt or polyethylene glycol (3,13; M. Zulauf, personal communication) but can be finely controlled using various detergent mixtures and by the addition of small amphiphiles (13; M. Garavito, unpublished observations; M. Zulauf, personal communication). The hybrid nature of a protein-detergent mixed micelle might imply that membrane protein crystallization would be dependant not only on protein-protein interactions but also on the micelle aggregation potential. The observation that crystal formation occurs at or near the phase boundary for demixing supports this speculation. The phenomenon of micelle clustering (2,16) would assist crystallization by allowing the close approach protein-containing micelles and encouraging the formation of crystal nuclei.

Finding the optimal conditions for crystallization may require then balancing several detergent factors while simultaneously encouraging protein-protein interaction. This sounds quite complicated, though increasing knowledge about the behavior of pure detergents in the presence of salts and small amphiphiles will help the selection of conditions. Nevertheless, even the straight forward application of the two methods described by Michel and Oesterhelt (12) and Garavito and Rosenbusch (7) have yielded microcrystals of photosynthetic reaction center (G. Feher, personal communication). In conclusion, membrane protein crystallization does occur in a systematic manner and promises to open up a new area to X-ray crystallography.

ACKNOWLEDGEMENTS

This work was supported by Swiss National Foundation grants 3.565.80 and 3.201.82.

REFERENCES

1. Benjamin, L. (1966) J. Colloid Interface Sci. 22, 386.

2. Corti, M. and Degiorgio, V. (1981) J. Phys. Chem. 85, 1442.

3. Doren, A. and Goldfarb, J. (1970) J. Colloid Interface Sci. 32, 67.

4. Garavito, R.M., Jenkins, J.A., Neuhaus, J.-M., Pugsley, A.P. and Rosenbusch, J.P. (1982) Ann. Microbiol. (Institut Pasteur) 133A, 37.

5. Garavito, R.M., Jenkins, J.A., Jansonius, J.N., Karlsson, R. and Rosenbusch, J.P. (1983) J. Mol. Biol. 164, 313.

6. Garavito, R.M., Hinz, U., and Neuhaus, J.-M. (1983) J. Biol. Chem., submitted.

7. Garavito R.M., and Rosenbusch, J.P. (1980) J. Cell Biol. 86, 327.

8. Hayter, J.B. and Zulauf, M. (1983) Colloid and Polymer Sci., in press.

9. Michel, H. (1982a) J. Mol. Biol. 158, 567.

10. Michel, H. (1982b) The EMBO Journal 1, 1267.

11. Michel, H. (1983) TIBS 8, 56.

12. Michel, H. and Oesterhelt, D. (1980) Proc. Nat. Acad. Sci. USA 77, 1283.

13. Rosen, M.J. (1978) Surfactants and Interfacial Phenomena, John Wiley and Sons, New York.

14. Rosenbusch, J.P., Garavito, R.M., Dorset, D.L. and Engel, A. (1982) in: Peeters, H. (Ed.), Protides of the Biological Fluids 29th Colloquium, Pergamon Press, Oxford, pp. 171-174.

15. Rosenbusch, J.P., Dorset, D.L., Garavito, R.M. and Zulauf, M. (1983) in: Palmieri, F. et al. (Eds.), Structure and Function of Membrane Proteins, Elsevier/North Holland Biomedical Press, Amsterdam, pp.

16. Zulauf, M. and Rosenbusch, J.P. (1983) J. Phys. Chem. 87, 856.

FACTORS INFLUENCING THE FORMATION OF TWO-DIMENSIONAL CRYSTALS OF
RENAL Na,K-ATPase.

ELISABETH SKRIVER AND ARVID B. MAUNSBACH.
Department of Cell Biology, Institute of Anatomy, University of Aarhus, DK-8000
Aarhus C, Denmark.

INTRODUCTION

Purified membrane-bound Na,K-ATPase can be directly demonstrated by electron microscopy after negative staining as surface particles, which show lateral mobility in the plane of the membrane (1,2). Incubation of purified membrane-bound Na,K-ATPase with vanadate or phosphate in the presence of magnesium induces the formation of two-dimensional enzyme crystals (3), which are suitable for further analysis by image processing of the structure and interaction of the protein units (4). The aim of this work was to further define the conditions for the formation of two-dimensional crystals of renal Na,K-ATPase.

METHODS

Na,K-ATPase was purified in membrane-bound form from the outer medulla of rabbit kidney by selective extraction of a microsomal fraction with sodium dodecyl sulphate (SDS) in the presence of ATP (5). The extracted membranes were purified by isopycnic centrifugation in continuous linear sucrose (15-45%) gradients, which were centrifuged at 108,000xg for 4 hours using a SW 27 rotor in a Beckman L2-65B Ultracentrifuge. The specific activity of the enzyme was 21-34 μmol $P_i \cdot min^{-1} \cdot mg$ protein^{-1}. The purified enzyme was incubated for 1-4 weeks at 4°C with one of the following ligand combinations: (a) without ligands in 10 mM imidazole buffer, pH 7.0-7.5 (control) (b) 150 mM KCl in 10 mM imidazole buffer, pH 6.1-7.5 (c) 150 mM NaCl in 10 mM Tris-HCl buffer, pH 6.1-8.7 (d) 1 mM sodium monovanadate and 3 mM $MgCl_2$ in 10 mM imidazole buffer, pH 6.3-8.1 (e) 1.0 mM sodium monovanadate and 1 mM $MgCl_2$ in Tris-HCl buffer, pH 6.3-8.1 (f) 1.0 mM ammonium monovanadate and 3 mM $MgCl_2$ in 10 mM imidazole buffer, pH 5.5-8.1. After 1-4 weeks samples were negatively stained with 1% uranyl acetate on hydrophilic carbon films and examined in a Jeol 100 CX electron microscope at magnifications of 50,000 and 66,000x.

RESULTS

The protein of membrane-bound Na,K-ATPase is observed by negative staining as 30-50 Å surface particles on the isolated membranes. In control preparations, in the absence of ligands, the enzyme protein has a random distribution in clusters and

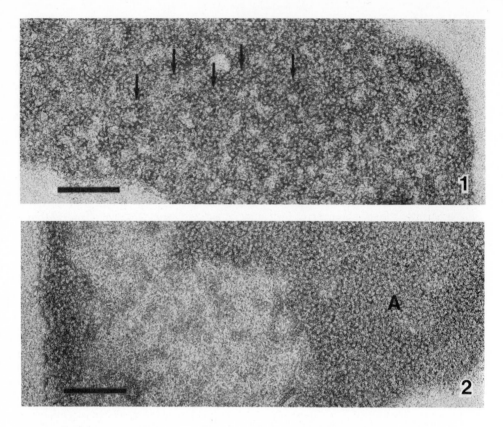

Fig. 1. Electron micrograph of membrane-bound Na,K-ATPase incubated under control conditions without ligands. The protein units (small light regions, arrows) have a random distribution in clusters and strands. Scale: 1000 Å.

Fig. 2. Membrane-bound Na,K-ATPase incubated with 150 mM NaCl, pH 8.7. Protein units are closely packed in large areas (A) without crystalline arrays. Scale: 1000 Å.

strands, even after weeks of storage (Fig. 1). Incubation of the enzyme in 150 mM NaCl produces a close packing of the enzyme protein, but regular crystalline arrays are not observed (Fig. 2). When the enzyme is incubated with 150 mM KCl the protein units likewise become densely packed, but in addition many membrane fragments show formation of crystalline arrays within two weeks (Fig. 3). The potassium-induced crystals appear similar to those induced by vanadate-magnesium (Fig. 4), but crystallisation induced by potassium is not as extensive as with vanadate-magnesium.

Induction of membrane crystals is also influenced by the pH of the incubation medium. Thus, potassium- as well as vanadate-induced crystals formed more rapidly

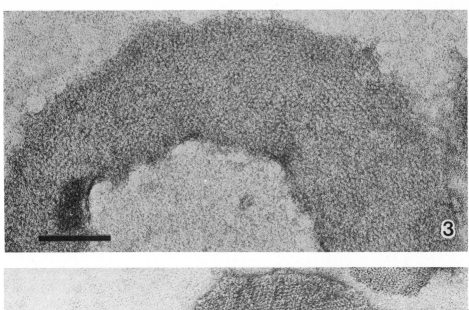

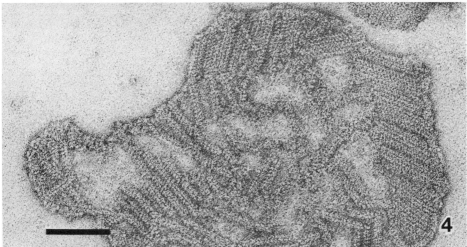

Fig. 3. Membrane-bound Na,K-ATPase incubated with 150 mM KCl, pH 7.5. The particles show extensive crystalline arrays, although some lattice lines are curved. Scale: 1000 Å.

Fig. 4. Membrane-bound Na,K-ATPase incubated in 1 mM $NaVO_3$ and 3 mM $MgCl_2$, pH 7.5. This membrane-fragment shows vanadate-induced two-dimensional crystals as well as linear, ladder-like arrays, that may represent a step in the assembly of the crystals (3). Scale: 1000 Å.

and in larger areas of the membrane fragments between pH 6.5 and 7.0 than at other pH values. There was no observable difference in the rate and extent of crystallization when ammonium monovanadate was used instead of sodium monovanadate. Crystallization with vanadate-magnesium occurred equally well in imidazole and in Tris-HCl buffers. Incubation of the enzyme in 10 mM Tris-HCl buffer pH 7.5 with 1 mM sodium monovanadate in the absence of magnesium did not induce the formation of crystals.

DISCUSSION

The present observations demonstrate that high concentrations of potassium, but not sodium, are capable of inducing two-dimensional crystals of Na,K-ATPase. In a previous study we found that incubation of Na,K-ATPase with vanadate-magnesium or phosphate-magnesium, which stabilize the protein in the E_2-conformation, induced the formation of two-dimensional crystals (3). Since potassium also induces the E_2-form of the enzyme, the present observations provide additional evidence that the E_2-conformation of the enzyme protein favours the assembly of the protein units in two-dimensional crystals. Furthermore, the observations that crystals form less extensively at or above pH 7.5 than at lower pH values, are also consistent with this interpretation since the E_2-conformation is favoured at low rather than high pH (6).

ACKNOWLEDGEMENTS

We wish to thank Einar Hansen and Elsebeth Thomsen for excellent technical assistance. This investigation was supported by the Danish Medical Research Council and the Research Fund of the University of Aarhus.

REFERENCES

1. Maunsbach, A.B. and Jørgensen, P.L. (1974) Proceedings of the Eight Internat. Congr. Electron Microscopy. Canbarra, vol. 2, 214-215.

2. Deguchi, N., Jørgensen, P.L., and Maunsbach, A.B. (1977) J. Cell Biol. 75, 619-634.

3. Skriver, E., Maunsbach, A.B., and Jørgensen, P.L. (1981) FEBS Lett. 131, 219-222.

4. Hebert, H., Jørgensen, P.L., Skriver, E., and Maunsbach, A.B. (1982) Biochim. Biophys. Acta 689, 571-474.

5. Jørgensen, P.L. (1974) Biochim. Biophys. Acta 356, 36-52.

6. Skou, J.C. and Esmann, M. (1981) Biochim. Biophys. Acta 647, 232-240.

THE STRUCTURE OF UBIQUINOL:CYTOCHROME REDUCTASE FROM NEUROSPORA MITOCHONDRIA

HANNS WEISS[1], STEPHEN J. PERKINS[1] AND KEVIN LEONARD[2]

[1]Institute of Biochemistry, University Düsseldorf, 4000 Düsseldorf, FRG

[2]European Molecular Biology Laboratory, 6900 Heidelberg, FRG

INTRODUCTION

Ubiquinol:cytochrome c reductase (Cytochrome reductase, EC 1.10.2.2.) is a cytochrome complex of mitochondrial oxidative phosphorylation. The enzyme transduces oxidative energy into the energy of an electrochemical proton gradient across the inner mitochondrial membrane and this energy can be used by the ATP synthetase to drive the synthesis of ATP from ADP and phosphate. Cytochrome reductase has a molecular weight of about 550 000, the enzyme is a dimer and the monomeric unit consists of probably nine different subunits. Three of the subunits carry redox centres, the cytochromes b and c_1 (subunit III and IV) and the iron-sulphur subunit (subunit V). Six subunits (subunit I, II and VI to IX) probably do not have prosthetic groups (for review see 1).

Our studies of the structure of cytochrome reductase from mitochondria of the fungus Neurospora crassa has followed three main courses. 1. The enzyme was cleaved in stages and the parts obtained (subunit complexes, subunits and subunit domains) were characterised as hydrophilic, amphiphilic or hydrophobic according to their solubility in aqueous buffers or detergent solution. This procedure assisted in the assignment of these parts to one of the two aqueous sections or the membraneous section of the enzyme. Furthermore, the parts were investigated for their affinity to cytochrome c a substrate of cytochrome reductase which is located on the outer surface of the mitochondrial inner membrane. This lead to an assignment of protein within either of the two aqueous sections (2 - 4). 2. Two-dimensional membrane crystals were prepared from the whole enzyme and a subunit complex of the enzyme by incorporation of the preparations into phospholipid bilayers (5 - 7). From tilted electron microscopic views of these membrane crystals, low resolution three-dimensional structures were calculated (4, 8). 3. Small angle neutron scattering experiments were performed with the whole enzyme and subunit complexes from the enzyme in protonated and deuterated alkyl polyoxyethylene detergents. The use of contrast variation permits low resolution structural information to be obtained separately for the protein and the protein bound detergent (9).

By combining the data now available a low resolution structure of cytochrome reductase has been obtained. The location of this structure in the phospholipid bilayer of the mitochondrial inner membrane and the topography of most of the subunits within this structure is known.

CLEAVAGE OF CYTOCHROME REDUCTASE AND PROPERTIES OF THE ISOLATED PARTS

Cytochrome reductase released from the membrane by Triton X-100 and purified by chromatographic procedures (10) readily dissociates when the ionic strength is increased. Three parts are obtained (Table). The largest part is a subunit complex that contains the cytochromes b and c_1 and the four smallest subunits without redox centres. The second part is a complex that contains the two largest subunits without redox centres. The smallest part is the single iron-sulphur subunit. These three parts can be separated easily. Gel filtration separates the cytochrome bc_1-subunit complex from the other two parts because of the different Stokes radii. Sucrose gradient centrifugation separates the subunit I and II complex, which is a pure protein complex, from the detergent-bound iron-sulphur subunit, because of their different buoyant densities (3,4). The isolated cytochrome bc_1-subunit complex dissociates when the protein-bound Triton is replaced by deoxycholate. So far the cytochromes b and c_1 have been purified by chromatographic steps as protein-deoxycholate complexes. After purification, the protein-bound deoxycholate can be exchanged for Triton X-100 by gel filtration (unpublished results).

Comparison of the molecular weights of the preparations determined from hydrodynamic properties (corrected for protein-bound detergent) with the molecular weight determined by SDS gel electrophoresis shows that the cytochrome reductase and the cytochrome bc_1 complex are in a dimeric state; the subunit I and II complex and the single subunits III, IV and V are in a monomeric state (Table, (4)).

The cleavage parts can be classified as hydrophilic, amphiphilic or hydrophobic (Table). The subunit I and II complex is hydrophilic because it is water soluble without detergent (4). The cytochrome c_1 and the iron-sulphur subunits are amphiphilic; these subunits are soluble only when a minor stretch of each is bound to a detergent micelle. After this stretch has been clipped off by proteolysis, the major part of each subunit, that carries the redox centre, is water soluble without detergent (2, 3). The cytochrome b subunit is hydrophobic; it is soluble only in detergent solution.

The binding of ferricytochrome c to cytochrome reductase or isolated parts of the enzyme was studied by gel filtration of the ^{14}C-leucine labelled preparations as a function of the concentration of free ^{3}H-leucine labelled cytochrome c and the ionic strength. One cytochrome c per cytochrome c_1 is bound to cytochrome reductase, the cytochrome bc_1 subunit complex, the 31 000 Mr membrane bound cytochrome c_1 and the 25 000 Mr water soluble cytochrome c_1 (Table) with similar affinities. The dissociation constants are 6×10^{-8} M at 25 mM Tris-Acetate pH 7.5 and 4×10^{-7} M at 40 mM. In the presence of the reductant ascorbate, only non-specific, low affinity binding was observed (2).

TABLE

PROPERTIES OF THE CLEAVAGE PRODUCTS FROM CYTOCHROME REDUCTASE

Preparation	Subunit composition	Triton binding (g/m mol)	Molecular weight of protein $(\times 10^{-3})$ determined from	
			Hydrodynamic properties	SDS-gel electro-phoresis
Cytochrome reductase	I to IX	120	550	280 - 330[a]
Cytochrome bc_1 subunit complex	III, V to IX	110	280	120 - 160[a]
Subunit I and II complex	I and II	0	170	140 - 190[a]
Cytochrome b	III	100		38
Cytochrome c_1	IV	90	30 - 40	31
Iron-sulphur subunit	V	80	20 - 40	25
water soluble cytochrome c_1	IV	0	24	23
water soluble iron-sulphur subunit	V	0	15	16

[a]The stoicheiometry of the subunits is still uncertain. The two extremes are 2:2:2:1:1:1:1:1:1 and 1:2:1:1:1:1:1:1:1; in the latter case two heme groups would be bound per apoprotein of Cytochrome b (4).

ELECTRON MICROSCOPY OF MEMBRANE CRYSTALS

Membrane crystals were prepared from cytochrome reductase and the cyto-chrome bc_1-complex by combining the proteins bound to Triton with mixed phospho-lipid Triton-micelles and subsequently removing the Triton (5, 6, 7). The dif-fraction pattern of the negatively stained cytochrome reductase crystals extends to the fifth order which corresponds to a resolution of about 2.5 nm. The pat-tern shows pgg symmetry which is consistent with the two-sided plane group $p22_12_1$. In this symmetry the alternate dimeric molecules are packed up-and-down across the bilayer. From sets of tilted electron microscopic views the

three-dimensional structure was calculated (8). The structure shows that the
monomeric units of the enzyme are related by a 2-fold axis perpendicular to the
membrane. They are elongated and extend approximately 15 nm across the membrane.
The protein is unequally distributed with about 30% of the total mass located in
the bilayer, 50% in a section which extends 7 nm from one side of the bilayer
and 20% in a section which extends 3 nm from the opposite side of the bilayer.
The two monomeric units are in contact essentially in the membraneous section
(Fig. 1).

In the membrane crystals prepared from the cytochrome bc_1 subunit complex, the
dimeric proteins also point up-and-down across the bilayer. The three-dimensional structure shows defined protein lobes correlated by a 2-fold axis that runs
perpendicular to the membrane plane (4). These lobes correspond in size and
shape to the smaller peripheral sections of the structure for the whole enzyme.
Below the lobes, the protein density merges into one less well defined region,
which is about 4 nm in thickness and corresponds roughly to the membraneous
section of the enzyme. The structure of the subunit complex shows no part
large enough to fit to the larger peripheral section of the enzyme (Fig. 1).

Fig. 1. Three-dimensional models of cytochrome reductase (right) and the cytochrome bc_1 subunit complex (left) determined by electron microscopy of membrane
crystals. The darker parts extend into the aqueous phases and the lighter
parts lie within the bilayer.

SMALL-ANGLE NEUTRON SCATTERING IN SOLUTION

The neutron scattering density of proteins is higher than that of nonionic detergent. By working with a deuterated detergent, this contrast relation can be reversed. Neutron scattering was applied in this way to cytochrome reductase and the cytochrome bc_1 subunit complex in either hydrogenated Triton X-100 (a tert.-$C_8 \emptyset E_{9.6}$ detergent) or deuterated Cemulsol LA 90 (a $C_{12}E_{8-10}$ detergent). Neutron scattering experiments were also carried out with the subunit I and II complex in detergent-free solution. The scattering data were quantitatively compared with the three-dimensional structures obtained by electron microscopy (9).

Molecular weights for the cytochrome reductase-detergent complex and the cytochrome bc_1-detergent complex (from I_o/c in H_2O) were about 660 000 and 380 000, respectively. Subtracting the amount of bound detergent, molecular weights consistent with the values derived from hydrodynamic properties result (Table). The molecular weight of the subunit I and II complex was about 340 000 at low strength indicating that the subunit complex was in a dimeric state as in the whole enzyme. At 200 mM NaCl the molecular weight was 190 000, indicating dissociation of the complex into the monomers.

The Stuhrmann plots (R_g^2 against $\Delta\rho^{-1}$) for the cytochrome reductase-detergent complex and cytochrome bc_1-detergent complex are parabolic indicating that the centres of protein and protein-bound detergent are displaced relative to one another. The displacement ranges between 3 - 5 nm for the cytochrome reductase-detergent complex and 1 - 3 nm for the cytochrome bc_1-detergent complex.

From the protein R_c values, which were 5.6 nm for cytochrome reductase, 4.3 nm for the cytochrome bc_1-complex and 4.9 nm for the subunit I and II complex, the separation between the centres of the subunit I and II complex and the cytochrome bc_1-complex was about 6.5 nm. This value is less than the half the estimated length of cytochrome reductase on its longest axis, which is between 7.5 nm and 8.5 nm. Therefore the greater part of the subunit complexes must be at their interface in the enzyme, i.e. there is no waist on the long axis of the enzyme.

For comparison of the neutron scattering data with the three dimensional models obtained from electron microscopy, the contour map of the enzyme was converted into 450 rectangular prisms (Fig. 2). The radius of gyration of the 450 units is 5.4 nm which is less than the experimental value 5.6 nm. By lengthening the model by 10% to a total length of 17 nm, the 450 units now give a radius of gyration of 5.7 nm. The electron microscopy model can thus account for the solution shape of cytochrome reductase.

Calculations were also made in which each of the above units was subdivided
into 8 equivalent units to give a total of 3594. Since 5 units in a line re-
sembles one detergent molecule, models for the bound detergent can be set up.
A total of 856 units arranged in a belt of height 10 units at a displacement of
3.6 nm was found to cover completely the protein surface (Fig. 2).

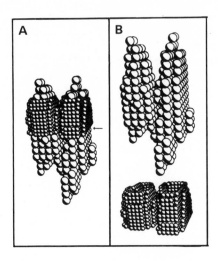

Fig. 2. Model for cytochrome reductase and the bound detergent in terms of
spheres used for simulation of the scattering data (A) cytochrome reductase-detergent complex,(B) cytochrome reductase and detergent separated. The cleavage
plane between the cytochrome bc_1 subunit complex and the subunit I and II complex is arrowed (9).

Further calculations showed that the electron microscopy model could be used
to confirm the experimental wide-angle scattering curve of cytochrome reductase
and the cytochrome bc_1 complex, and account for the R_g values of the cytochrome bc_1 complex and the subunit I and II complex (9).

CONCLUSION

A first assignment of the positions of the subunit complexes, subunits and
subunit domains to within either the two peripheral sections or the membraneous
section of cytochrome reductase can be made on the basis of their hydrophilic,
amphiphilic or hydrophobic character. The hydrophilic subunit I and II complex
is assumed to extend completely from the membrane into the aqueous phase. The
amphiphilic cytochrome c_1- and iron-sulphur-subunits are assumed to extend into
the aqueous phase and to be anchored to the membrane only by small protein
stretches. The hydrophobic cytochrome b subunit is assumed to lie mainly
within the membrane.

The assignment of subunits within either of the two peripheral sections of cytochrome reductase can be made by comparing the structures obtained for the whole enzyme to that of the cytochrome bc_1 subunit complex. A part corresponding to the larger peripheral section of cytochrome reductase is missing from the structure of the subunit complex and the overall size of the complex fits well with that part of cytochrome reductase taken by the smaller peripheral section and the membrane section. The larger peripheral section of cytochrome reductase must therefore be assumed to be contributed by the subunits I and II since only these two subunits can account for the protein mass and the protein R_g (3). The question of where to locate the iron-sulphur subunit is more difficult to answer. Since the resolution limit of the reconstruction is of the same order as the expected diameter of the subunit. The comparison of the smaller peripheral section of cytochrome reductase with the (only) peripheral section of the cytochrome bc_1 complex shows that the distance between the centre of mass of the latter is 10% less than that of the former. This difference suggests that protein is missing from the external faces of the cytochrome bc_1 complex as compared with cytochrome reductase thus shifting inwards the mass centres of protein. The iron-sulphur protein could thus tentatively be located on the outside of the smaller peripheral section of cytochrome reductase (3, 4).

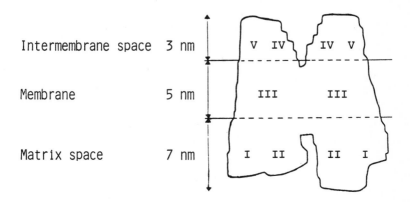

Fig. 3. Schematic drawing of the location of cytochrome reductase in the mitochondrial inner membrane and the topography of the subunits in the enzyme. The subunits VI to IX have not been included into the scheme because we have no information about their location.

The orientation of the structure in the mitochondrial membrane results from the topography of the cytochrome c_1 subunit. This subunit interacts with cytochrome c at the outer surface of the mitochondrial inner membrane. The sub-unit is present in the cytochrome bc_1 subunit complex and because of its amphi-philic nature must be located mainly in the peripheral part of this subunit complex. Therefore, the smaller peripheral section of the enzyme is assumed to extend into the intermembrane space of mitochondria and the larger peripheral section inside of mitochondria into the matrix space (Fig. 3).

REFERENCES

1. Wikström, M., Krab, K. and Saraste, M. (1981) Annu. Rev. Biochemistry, 50, 623 - 655

2. Li, Y., Leonard, K. and Weiss, H. (1981) Eur. J. Biochem., 116, 199 - 205

3. Li, Y., De Vries, S., Leonard, K. and Weiss, H. (1981) FEBS Lett., 135, 277 - 280

4. Karlsson, B., Hovmöller, S., Weiss, H. and Leonard, K. (1983) J. Mol. Biol., 165, 287 - 302

5. Wingfield, P., Arad, T., Leonard, K. and Weiss, H. (1979) Nature, 230, 696 - 697

6. Hovmöller, S., Leonard, K. and Weiss, H. (1981) FEBS Lett., 123, 118 - 122

7. Hovmöller, S., Slaughter, M., Berriman, J., Karlsson, B., Weiss, H. and Leonard, K. (1983), J. Mol. Biol., 165, 401 - 406

8. Leonard, K., Wingfield, P., Arad, T. and Weiss, H. (1981) J. Mol. Biol., 149, 259 - 274

9. Perkins, S. J. and Weiss, H. (1983) J. Mol. Biol., in press

10. Weiss, H. and Kolb, H. J. (1979) Eur. Biochem., 99, 139 - 149

© 1983 Elsevier Science Publishers B.V.
Structure and Function of Membrane Proteins,
E. Quagliariello and F. Palmieri editors.

STRUCTURAL STUDIES ON BEEF HEART CYTOCHROME C OXIDASE FROM WHICH SUBUNIT III HAS BEEN REMOVED BY CHYMOTRYPSIN TREATMENT

FRANCESCO MALATESTA, GRADIMIR GEORGEVICH AND RODERICK A. CAPALDI
Institute of Molecular Biology, University of Oregon, Eugene, Oregon 97403

INTRODUCTION

Cytochrome c oxidase is the terminal enzyme in the mitochondrial electron transfer chain, catalyzing the four electron reduction of molecular oxygen and coupling this reaction to the generation of a proton gradient across the mitochondrial inner membrane (1-3).

The enzyme from beef heart mitochondria has been established to contain at least seven different subunits (4-6). These have all been sequenced either from the polypeptide directly (e.g., 7-8) or by sequencing of the appropriate genes in the case of the three largest subunits which are coded for on mtDNA (9). Beef heart cytochrome c oxidase has been obtained in two dimensional crystals and these have been studied by electron microscopy and image analysis to give a low resolution (25 Å) structure of the protein (10-13). It is a Y-shaped enzyme with the two arms of the Y, the M_1 and M_2 domains, each spanning the mitochondrial inner membrane and with the stalk of the Y, the C domain, extending into the intracristal space between the inner and outer mitochondrial membranes (3,14).

The arrangement of different subunits within the Y-shaped cytochrome c oxidase has been explored by chemical labeling and cross linking experiments (15-20). These studies have provided information on which subunits are exposed at the matrix side and which on the cytoplasmic side of the complex in the inner membrane, on which subunits contribute to the bilayer intercalated part and on near neighbor relationships of subunits within the complex (3).

Recently, there have been several reports of cytochrome c oxidase preparations missing subunit III, one of the mitochondrially synthesized components (21-23). Such preparations are of importance for functional studies because subunit III has been implicated in the proton pumping function of the enzyme. Here we describe the use of chymotrypsin to remove subunit III from beef heart cytochrome c oxidase. Enzyme missing subunit III has been compared structurally with that containing a full complement of subunits. The results suggest that subunit III provides the M_2 domain of cytochrome c oxidase.
(Abbreviations used: AD, adamantane diazirine; DCCD, dicyclohexylcarbodiimide; DABS diazobenzene sulfonate; DSP, disuccinimidylpropionimidate; NaDodSO4, sodium dodecyl sulfate; PMSF, phenylmethyl-sulfonyl flouride; TLCK, tosyl-lysine chloromethyl ketone.)

EXPERIMENTAL PROCEDURES

Enzyme Preparation and Assays. Beef heart cytochrome c oxidase was prepared as described by Capaldi and Hayashi (24). Electron transport activity was measured polarographically by the procedure of Vik and Capaldi (25). Protein concentrations were determined according to Lowry et al. (26).

Chymotrypsin Digestion and Removal of Subunit III. Purified cytochrome c oxidase dissolved in 1% Triton X-100, 20 mM Tris-HCl pH 8.2 at 40 μM heme a was incubated with α chymotrypsin (TLCK treated; Worthington) at a ratio of 1:20 (wt/wt) for 2 hr at room temp. The reaction was terminated by adding PMSF (0.5 mM).

Peptide fragments released by the protease digestion were separated from the remaining cytochrome c oxidase complex by chromatography on DEAE agarose (BioRad) using a 1 x 2 cm column. Protein was bound to the column and fragments washed off in 100 ml of 1% Triton X-100, 20 mM Tris-HCl pH 8.2. The cytochrome c oxidase complex was then eluted in the same buffer with 200 mM NaCl added. Removal of subunit III was monitored by loss of DCCD or AD-labeled subunit as described in the Results section.

Labeling of Cytochrome c Oxidase with Different Reagents. [^{35}S] DABS (5-9 Ci/mmol) was prepared from [^{35}S] sulfanilic acid (Amersham) according to Tinberg et al. (27). [^{3}H] adamantane diazirine was the generous gift of Dr. J. Kite, Department of Chemistry, University of California at San Diego. [^{14}C] iodoacetamide and [^{14}C] DCCD were purchased from New England Nuclear and Amersham respectively.

The labeling of cytochrome c oxidase by [^{15}S] DABS, [^{3}H] AD, [^{14}C] iodoacetamide and [^{14}C] DCCD was conducted as described previously in Ludwig et al. (16), Georgevich and Capaldi (28), Darley-Usmar et al. (29), and Prochaska et al. (30) respectively. For examination of the extent of protease digestion, chymotrypsin-cleaved, radiolabeled enzyme samples were electrophoresed after quenching the labeling reaction and unincorporated counts were removed by the gel staining and destaining procedures.

NaDodSO$_4$ polyacrylamide gel electrophoresis was conducted as described by Fuller et al. (31). Staining with Coomassie Brilliant blue and destaining of gels suspended in this dye solution were performed according to Downer et al. (5). Silver staining was performed according to Oakley et al. (32). Gels were scanned at 560 nm on a Gilford Spectrophotometer with a linear scanning attachment and sliced into 1mm pieces with a Mickle gel slicer. Radioactivity was measured by liquid scintillation counting in a packard Tri Carb Model 3385 liquid scintillation counter.

Cross-Linking of Cytochrome c Oxidase Samples with DSP. Stock enzyme and subunit III-less cytochrome c oxidase (4 μm heme a) dissolved in 1% Triton X-100, 0.2 M NaCl, 10 mM sodium phosphate (pH 8.0) were reacted with DSP (1 mg/ml) for 15 min at room temp and the cross linking reaction then terminated by adding 50 mM ammonium acetate. Samples were denatured in NaDodSO$_4$ without reducing agents and polypeptides resolved by NaDodSO$_4$ polyacrylamide gel electrophoresis under the conditions of Fuller et al. (31). The second dimensional analysis was conducted as described by Briggs and Capaldi (18).

Molecular Weight Determinations. The molecular weight of subunit III-less cytochrome c oxidase was determined by sedimentation equilibrium analysis as described by Tanford et al. (33), using a Beckman-Spinco Model E Analytical Ultracentrifuge equipped with a photoelectric scanner. The buffer conditions for both sedimentation velocity and sedimentation equilibrium experiments were as described in Georgevich et al. (34). Molecular weights were calculated by correcting for bound Triton X-100 (partial specific volume 0.908) and bound lipid as described by Tanford et al. (33). Bound detergent was measured as described in Georgevich et al. (34), using ε_{280} = 213 mM^{-1} cm^{-1} per heme a for subunit III-less enzyme as determined from enzyme suspended in sodium cholate (1%).

RESULTS

Cleavage of Cytochrome c Oxidase by Chymotrypsin. The polypeptide composition of beef heart cytochrome c oxidase incubated with chymotrypsin (1:20) for 2 hr at room temp in 1% Triton X-100, 20 mM Tris HCl (pH 8.2) is compared with that of untreated enzyme in Figure 1A. As first reported by Carroll and Racker (35), chymotrypsin cleaves subunit III (subunit IIb in their terminology) and small molecular weight polypeptides (here identified as components b and c). These are the only polypeptides whose migration on gels is visibly altered either when the protease digestion is done in Triton X-100 (this study) or cholate (this study, results not shown; also in Carroll and Racker, 35). Figure 1B shows the rate of disappearance of polypeptides b, c and subunit III from the gel profile as a function of time and the effect of the cleavage of these polypeptides on electron transfer activity. Essentially complete digestion of subunit III results in less than 10% loss of activity when the reaction is done at pH 8.2.

Identification of the Fragments of Subunit III Generated by the Protease Cleavage. The gel profile in Figure 1 is of chymotrypsin-treated cytochrome c oxidase applied directly to the gel (the protease having been inhibited by PMSF prior to denaturation in SDS) without prior separation of fragments released from the core of the complex during digestion.

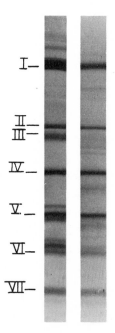

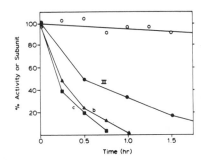

Figure 1. Effect of chymotrypsin on digestion of cytochrome c oxidase activity and polypeptide composition.
A. SDS polyacrylamide gels of chymotrypsin treated and
control beef heart cytochrome c oxidase. Gels were
stained with Coomassie brilliant blue. Polypeptides are
numbered using the nomenclature of Downer et al. [21].
B. Time course of chymotrypsin digestion of subunit III
(● - ● - ●) and polypeptides b (Δ - Δ) and c (■ - ■)
measured by following the decrease in area of peaks in
scans of the stained gels. The effect of digestion on
electron transfer activity (o - o - o) was monitored
polarographically.

The only new bands visible on the gel are those of polypeptides of chymotrypsin; fragments of subunit III were not visible. This was because they stained poorly with Coomassie blue (as does the intact subunit III) and because they comigrated on gels with the smaller subunits of the enzyme (see later).

An obvious approach to identifying fragments of subunit III was to use antibody to this subunit in combination with the blotting technique of Towbin et al. (36). However, we have been unable to generate a good antibody against this very hydrophobic polypeptide. The alternative has been to use labeling procedures employing [^{3}H] adamantane diazirine to label multiple sites of subunit III, [^{14}C] DCCD to label Glu 90 and [^{14}C] iodoacetamide to label Cys 115 prior to the cleavage step.

Adamantane diazirine is a lipophilic reagent, which when illuminated by UV light, is converted to a highly reactive carbene that can insert into covalent bonds (Richards and Brunner, 1980). Our previous studies have established that adamantane diazirine reacts with cytochrome c oxidase to label predominantly subunit I and III (28), in a pattern very similar to the labeling by arylazidophospholipids (17,20). A typical labeling pattern with the reagent is shown in Figure 2B; also shown are the distribution of counts after cleaving adamantane diazirine-labeled-cytochrome c oxidase with chymotrypsin (Figure 2A). In this experiment there was approximately a 60% reduction in the amount of subunit III by chymotrypsin cleavage as judged both by the change in stain intensity and the loss of counts from the subunit III peak. One major new peak of radioactivity comigrated with subunit IV and accounted for more than half of the counts lost from subunit III. The remaining counts appeared in two broad peaks, one running around the position of subunit V, the other migrating in the position of uncleaved polypeptides b and c.

The labeling of beef heart cytochrome c oxidase by ^{14}C DCCD is shown in Figure 3B. As reported previously, this reagent labels subunits II, III and IV about equally in detergent dispersed enzyme as opposed to the labeling of subunit III almost exclusively in membrane preparations of the enzyme (Prochaska et al., 1981). The site of incorporation of DCCD into subunit III has been localized to Glu 90 (30). Cleavage of DCCD-labeled cytochrome c oxidase (Figure 3A) shows that digestion of subunit III generates two broad peaks of DCCD-labeled fragments one comigrating with subunit V, and the other with polypeptides b and c. These peaks of radioactivity migrate on gels at the same position as the new peaks of adamantane diazirine labeling. Each peak must contain fragments including Glu 90, with the larger of the two concatamers including the smaller fragment. The absence of DCCD labeling from the fragment comigrating with subunit IV, identifies this as the C-terminal part of subunit III.

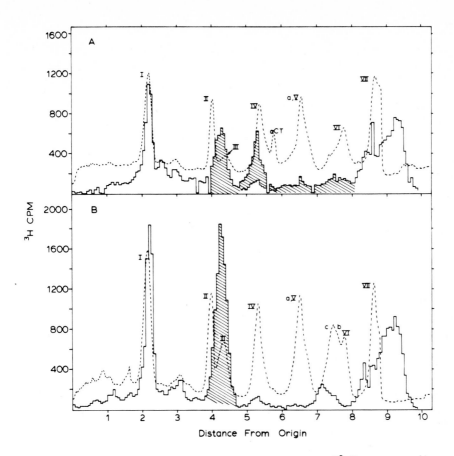

Figure 2. Labeling of Cytochrome c Oxidase with [³H] Adamantane Diazirine. A. Enzyme treated with chymotrypsin after labeling. B. Control cytochrome c oxidase. Dashed areas identify subunit III and its fragments generated by chymotrypsin digestion.

Preparation of Subunit III-less Enzyme. Chymotrypsin cleavage offers an approach to preparing subunit III-less enzyme for structural and functional studies provided that the fragments generated by the cleavage procedure can be removed under mild conditions. Several approaches were tested. The one chosen involved chromatography on DEAE agarose. Samples of enzyme that had been cleaved with chymotrypsin were bound to DEAE agarose in the buffer system used in the cleavage step.

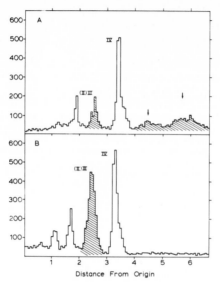

Figure 3. Labeling of Cytochrome c Oxidase with (^{14}C) DCCD.
A. Enzyme treated with chymotrypsin after labeling. B. Control cytochrome c
oxidase. Dashed areas identify subunit III and its fragments generated by chymotrypsin digestion.

The preparation was washed in detergent containing buffer (0.1 Triton X-100;
0.1% cholate) at low ionic strength to remove fragments of subunit (as monitored
by radioactivity) and then the remainder of the cytochrome c oxidase complex was
eluted in the same buffer at high ionic strength. The column treatment by
itself removes a large part of polypeptide b and c and a small amount of subunit
III. However, subunit III-free enzyme was obtained only by the combination of
chymotrypsin cleavage and DEAE chromatography. These two steps together
increased the heme to protein ratio of cytochrome c oxidase preparations from
8.5-9.7 to 13.5-14.2 nmoles heme a/mg protein. Subunit III-less enzyme had an
electron transfer activity that was 90% of that of an untreated control assayed
under the same conditons. Studies are in progress to examine the effect of
removing subunit III on the proton pumping function of the enzyme.

Molecular Weight of Subunit III-less Enzyme. Beef heart cytochrome c
oxidase has been found to exist as a dimer in low concentrations of Triton X-100
or Tween 80 at pH's in the range of 7.0-7.8 (37) but is dissociated into
monomers in Triton X-100 (but not Tween 80) at high ionic strength, pH 8.5 (34).
The enzyme is also monomeric after prolonged incubation with DOC at pH 8.0, as
in the preparation of two dimensional crystals in this detergent (11).
Sedimentation equilibrium studies of a typical subunit III-less enzyme that was

eluted from the DEAE agarose column in 0.1% Triton X-100 pH 7.4, gave a
molecular weight of 139,000 after correction for detergent (180 ± 20 moles
Triton X-100 bound per mole heme aa_3). This compares with a molecular weight of
126,000 calculated from sequence data for a monomer containing one copy each of
subunits I, II, IV, V, VI and VIIs Ser and Ile.

Cytochrome c oxidase which has been dissociated into monomers in the presence
of Triton X100 by a combination of high pH and high ionic strength, reassociates
into dimers when the ionic strength and pH (7.4) are reduced. This is shown in
Figure 4 with the monomer and dimer being separated by sucrose gradient centrifugation. In contrast, subunit III-less enzyme remains fully monomeric in
Triton X-100 or Tween 80 at low pH, indicating an important difference in aggregation properties depending on the presence or absence of subunit III.

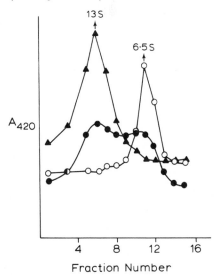

Figure 4. Density gradient centrifugation of cytochrome c oxidase. The
enzyme was sedimented through 10-40% linear sucrose gradients containing 0.1%
Triton X-100, and 20 mM Tris-HCl at pH 7.4 for 1.5 hours at 80,000 rpm in the
Beckman UTi 80 rotor (5°C).
Monomeric enzyme containing subunit III was generated by the high pH, high
Triton X-100 incubation [34] and dialyzed against 20 mM Tris HCl, pH 7.4 and
0.1% Triton X-100 and loaded on the gradients (● - ●). The untreated control
enzyme (Δ - Δ) and the chymotrypsin digested subunit III-less enzyme (o - o)
were dialyzed against the same buffers before centrifugation.

Chemical Labeling of Subunit III-less Enzyme. The removal of subunit III from
the cytochrome c oxidase complex might be expected to expose sites for labeling
by protein modifying reagents that were previously shielded by interaction with
subunit III. Identification of such sites would then provide important clues

230

about the association of subunits in the enzyme complex. This possibility was examined by comparing the labeling of subunit III-less and subunit III-containing enzyme with [^{35}S] DABS and [^{14}C] iodoacetamide as reagents for the water exposed parts of the protein and [^{3}H] adamantane diazirine as a probe of the bilayer-intercalated part of the complex.

Within experimental error, the labeling of subunit I, II, IV, V, VI and VIIs by any of these reagents was the same whether subunit III was present or not. This is shown for [^{3}H] adamantane diazirine-labeling in Figure 5 (the number of counts on the gel migrating with subunit I is less in the chymotrypsin-treated sample because of removal of a comigrating impurity). The almost identical labeling pattern with and without subunit III present is in marked contrast to denaturing cytochrome c oxidase with SDS which leads to heavy labeling of all polypeptides with AD (G. Georgevich, unpublished results).

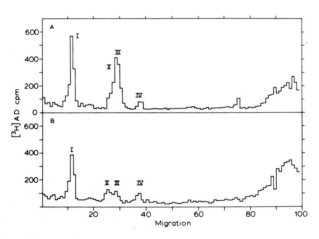

Figure 5. Labeling of Subunit III-less Cytochrome c Oxidase with [^{3}H] Adamantane Diazirine. A. Control enzyme. B. Subunit III-less preparation.

The only clear difference between subunit III-less and subunit III-containing cytochrome c oxidase was seen in cross-linking experiments. Figure 6 shows a two dimensional gel of the cross linking of near neighbor polypeptides in subunit III-less enzyme by disuccinimidylpropionimidate (DSP). The major cross linked product in the subunit III-less enzyme were I + V, II + V, IV + VI, IV + VII and VII + VII. These same products were identified in a sample of enzyme retaining subunit III and made monomeric for the cross linking reaction. The difference was that the cross linked product of III + V seen in the control sample (and see Briggs et al., 1977) was missing from the gel of the subunit III-less enzyme.

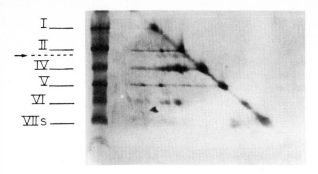

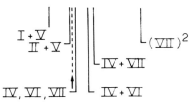

Figure 6. Cross Linking of Subunit III-less Cytochrome c Oxidase with DSP. The arrow shows the position of subunit III on this gel system.

DISCUSSION

The studies described here were undertaken in order to locate subunit III within the three dimensional, Y-shaped structure of cytochrome c oxidase. Firstly, a method has been devised to remove subunit III under mild conditions. Second, subunit III-less enzyme has been studied with respect to molecular weight and lastly, the arrangement of subunit III in the complex has been explored by chemical labeling and by cross linking experiments.

There have been several recent reports of the removal of subunit III from beef heart (21,23) and rat liver cytochrome c oxidase (38). In this study we have used chymotrypsin cleavage followed by DEAE agarose chromatography to cleave and then remove this subunit. The removal of fragments after cleavage was monitored by selective labeling of subunit III with iodoacetamide, DCCD and through labeling with adamantane diazirine.

The heme a content of subunit III-less enzyme prepared by Saraste et al. (21), Bill and Azzi (23), and the chymotrypsin-generated preparation described here is approximately the same, i.e. 14-15 nmoles/mg protein. In each case, the enzyme retains 90% of the electron transfer activity of untreated, subunit III-containing preparation. The proton pumping function has only been examined in the one case and was lost upon removal of subunit III (21). An advantage of the chymotrypsin method is speed; it takes less than 4 hrs to complete. Also it can be done in detergents other than Triton X-100, e.g. cholate, which may be important for reconstitution experiments where it is an advantage to be able to remove detergent easily and fully by dialysis.

The major emphasis of the present studies was a structural one. Cleavage of subunit III with chymotrypsin was shown to occur in at least two places in the sequence, one near the center of the polypeptide and to the C-terminus of Glu 90 and close to Cys 115. This latter residue has been established as the site of disulfide bond formation with Cys 102 in yeast cytochrome c (39), placing it in the C domain. Preliminary studies with oriented membranes or cytochrome c oxidase confirm that chymotrypsin cleaves subunit III from the C side of the membrane (Zhang Yu-Zhong and R.A. Capaldi, unpublished studies) and experiments in progress should indicate whether one or both cleavage sites are present in the C domain. The key results of the structural analysis on the subunit III-less enzyme are as follows:

i) Cytochrome c oxidase missing subunit III is monomeric under conditions where the enzyme retaining this subunit is dimeric.

ii) Removal of subunit III does not expose new sites on the bilayer-intercalated subunits I, II, IV and VII's for labeling by adamantane diazirine.

iii) Removal of subunit III has a minimal effect on the water exposed part of the complex. The electron transfer activity of the enzyme and thus the cytochrome c binding site on subunit II is unaltered. The spectral properties of the hemes are altered to a small extent but apparently without functional significance. No major new surface for reaction with iodoacetamide or DABS is exposed and the only structural alteration detected is the absence of cross linking between subunit V and subunit III.

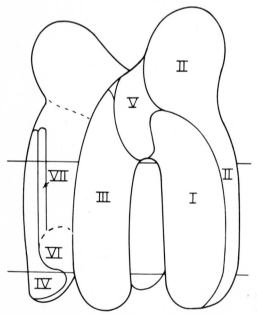

Figure 7. A Model for the Arrangement of Subunits in Cytochrome c Oxidase.

To integrate and interpret these findings it is necessary to review the emerging picture of the three dimensional structure of cytochrome $\underline{c}$ oxidase. Electron microscopy and imaging procedures have shown that the enzyme complex is Y-shaped and exists as a dimer in the lipid bilayer (10-13). Figure 7 presents a model for the arrangement of subunits in this Y-shaped complex based on arguments presented below and consistent with the results of experiments discussed here. In this model subunit III is placed in the M_2C_2 domain and the larger subunits I, II, IV, V and VI are all placed in the $C_1C_3M_1$ parts of the molecule. The location of subunit III in the M_2C_2 domain is consistent with considerations of the number of likely transmembrane segments in the various subunits. Briefly, the areas of the M_1 and M_2 domain in projection (at 12 Å resolution) are large enough to contain between 9 and 13 and 6-8 helices respectively if these helical segments are packed the same way as in bacteriorhodopsin (see ref. 13). Examination of the published sequences of the cytochrome $\underline{c}$ oxidase subunits identify 10 likely transmembrane segments in subunit I, 2 in II, 6 in III, 1 in VI and 2 in VIIs for 21 in all (3). Given these numbers of helices in each subunit, it would be impossible to fit both subunits I and III in the same M domain (16 helices in all). Moreover, the M_2 domain would be too small to contain subunit I arguing that it is the smaller of the membrane spanning domains which contains subunit III.

The separation of the bilayer-intercalated part of subunit III from the bilayer intercalated parts of the other subunits would explain why removing subunit III does not expose new sites for labeling by adamantane diazirine.

Inspection of Figures 2 and 3 in Deatherage et al. (12) shows that the connection between the M_2C_2 part and the rest of the structure is small and limited to between the C_2 and C_3 regions of the molecule. This small area of contact when disrupted, could well be devoid of sites for reaction with DABS or iodoacetamide. Based on the cross linking data, the contact of subunit III with the rest of the complex involves subunit V. Subunit V is also a near neighbor of subunits I and II (18,19). A limited area of contact between subunit III and the C_2M_2 domain and the core of the complex would explain the easy and selective dissociation of subunit III from the molecule by the relatively mild, non-denaturing detergent Triton X-100. Finally, it is significant for the model proposed in Figure 7 that the subunit III-less enzyme is monomeric. The interaction between monomers has been shown by Deatherage et al. (12,13) to involve mainly contacts between the C_2 parts of the C domain. Removal of subunit III as the C_2M_2 part of the molecule would thus be expected to dissociate the protein into monomers.

A structural prediction of the model for cytochrome $\underline{c}$ oxidase in Figure 7 is that subunit III-less enzyme should no longer be Y-shaped as one of the arms of the Y would be missing. Efforts are now underway to crystallize chymotrypsin cleaved and subunit III-less cytochrome $\underline{c}$ oxidase so that this prediction can be tested.

ACKNOWLEDGMENTS

This work was supported by PHS Grant No. HL22050.

REFERENCES
1. Azzi, A. (1980) Biochim. Biophys. Acta. 594, 231-252.
2. Wickstrom, M., Krab, K. and Saraste, M. (1981) Ann. Rev. Biochem. 50, 623-655.
3. Capaldi, R.A., Malatesta, F. and Darley-Usmar, V.M. (1983) Biochim. Biophys. Acta., in press.
4. Steffens, G.J. and Buse, G. (1976) Hoppe Seylers Z. Physiol. Chem. 357, 1125-1137.
5. Downer, N.W., Robinson, N.C. and Capaldi, R.A. (1976) Biochemistry 15, 2930-2936.
6. Kadenbach, B. and Merle, P. (1981) FEBS Lett. 135, 1-11.
7. Steffens, G.C.M., Steffens, G.J. and Buse, G. (1979) Hoppe Seyler's Z. Physiol. Chem. 360, 1641-1650.
8. Tanaka, M., Haniu, M., Yasunobu, K.T., Yu, C.A., Yu, L. Wei, Y-H and King, T.E. (1975) J. Biol. Chem. 254, 3879-3885.
9. Anderson, S., de Bruijn, M.H.L., Carlson, A.R., Eperon, I.C., Sanger, F. and Young, I.G. (1982) J. Mol. Biol. 156, 683-718.
10. Henderson, R., Capaldi, R.A. and Leigh, J.S. (1977) J. Mol. Biol. 112, 631-648.
11. Fuller, S.D., Capaldi, R.A. and Henderson, R. (1979) J. Mol. Biol. 134, 305-327.
12. Deatherage, J.F., Henderson, R. and Capaldi, R.A. (1982a) J. Mol. Biol. 158, 487-499.
13. Deatherage, J.F., Henderson, R. and Capaldi, R.A. (1982b) J. Mol. Biol. 158, 500-514.
14. Frey, T.G., Chan, S.H.P. and Schatz, G. (1978) J. Biol. Chem. 253, 4389-4395.
15. Eytan, G.D., Carroll, R.C., Schatz, G. and Racker, E. (1975) J. Biol. Chem. 250, 8598-8603.
16. Ludwig, B., Downer, N.W. and Capaldi, R.A. (1979) Biochemistry 18, 1401-1407.
17. Bisson, R., Montecucco, C., Gutweniger, H. and Azzi, A. (1979) J. Biol. Chem. 254, 9962-9965.
18. Briggs, M.M. and Capaldi, R.A. (1977) Biochemistry 16, 73-77.
19. Briggs, M.M. and Capaldi, R.A. (1978) Biochem. Biophys. Res. Commun. 80, 553-559.
20. Prochaska, L., Bisson, R. and Capaldi, R.A. (1980) Biochemistry 19, 3174-3179.
21. Saraste, M., Penttila, T. and Wickstrom, M. (1981) Eur. J. Biochem. 115, 261-268.
22. Penttila, T., Saraste, M. and Wikstrom, M. (1979) FEBS Lett. 101, 295-300.
23. Bill, K. and Azzi, A. (1982) Biochem. Biophys. Res. Commun. 106, 1203-1209.
24. Capaldi, R.A. and Hayashi, H. (1972) FEBS Lett. 26, 261-263.

25. Vik, S.B. and Capaldi, R.A. (1980) Biochem. Biophys. Res. Commun. 94, 348-354.
26. Lowry, O.H., Rosebrough, N.J., Farr, A.L. and Randall, R.J. (1951) J. Biol. Chem. 193, 265-270.
27. Tinberg, H.M., Melnick, R.L., Maguire, J. and Packer, L. (1974) Biochim. Biophys. Acata. 345, 118-128.
28. Georgevich, G. and Capaldi, R.A. (1982) Biophys. J. 37, 66-67.
29. Darley-Usmar, V.M., Capaldi, R.A. and Wilson, M.T. (1981) Biochem. Biophys. Res. Commun. 103, 1223-1330.
30. Prochaska, L.J., Bisson, R., Capaldi, R.A., Steffens, G.C.M. and Buse, G. (1981) Biochim. Biophys. Acta. 637, 360-373.
31. Fuller, S.D., Darley-Usmar, V.M. and Capaldi, R.A. (1981) Biochemistry 20, 7046-7053.
32. Oakley, B.R., Kirsch, D.R. and Morris, N.R. (1980) Anal. Biochem. 105, 361-362.
33. Tanford, C., Nozaki, Y., Reynolds, J.A. and Makino, S. (1974) Biochem. 13, 2369-2374.
34. Georgevich, G., Darley-Usmar, V.M., Malatesta, F. and Capaldi, R.A. (1983) Biochemistry 22, 1317-1322.
35. Carroll, R.C. and Racker, E. (1977) J. Biol. Chem. 252, 6981-6990.
36. Towbin, H., Staehlin, T. and Gordon, J. (1979) Proc. Natl. Acad. Sci. USA 76, 4350-4354.
37. Robinson, N.C. and Capaldi, R.A. (1977) Biochemistry 16, 375-381.
38. Thompson, D.A., Suarez-Villafane, M. and Ferguson-Miller, S. (1982) Biophys. J. 37, 227-232.
39. Malatesta, F. and Capaldi, R.A. (1982) Biochem. Biophys. Res. Commun. 118, 1180-1185.

STRUCTURE OF BOVINE MITOCHONDRIAL NADH: UBIQUINONE OXIDOREDUCTASE STUDIED BY ELECTRON MICROSCOPY

EGBERT J. BOEKEMA AND ERNST F.J. VAN BRUGGEN
Biochemisch Laboratorium, Rijksuniversiteit Groningen, Nijenborgh 16, 9747 AG
Groningen (The Netherlands)

INTRODUCTION

NADH:Ubiquinone(Q) oxidoreductase (E.C. 1.6.99.3) is a transmembraneous lipoprotein, localized in the inner mitochondrial membrane. It catalyzes the first step in the respiratory chain: proton translocation coupled to the oxidation of NADH. A purified form of the enzyme, that has been studied by several research groups, is the Complex I from bovine heart (1). It is probably the most complex protein of the mitochondrial electron transfer chain. The number of subunits of which it is composed has been reported to vary up to 26 (2). A flavoprotein fragment consisting of three subunits and FMN as prosthetic group, forms the core of the enzyme, buried within a shell of more hydrophobic subunits (2).

In Complex I preparations the NADH:Q oxidoreductase is solubilized with the detergent cholate. When the cholate is removed by dialysis against buffers containing 1.0 - 1.5 M ammonium sulphate at pH 5.6 - 7.4, two-dimensional sheets are formed that are very suitable for low-dose electron microscopy (3).

We have investigated the structure of NADH:Q oxidoreductase by electron microscopy and computer image analysis of micrographs of these two-dimensional crystals. The insufficient statistical significance of an image of a single molecule must be improved by averaging over many molecules before a trustworthy interpretation of the studied structure is possible. To obtain noise-free images of two-dimensional crystals, Fourier peak filtering is an easy and well-known technique. The resolution of averaged projections by this method is generally limited by long-range distortions of the crystal lattice and by small rotational and translational shifts of the repeating unit with respect to its ideal position. To correct for these lattice imperfections, correlation averaging techniques have been developed (4,5). We will describe our application of correlation averaging on the projected stucture of NADH:Q oxidoreductase.

MATERIALS AND METHODS

Complex I was isolated from bovine heart muscle particles, according to the method of Hatefi et al. (1) and was a gift from Dr. S.P.J. Albracht (University of Amsterdam). It was crystallized by means of equilibrium dialysis in glass capillaries (3).

Electron microscopy was carried out with negatively stained specimens (1% unbuffered uranyl acetate) in a Philips EM 400 at 80 kV, using a low-dose unit for off-specimen focussing and astigmatism correction.

Micrographs were selected for computer processing on the basis of the quality of their optical diffraction pattern. Negatives were scanned on a Joyce-Loebl Scandig 3 rotating drum densitometer using a 25x25 μm sampling aperture and sampling grid, corresponding to 0.54 nm at the specimen level. The digitized images were processed using the IMAGIC image processing system (6) on a Norsk Data Nord-10 minicomputer.

RESULTS AND DISCUSSION

To resolve the structure of an intrinsic membrane protein like NADH:Q oxido-reductase at high resolution, crystallization in two or three dimensions is an essential step. In general, averaging of single molecules after alignment procedures may result in noise-reduced projections with reliable details in the range of 2 nm (7), but in our case single molecules (fig. 1) are difficult to prepare, and close examination in the computer shows that their images lack structural detail below 3-4 nm (8).

With the microdialysis technique two forms of two-dimensional crystals could be obtained. By dialysis against a 50 mM Tris.HCl buffer of pH 7.2, containing 1.5 M ammonium sulphate, the compact clots of the original detergent-lipo-protein complex could be forced to form partly aligned areas. The packing is however very difficult to interpret. The spacings a and b of the smaller stain-excluding rows are 17.4 $\pm$ 1.7 nm for a and 14.9 $\pm$ 1.3 nm for b (standard error of the mean for 12 crystals). A highly-ordered crystal-type could be obtained by slight alterations of the dialysis-medium within 14 days. Large monolayer-crystals were obtained by dialysis against 1.0 M ammonium sulphate in 50 mM sodium acetate buffer pH 5.9 plus 5 mM NADH, with repeated refresh-ments every two days (fig. 2). The crystal has a square unit cell of 15.2x15.2 nm (fig. 3) and is built up from the tetrameric molecules seen in figure 1.

These crystals were incubated with Phospholipase A2 from the Forest cobra Naja melanoleuca. This enzyme is capable of breaking down the lipid bilayer (9). Treatment of crystallized Complex I with a relatively large amount of

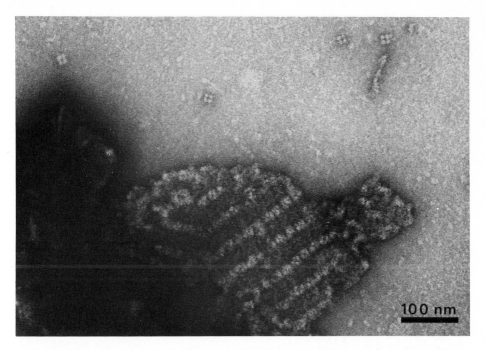

Fig. 1. An electron micrograph of a compact clot of lipid and NADH:Q oxido-reductase of a Complex I preparation after microdialysis against 1.5 M ammonium sulphate, surrounded by some single tetrameric molecules. Here and subsequently, protein appears white.

Phospholipase A2 (1 mg/ml in Tris.HCl pH 7.4/5 mM CaCl2/18 hr) resulted in the crushing of all types of lipid-protein clots. The highly-ordered crystals however were still intact and, as judged by their optical diffraction pattern, unaltered by treatment.

Figure 2 shows a large crystal that was used to obtain a noise-free projection of the NADH:Q oxidoreductase structure at high resolution by correlation averaging. The first step in the alignment procedure was to set up a reference by Fourier peak filtering of an area of 512x512 pixels (image elements) selected from the crystal. The central part of this filtering (128x128 pixels) was used for the alignment of 200 noisy equally sized fragments from the complete crystal. The fragments overlapped some 10% with their neighbours to compensate for the loss of cut unit cells in any of the fragments. The position in which a fragment was exactly equivalent to the reference could

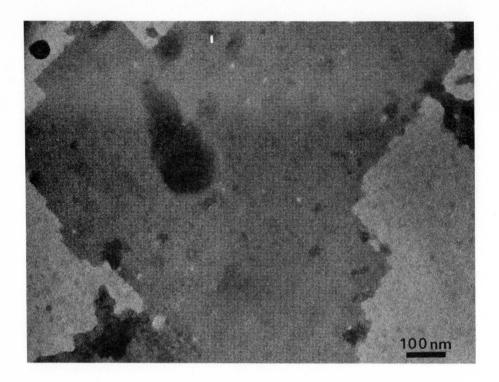

100 nm

Fig. 2. Typical appearance of a large two-dimensional NADH:Q oxidoreductase crystal, slightly stained with 1% uranyl acetate. It was prepared after microdialysis against a buffer with 1.0 M ammonium sulphate plus 5 mM NADH.

be determined by calculating the two-dimensional cross-correlation function between the reference and each fragment. This resulted in small shifts of the aligned areas (fig. 4a). Since local deviations of the repeating motive are compensated for by tanslation, the resolution of the average of 200 fragments is necessarily better than the resolution obtained by computer Fourier filtration (compare the Fourier test-pattern of figure 4b with the optical diffraction pattern of the uncompensated lattice of figure 3).

Averaging of the 200 fragments resulted in a summation in which 9 whole, individual molecules were present. They were also selected, aligned and summed, resulting in an average of 1800 molecules (fig. 4c).

We investigated if further improvement of resolution could be obtained by alignment (rotational and translational) of fragments of 48x48 pixels,

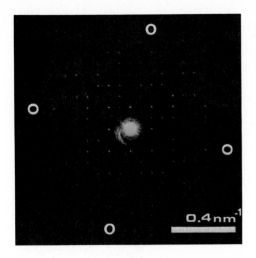

Fig. 3. The optical diffraction pattern of the image of figure 2 shows the preserved periodicity in the crystal. The 9th orders of reflection, equivalent to a spacing of 1.7 nm, are indicated.

regions that have the dimensions of figure 4c. A trial alignment of 500 of these regions resulted in a mean rotational shift of 1.31 ($\pm$ 1.05 standard deviation) degrees. In a reconstruction such small rotational shifts will contribute to details well below 1 nm. The transfer-function shown by this micrograph, however, fades rapidly below 1.2 nm (fig. 3), so almost no structural detail below this value was recorded. Hence it is not surprising that the summation over 500 aligned regions differed only marginally from the picture of 1800 molecules. The resolution obtained by the correlation averaging was about 1.3 nm, determined by calculating the centro-correlation (7) between two individual subsets of 900 regions.

Until correlation averaging on a larger scale has been worked out it is not possible to determine to what extent the preparation method limits the final resolution. At the moment we can see a two-fold rotational symmetry for the NADH:Q oxidoreductase molecule from close examination of the averaged projection (fig. 4c). This implies that only two pairs of two monomers are identical, although all four monomers look quite similar. So the NADH:Q oxidoreductase must be a functional dimer. This conclusion was supported by determination of the molecular mass by scanning transmission of electron microscopy

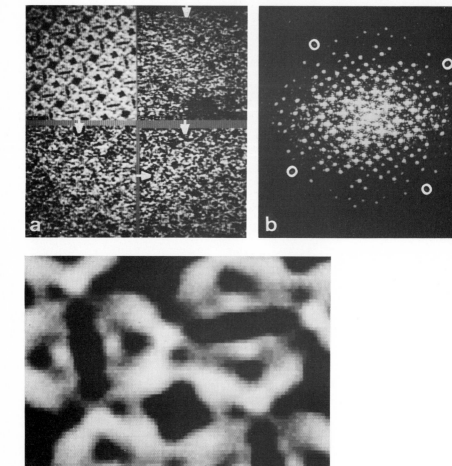

Fig. 4. Correlation averaging applied to the crystal of figure 2. (A) Fourier peak filtering was applied to make a low-noise reference (top left) for the alignment of 200 noisy fragments of the crystal. The direction of the shifts to an optimal match with the reference is indicated for three aligned fragments. (B) A calculated Fourier pattern from 200 summed crystal fragments showing an improvement in resolution. The 10th orders of reflection on both crystallographic axes are indicated. (C) The final result from the correlation averaging: summation of 9 individual unit cells from the sum of the 200 fragments results in an average of 1800 NADH:Q oxidoreductase molecules.

as 1.6 $\pm$ 0.2x10^6 daltons (8). This figure indicates a dimeric molecule since a ratio of FMN to protein of 1 : 700 000 has been found (10) and a minimal unit of NADH:Q oxidoreductase must contain two clusters 2 and two FMN molecules in order to accommodate one NADH-reducible cluster 1 (10).

The results obtained by correlation averaging on the projected structure can be compared with a three-dimensional model calculated form a tilt-series of a two-dimensional crystal (8). The most interesting features of this model are the pores through the centre of each of the four monomers, extending from the top to the bottom of the molecule (fig. 5). Their diameter is about 1.5-2 nm, measured from the averaged projection. Four smaller pores that seem to be present in the average of figure 4c can also be localized in the model. They all begin on one side but are only 5 nm long, about half of the thickness of the model (10-11 nm). The interpretation of these pores in terms of biochemical relevance is however still a long way off.

Fig. 5. View of a model of NADH:Q oxidoreductase, consisting of 20 sections parallel to the plane of the crystal, each with a thickness of 0.54 nm. Remarkable features are the pores through the centre of the four monomers, extending from the top to the bottom. Four smaller pores (see *) all begin on one side, but are only 5 nm long, about half of the thickness of the model.

ACKNOWLEDGEMENTS

We thank Dr. W. Keegstra for his help with computer facilities, Mrs. K. Gilissen and J. Haker for excellent technical assistance and Mrs. I. Bont for typing the manuscript. This work was supported by the Netherlands Organization for Chemical Research (SON) with financial aid from the Netherlands Organization for the Advancement of Pure Research (ZWO).

REFERENCES

1. Hatefi, Y., Haavik, A.G. and Griffiths, D.E. (1962) J. Biol. Chem., 237, 1676.

2. Ragan, C.I. (1980) in: Roodyn, D.B. (Ed.), Subcellular Biochemistry, Vol. 7, pp. 267-307.

3. Boekema, E.J., Van Breemen, J.F.L., Keegstra, W., Van Bruggen, E.F.J. and Albracht, S.P.J. (1982) Biochim. Biophys. Acta, 679, 7.

4. Frank, J. (1982) Optik, 63, 67.

5. Saxton, W.O. and Baumeister, W. (1982) J. Microscopy, 127, 127.

6. Van Heel, M.G. and Keegstra, W. (1981) Ultramicroscopy, 7, 113.

7. Frank, J., Verschoor, A. and Boublik, M. (1981) Science, 21, 1353.

8. Boekema, E.J., Van Heel, M.G. and Van Bruggen, E.F.J. submitted for publication.

9. Verheij, H.M., Slotboom, A.J. and De Haas, G.H. (1981) Rev. Physiol. Biochem. Pharmacol., 91, 91.

10. Galante, Y.M. and Hatefi, Y. (1979) Arch. Biochem. Biophys., 192, 559.

11. Beinert, H. and Albracht, S.P.J. (1982) Biochim. Biophys. Acta, 683, 245.

NATURE OF CONFORMATIONAL TRANSITIONS ASSOCIATED WITH CATION TRANSPORT BY PURE Na,K-ATPase.

PETER L. JØRGENSEN, Institute of Physiology, Aarhus University, 8000 Aarhus C, Denmark.

INTRODUCTION

The characteristics of the Na,K-transport reaction lead one to assume that the translocation process involves motion of protein components in the pump; but it has been a problem to obtain direct evidence for coupling of conformational changes to ion translocation. Pure Na,K-ATPase consists of two proteins in a molar ratio of 1:1. The α-subunit with M_r 94-106,000 forms binding sites for nucleotides and cardiac glycosides, and is phosphorylated from ATP. The β-subunit, a sialoglycoprotein with protein M_r 32-38,000 forms a part of the extracellular binding area for cardiac glycosides, but other catalytic functions have not been associated with this protein (for ref. see 1.). Two principal conformations of the α-subunit, E_1 and E_2, are defined by two patterns of tryptic cleavage (2) and by different intensities of fluorescence from tryptophan (3) or extrinsic probes (4-6). In absence of complete aminoacid sequences and detailed structure models of Na,K-ATPase, the transitions can be described in terms of differences between spatial arrangements of aminoacid residues in states of the pump with different ligand affinities. In addition to the trypsin sensitive bonds and the tryptophan residues there is evidence that the E_1 to E_2 transition involves sulfhydryl groups, ionizable groups and intramembrane segments. The two forms have different affinities for nucleotides and phosphate and for the specific inhibitors, ouabain and vanadate (For ref. see 1). Explanations of the relationship of these transitions to ATP hydrolysis and ion translocation will be essential for understanding the transport process.

The first question to be addressed is whether the transitions between E_1 and E_2 forms consist of rearrangement within one α-subunit or if interaction between protomer $\alpha\beta$-units is required. For this we have examined the conformational responses in preparations of soluble and fully active $\alpha\beta$-units in dodecyl-octa-ethylen-glycol-monoether ($C_{12}E_8$). Next, we have used selective proteolytic mofification of the α-subunit to identify functional domains of the protein and to study the relationship between structural changes in the protein, saturation of high affinity ligand binding sites and flipping of cation sites between inside exposed and outside exposed states.

MINIMUM ACTIVE PROTEIN UNIT OF SOLUBLE Na,K-ATPase.

Preparation of soluble and fully active $\alpha\beta$-units of Na,K-ATPase requires that the period of exposure to the non-ionic detergent $C_{12}E_8$ is as brief as possible. The pure membrane-bound Na,K-ATPase is solubilized in $C_{12}E_8$ immediately prior to enzyme assay and analytical ultracentrifugation (7,8). This procedure yields a solution of predominantly protomeric αβ-units with maximum M_r 141,000-170.000 and specific activity of Na,K-ATPase equal to that of pure membrane bound enzyme. This fully active preparation of αβ-units in $C_{12}E_8$ reconstitutes directly by the freeze thaw sonication procedure to catalyze Na,K-transport at high rates (8). Secondary aggregation to dimers or trimers was observed after chromatography or storage overnight of the solublized preparations of both Na,K-ATPase and Ca-ATPase (7). This aggregation is not due to specific interaction between protein units, but can be explained by time dependent aggregation of detergent micelles. Polyether-derived detergents are known to form oxidizing degradation products (9) that may catalyse artefactual time-dependent aggregation of the soluble protein (10). Purification of soluble Na,K-ATPase from shark rectal gland with M_r 257,000-380,000 involved chromatography with prolonged periods of contact with $C_{12}E_8$ or Lubrol WX (11,12) and it is probable that these higher molecular weights are explained by secondary aggregation of detergent micelles.

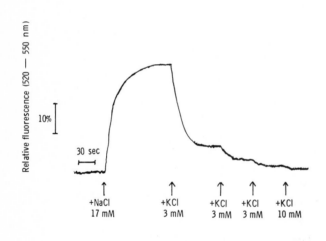

Fig.1. Changes in Eosin fluorescence (6) accompanying transitions between E_2K and E_1Na forms of soluble Na,K-ATPase in $C_{12}E_8$ (7,8). 150 μg soluble Na,K-ATPase was suspended in 25 nM Eosin, 25 mM imidazole, 0.1 mM EDTA, 0.4 mM KCl, pH 7.5 in a stirred cuvette at 10oC. Fluorescence was recorded in a Perkin Elmer MPF 44A fluorimeter. NaCl and KCl were added from Hamilton syringes.

The soluble Na,K-ATPase binds nucleotides with the same capacity as the membrane-bound enzyme and analysis of the effects of K^+ on the binding suggests that cooperative interactions between binding sites are relieved on solubilization (13). With one site for ATP binding per αβ-unit (5), the data on our soluble preparation thus support the notion that a single αβ-unit performs the entire Na,K-ATPase cycle and that a permanent association to $(\alpha\beta)_2$-units is not necessary for the Na,K-ATPase activity.

One of the requirements for accepting the $\alpha\beta$-unit as the transport system is that it can undergo the transitions between E_1 and E_2 forms of the α-subunit that are assumed to be associated with cation translocation. In dilute suspensions of $\alpha\beta$-units temporary associations between $\alpha\beta$-units can be excluded. Binding data and fluorescence studies show that both the cation induced (E_1Na -- E_2K) and the ATP or P_i dependent (E_1P --E_2P) transitions can be demonstrated in preparations consisting of of soluble and fully active $\alpha\beta$-units. Similar fluorescence changes are observed for monomeric Ca-ATPase in $C_{12}E_8$ solution (14). The example Fig. 1 demonstrates fluorescence changes accompanying trnsition between E_1Na and E_2K forms in pure soluble Na,K-ATPase consisting predominantly of $\alpha\beta$-units. This is consistent with the notion that a single $\alpha\beta$-unit can catalyze the whole series of intermediary reactions normally occurring in the membrane.

It is questionable to what extent identification of the $\alpha\beta$-unit as the minimum active protein unit of soluble Na,K-ATPase is valid for the membrane-bound Na,K-pump. Transport across phase boundaries cannot be demonstrated in detergent solution and it is not apriori given that the active unit of Na,K-ATPase is identical to the protein complex required for active transport. This question also concerns the pathway for cations across the membrane. In a pump consisting of one $\alpha\beta$-unit the pathway for ions must pass through the protein. In an oligomeric system the transport pathway can be formed in the space between subunits. These problems and the significance of crystallization of the pure membrane-bound Na,K-ATPase in two patterns (15-17) have previously been discussed (1,20).

DISPOSITION OF PROTEINS IN THE MEMBRANE AND PROTEOLYTIC CLEAVAGE OF THE α-SUBUNIT

The α-subunit possesses several transmembrane segments and exposes residues on both the cytoplasmic and extracellular membrane surfaces. Recent data show that also the β-subunit is a transmembrane protein. The amino terminus of the α-subunit and the three trypsin sensitive bonds are exposed at the cytoplasmic surface, (for ref. see 1) but the sidedness of the carboxyl terminus remains unknown. The path of the α-subunit in the membrane has been traced by determining the sidedness of primary proteolytic splits and the localization of membrane embedded segments relative to residues involved in formation of sites for ouabain and ATP at the two membrane surfaces (18).

The bonds exposed to primary tryptic or chymotryptic cleavage at the cytoplasmic membrane surface are unique and well-defined points of reference in the α-subunit (2, 19,20). At any time only one or two bonds are exposed to cleavage. Transition from E_1Na to E_2K or from E_1P to E_2P involves protection of bond 3 and exposure of bond 1 near the middle of the α-subunit to trypsin while the position of bond 2 near the amino terminus is altered so that cleavage of bond 2 becomes secondary to cleavage of bond 1 within the same α-subunit. In NaCl medium of low ionic strength, chymotrypsin cleaves bond 3 about 250 residues away from the amino terminus and the α-subunit can be converted

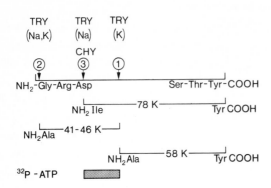

Fig. 2. Linear map of α-subunit with position of sites of primary tryptic cleavage (TRY) in presence of NaCl or KCl (2,19,20) and chymotryptic cleavage (CHY), (19,20). amino and carboxyl terminal residues from Castro and Farley (19). Labelling with 32P-ATP as described before (19,20).

almost quantitatively to the 78 000 fragment because secondary cleavage is negligible (20). Bond 1 and bond 3 in the α-subunit are separated by intramembrane segments and may be located in loops between domains that are protruding at the cytoplasmic surface. The changes in structure can be due to altered positions of side chains within the loops or to rearrangement of the cytoplasmic domains relative to each other.

EFFECT OF CLEAVAGE OF BOND 3 OR BOND 1 ON CONFORMATIONAL TRANSITIONS AND BINDING OF CATIONS AND ATP

Selective cleavage of bond 1 with trypsin in KCl medium or of bond 3 with chymotrypsin in NaCl medium are powerful tools for examination of the function of the α-subunit because the two bonds are located at either side of the aspartyl phosphate residue. Both split 1 and 3 block Na,K-ATPase and active ATP-dependent Na/K-exchange but the resulting 78 000 and 46 000 fragments can both be phosphorylated from $[\gamma^{32}P]ATP$ at 20-25 µM concentration (20) and the two splits have widely different effects on ligand binding and the passive cation exchange reactions that are catalyzed by the Na,K-pump. The study of seclectively cleaved enzymes therefore leads to identification of functional domains of the pump and to a better understanding of the transport process.

Cleavage of bond 3 abolishes both cation induced and phosphate dependent transitions between E_1 and E_2 as monitored by fluorescence from tryptophan or the extrinsic probes, fluorescein-isothiocyanate and eosin. As an example, the absence of cation induced fluorescence responses after cleavage of bond 3 is shown in Fig. 3. It is seen that the responses are retained after cleavage of bond 1 by trypsin in KCl medium.

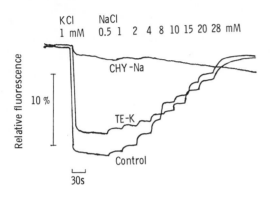

Fig. 3. Cation induced changes in emission from fluorescein attached to the α-subunit (4) in untreated Na,K-ATPase (Control); after chymotryptic cleavage in NaCl (CHY-Na) or after tryptic cleavage in KCl medium (TE-K). The cuvettes contained 2.5 ml 25 mM Tris-HCl, 0.1 mM EDTA, pH 7.5 and 25 μg FITC-labeled Na,K-ATPase protein. KCl or NaCl were added from Hamilton syringes. Emission from fluorescein was recorded at 519 nm with exitation wavelength 495 nm using 10 nm slits on both monochromators.

The absence of fluorescence changes could be due to interference with cation binding, but this is not the case. Fig. 4 shows that neither split 1 nor split 3 affect high affinity binding of two Rb^+-ions per α-subunit (1). Chymotryptic cleavage of bond 3 stabilizes the protein in the E_1 form without affecting the affinity or the capacity of ^{86}Rb binding. This means that the pump exposes high affinity binding sites for Rb^+ at the

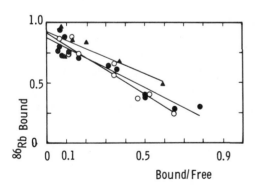

Fig. 4. Equilibrium binding of ^{86}Rb to native Na,K-ATPase (o) and after cleavage with chymotrypsin in NaCl (o) or with trypsin in KCl ( ). The binding capacities fell in the range 10.4–12.1 nmol/mg protein and the dissociation constants were -12.2 μM (o), -9.9 μM (o), and -9 μM ( ), respectively.

cytoplasmic surface. Selective cleavage of the two bonds thus demonstrates that the structural changes accompanying transitions between E_1 and E_2 forms of the protein are coupled to events that occur after saturation of cation binding sites.

Cytoplasmic sites for high affinity binding of Rb^+ or K^+ have previously been inferred from data on enzyme treated with Thimerosal (1). It has also been shown that Rb^+ can bind to Na,K-ATPase with relatively high affinity, K_D 60-70 μM, in presence of 2 mM ATP which stabilizes the E_1 conformation even at high concentrations of K^+ (13). The high apparent affinity of K^+ binding can therefore not be explained by weak binding of K^+ to the E_1-form in coupling with the conformational equilibrium between E_1K and E_2K as proposed by Karlish et al.(3).

Physiologically, the site for high affinity binding of K^+ at the cytoplasmic surface can be very important. Release of K^+ from these sites requires the combined presence of ATP and Na^+. In physiological conditions with a high concentration of K^+ in the cytosol, the slow rate of release of K^+ from the cytoplasmic sites will be rate limiting and thus control the overall rate of the Na,K-transport reaction.

Another surprising result is that ATP and Rb^+ or K^+ bind simultaneously with high affinity after cleavage of bond 3. The data in Fig. 5 demonstrates high affinity binding of ATP both in 150 mM KCl and in 150 mM NaCl after cleavage with chymotrypsin, while ATP, as usual, binds with very low affinity to native Na,K-ATPase in KCl medium.

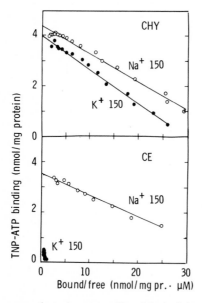

Fig. 5. Fluorescence titration of trinitrophenyl-ATP binding (5) to (CE) native Na,K-ATPase and (CHY) enzyme digested with chymotrypsin in NaCl medium. The dissociation constants were for native Na,K-ATPase (CE): - 0.081 μM in 150 mM NaCl and - 3.4 μM in 150 mM KCl and for chymotrypsin-treated enzyme (CHY): -0.11 μM in 150 mM NaCl and -0.13 μM in 150 mM KCl.

This means that cleavage of bond 3 abolishes the K-ATP antagonism. This antagonism is usually ascribed to the stabilization by K^+ of the E_2K form in which the nucleotide site is adapted for weak binding of ATP. In agreement with this we observe that cleavage of bond 3 prevents this conformational change, but this cannot explain that ATP is without effect on the affinity for Rb^+ after cleavage of bond 3. To explain this it is necessary to assume that cleavage of bond 3 disrupts the interaction between a part of the protein (Amino terminus -- bond 3) engaged in formation of sites for binding of Rb+ or K+ at the cytoplasmic surface and a domain (Bond 3 -- carboxyl terminus) engaged in formation of the nucleotide binding area (Cf. Fig. 7). The disruption of interaction between the two domains may explain that both the K-ATP antagonism and the fluorescence changes are abolished.

Binding of Na+ could not be measured at equilibrium with our technique, but examination of the Na-dependence of phosphorylation after cleavage of bond 3 shows that the 78 000 fragment has a higher apparent affinity for Na with $K_{\frac{1}{2}}=0.7$ mM as compared with $K_{\frac{1}{2}}=1.2$ mM for the intact α-subunit (20). This stimulation of the transfer of γ-phosphate from ATP to the protein means that at least one Na binding site is intact at the cytoplasmic surface after cleavage of bond 3. Recently, it was also shown that E_1P formed when chymotrypsin cleaved enzyme is phosphorylated by ATP contans three occluded Na+-ions per phosphorylation site. Occluded ions do not exchange with medium cation when the enzyme is forced rapidly down a cation exchange column (21).

Selective clevage of bond 3 thus offers an unique opportunity to study the cytoplasmic cation binding sites in conditions where transition to the E_2-form is blocked. Two Rb+-ions can bind per α-subunit with high affinity both in the native Na,K-ATPase and after cleavage of bond 3. These Rb+-ions can be displaced by Na+ in the cleaved enzyme, but not by ATP or ADP as in the native Na,K-ATPase. Cleavage of bond 3 does not interfere with binding of ATP and Na-stimulated transfer of γ-phosphate to the protein and the E_1P-form of the cleaved enzyme occludes 3 Na+ ions per EP from the cytoplasmic surface. As cleavage of bond 3 stabilizes the enzyme in the E_1-form, it appears that none of these reactions are associated with transition from E_1 to E_2.

CATION TRANSPORT AFTER CLEAVAGE OF BOND 3 OR BOND 1

Active Na,K-exchange in reconstituted vesicles is abolished after cleavage of bond 1 or bond 3, but the two splits have widely different effects on the Na/Na and Rb/Rb exchange reactions which are coupled to transition betweeen E_1P and E_2P. Transitions between the two phosphoforms are possible after cleavage of bond 1 and this split does not interfere with either of the exchange reactions (20).

In contrast, cleavage of bond 3 effectively disrupts the coupling of phosphoryl transfer with cation exchange. ATP- and P_i-dependent transport, active Na/K-exchange as well as Na/Na or Rb/Rb exchange reactions are blocked and split 3 prevents the cation effects on the conformational equilibrium between the phosphoenzyme forms (20). The disruption of continuity between a cation binding domain (Amino terminus -- bond 3) and the segment containing the aspartyl phosphate residue (Bond 3 -- bond 1) offers a straightforward explanation for these results, but it remains uncertain to what extent the defective cation exchange is secondary to effects on the conformational equilibrium between E_1 and E_2.

While split 3 effectively blocks cation exchange coupled to transitions between phosphoforms of the protein, it remains possible that the cations bound at the cytoplasmic surface can exchange with cations from the extracellular surface in a passive process without coupling to phosphorylation reactions (22). We therefore examined the vanadate sensitive exchange of ^{86}Rb with Na$^+$ in vesicles reconstituted with chymotrypsin treated enzyme. The results were essentially the same whether the protein was cleaved before or after reconstitution. It is seen from Fig. 6 that reduction of the rate of this passive cation exchange process is moderate, while the Na/Na or Rb/Rb exchange processes were completely blocked after cleavage of bond 3. This suggests that flipping of cation sites between inside and outside exposed states is possible after cleavage of bond 3, but this

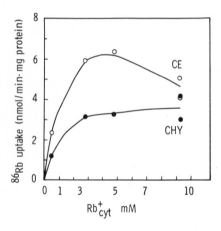

Fig. 6. Vanadate sensitive passive fluxes of ^{86}Rb into vesicles reconstituted with pure native Na,K-ATPase (CE) (o) or with chymotrypsin- cleaved enzyme (CHY) (o). Vesicles were formed (22) with 20 mM NaCl and 10 mM Tris-HCl, pH 7.0 in the inside medium and varying concentrations of ^{86}Rb and Tris-HCl in the outside medium. Incubation for 5, 10, and 20 min without and with 3 mM MgCl$_2$, 3 mM NaVO$_3$.

exchange process is not accompanied by changes in levels of fluorescence intensity. The explanation can be that the changes in fluorescence intensity requires that flipping of the cation sites is coupled to movement of the segment containing the aspartyl phosphate group (Bond 3 -bond 1) and that this coupling is disrupted by cleavage of bond 3.

FUNCTIONAL DOMAINS OF THE α-SUBUNIT

Cleavage of bond 1 or bond 3 at either at either side of the aspartyl phosphate residue blocks different aspects of the mutual interaction between ligand binding sites. Cleavage of bond 1 interferes with the proper orientation between the aspartyl phosphate residue in the 46 000 fragment and a domain in the 58 000 fragment which is engaged in nucleotide binding (20). Cleavage of bond 3 prevents interaction between a cation binding domain (Amino terminus -- bond 3) and the phosphorylated segment between bond 3 and bond 1. This split abolishes the K+-induced changes in conformation of the nucleotide binding area and phosphorylation sites that are observed in the native Na,K-ATPase.

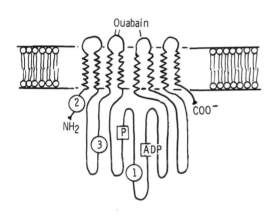

Fig. 7. Model for arrangement of α-subunit of Na,K-ATPase in the membrane. The circled numbers mark the sites of primary tryptic and chymotryptic cleavage. Based on specific labeling of α-subunit (18), but the data on Na,K-ATPase are accomodated in a model based on the distribution of intramembrane and surface domains in Ca-ATPase from sarcoplasmic reticulum as described by Allen, Trinnaman and Green (24).

The structural changes acompanying transitons between E_1 and E_2 forms of the protein are tightly coupled to flipping of cation sites both in native Na,K-ATPase and after cleavage of bond 1 (20). Only after cleavage of bond 3 can passive exchange along cation gradient be demonstrated in absence of fluorescence responses. This suggests that the fluorescence changes are associated with a shift in position of the phosphorylated segment (Bond 3 -- bond 1) relative to neighbouring surface domains and transmembrane segments within the same α-subunit. In support of this, analysis of tryptophan fluorescence (3) and labeling with a hydrophobic probes (18) suggest that residues within the α-subunit move from hydrophilic to a relatively hydrophobic environment during transition from E_1 to E_2 and there is evidence that the aspartyl phosphate residue moves from a hydrophilic environment in E_1P to a more hydrophobic environment in E_2P. (20).

The identification by controlled proteoysis of a segment (Amino terminus -- bond 3) involved in binding of Rb+ or K+ from the cytoplasmic surface is in agreement with the observaton by radiation inactivation that the target size for the part of the Na,K-ATPase responsible for occlusion of Rb+ is only 39 000 dalton (23). The identification of this domain does not mean that cation binding is restricted to this protein segment. On the contrary the present data suggest that a Na+-binding site is formed by the protein segment between bond 3 and bond 1 since Na-stimumlation of phosphorylation from ATP is unaffected by cleavage of bond 3.

REFERENCES

1. Jørgensen, P.L. (1982) Biochim. Biophys. Acta, 694, 27-68.

2. Jørgensen, P.L. (1975) Biochim. Biophys. Acta, 401, 399-415.

3. Karlish, S.J.D. and Yates, D.W. (1978) Biochim. Biophys. Acta, 527, 115-130.

4. Karlish, S.J.D. (1980) J. Bioenerg. Biomembr., 12, 111-136.

5. Moczydlowsky, E. and Fortes, P.A.G. (1981) J. Biol. Chem., 256, 2346-2356.

6. Skou, J.C. and Esman , M. (1981) Biochim. Biophys. Acta, 647, 232-240.

7. Brotherus, J.B., Møller, J.V. and Jørgensen, P.L. (1981) Biochem. Biophys. Res. Commun.,100, 146-154.

8. Brotherus, J.B., Jacobsen, L. and Jørgensen, P.L. (1983) Biochim. Biophys. Acta, in press.

9. Lever, M. (1977) Anal. Biochem., 83, 273-284.

10. Papper, G. and Schubert, D. (1982) Hoppe Seylers Z. für Physiol. Chemie., 363, 904.

11. Hastings, D.F. and Reynolds, J.A. (1979) Biochemistry, 18, 817-821.

12. Esmann, M., Christiansen, C., Karlsson, K.A.,Hansson, G.C. and Skou, J.C. (1980). Biochim. Biophys. Acta, 603, 1-12.

13. Jensen, J. and Ottolenghi, P. (1983) Biochim. Biophys. Acta, in press.

14. Andersen, J.P., Møller, J.V. and Jørgensen, P.L. (1982) J.Biol. Chem., 257, 8300-8307

15. Deguchi, N., Jørgensen, P.L. and Maunsbach, A.B. (1977) J. Cell. Biol., 75, 619-634.

16. Hebert, H., Jørgensen, P.L., E., Skriver, E. and Maunsbach, A.B. (1982) Biochim. Biophys. Acta, 689, 571-574

17. Skriver, E., Maunsbach, A.B. and Jørgensen, P.L. (1981) FEBS Lett. 131: 219-222.

18. Jørgensen, P.L., Karlish, S.J.D. and Gitler, C. (1982) J. Biol. Chem. 257, 7435-7442.

19. Castro, J. and Farley, R.A. (1979) J. Biol. Chem. 254: 2221-2228.

20. Jørgensen, P.L., Skriver, E., Hebert, H. and Maunsbach, A.B. (1982) Ann. N.Y. Acad. Sci., 402, 207-225.

21. Glynn, I.M., Hara, Y. and Richards, E.E. (1983) J. Physiol., in press.

22. Karlish, S.J.D. and Stein, W.D. (1982) J. Physiol., 328, 295-316.

23. Richards, D.E., Ellory, J.C. and Glynn, I.M. (1981) Biochim. Biophys Acta, 648, 284-286.

24. Allen, G., Trinnaman, B.J. and Green, N.M. (1980) Biochem. J., 187, 591-616.

ACKNOWLEDGEMENT

Work in the authors laboratory is supported by the Danish Medical Research Council

IMAGE RECONSTRUCTION OF TWO-DIMENSIONAL CRYSTALS OF MEMBRANE-BOUND Na,K-ATPase.

HANS HEBERT[1], ELISABETH SKRIVER[2], PETER L. JØRGENSEN[3] AND ARVID B. MAUNSBACH[2].

[1]Department of Medical Biophysics, Karolinska Institutet, Box 60400, S-104 01 Stockholm, Sweden, [2]Deparment of Cell Biology at the Institute of Anatomy and [3]Institute of Physiology, University of Aarhus, DK-8000 Aarhus C., Denmark.

INTRODUCTION

The protein units of Na,K-ATPase can be directly demonstrated by electron microscopy in preparations of pure, membrane-bound enzyme. Stabilization of the protein in the E_2 conformation with vanadate and phosphate induces formation of two-dimensional crystals in the pure membrane-bound Na,K-ATPase (1,2). The assembly of a protein into a crystalline array gives information about the interaction between protein units. Moreover, Fourier methods can be applied for structure analysis by image processing of electron micrographs. In this paper the projection structure of Na,K-ATPase crystals obtained by incubation in magnesium-vanadate or magnesium-phosphate medium is presented.

METHODS

Na,K-ATPase was purified in membrane-bound form from pig kidney outer medulla by selective extraction of plasma membranes with SDS in presence of ATP, followed by isopycnic zonal centrifugation in a Ti-14 Beckman zonal rotor (3). Two-dimensional crystals were formed during incubation for 4 weeks at $4°$ C in 10 mM Tris-HCl (pH 7.5) with either 0.25 mM sodium monovanadate, 1 mM magnesium chloride or with 12.5 mM Pi, 3 mM magnesium chloride. The membranes were negatively stained with 0,5 % uranyl acetate and micrographs were obtained at magnifications of 54,700 or 65,000 with a Jeol 100 CX electron microscope. Electron micrographs suitable for analysis were selected by optical diffraction and densitometered at 20 µm intervals. The scanned area was 512 x 512 steps. Projection maps were calculated using the Fourier transform amplitudes and phases collected at the reciprocal lattice points. For the phosphate-induced crystals two-fold rotational symmetry was included in the calculation restricting the phase values to $0°$ and $180°$ with a phase residual of about $15°$. The processing of the electron micrographs was made on VAX 11/780 using the image analysis system EM (4).

RESULTS

Incubation of membrane-bound Na,K-ATPase in vanadate or phosphate medium induced formation of two-dimensional crystals (Fig. 1). The degree of order can be measured from the optical diffraction patterns which show sharp peaks to 25 Å. The crystals are limited in size by the dimensions (100-500 nm) of the membrane fragments in which they grow. The result of reconstructions from two vanadate induced and one phosphate-induced crystal are shown in Fig. 2. The first two crystals show no symmetry element but the unit cell dimensions are different. However, in both cases one large positive peak, corresponding to protein in the structure, occupies the unit cells. Both peaks have a triangular shape. In the most tightly packed crystal this local three-fold symmetry corresponds to an approximate crystallographic symmetry. The crystal grown in the phosphate medium shows a two-fold rotational symmetry. Two identical protein regions occupy the unit cell. Each region is subdivided into one large and one small peak. The large peak has a triangular shape similar to the peaks in the maps of the vanadate induced crystals.

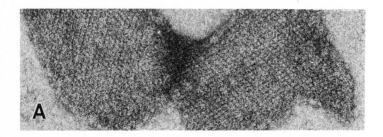

Fig. 1. Electron micrographs of two-dimensional crystals induced in vanadate-magnesium (A) and phosphate-magnesium (C). Magnification 300,000.

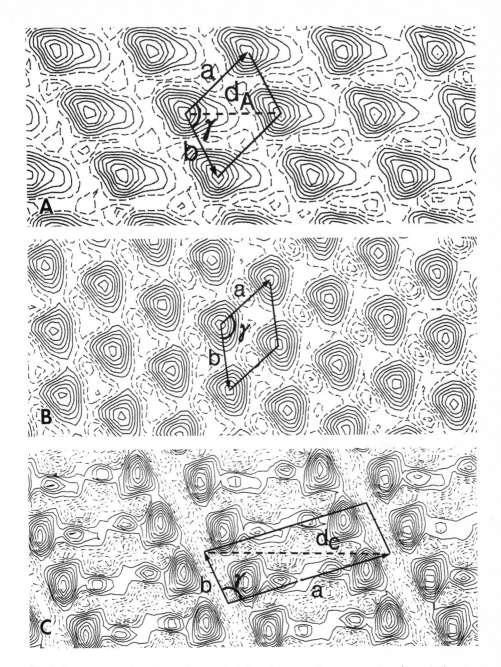

Fig.2. Image-reconstructions of crystals induced in vanadate-magnesium (A and B) and phosphate-magnesium (C). The unit cell dimensions are: (A) a=69 Å, b=53 Å, γ=105°; (B) a=53 Å, b=51 Å, γ=120°; (C) a=135 Å, b=44 Å, γ=101°.

DISCUSSION

Electron microscopy studies (5,6) suggest that the size of an αβ-unit (protomer) of Na,K-ATPase when viewed along an axis perpendicular to the membrane surface is about 50 Å. In the reconstructed images presented here it is assumed that the region of highest contrast variation is the approximate border of the protein. Using this assumption the size of the predominant peaks in the maps is close to the value above. It is thus suggested that the peaks correspond to protomers of the protein. Consequently, the unit cells in the crystals formed in vanadate medium contain one single protomer while the phosphate induced crystal contains two identical copies of the protomer. In the latter case these are related by a two-fold rotation and connected along a stain deficient region. Moreover, the phosphate induced crystal has one large and one small well resolved peak in the protein region. These may correspond to the α- and β-subunits, respectively. The large peak is similar in shape to the peaks in the vanadate induced crystals. In the more tightly packed vanadate crystals it is thus possible that the β subunit is buried in negative stain.

Our observations are consistent with the possibility that the proteins exist in equilibrium between protomeric and oligomeric forms in the membrane. The functional significance of transitions between the two forms is not clear at the present, but it remains possible that active Na,K-transport requires formation of oligomeric $(\alpha\beta)_2$-units, despite the fact that the minimum active protein unit of Na,K-ATPase is an αβ protomer (7).

REFERENCES

1. Skriver,E., Maunsbach, A.B., and Jørgensen, P.L. (1981) FEBS Lett. 131, 219-222

2. Hebert, H., Jørgensen, P.L., Skriver, E., and Maunsbach, A.B. (1982) Biochim. Biophys. Acta 689, 571-474

3. Jørgensen, P.L. (1974) Biochim. Biophys. Acta 356, 36-52.

4. Hegerl, R. and Altbauer, A. (1982) Ultramicroscopy 9, 109-116.

5. Deguchi, N., Jørgensen, P.L., and Maunsbach, A.B. (1977) J. Cell Biol. 75, 619-634.

6. Maunsbach, A.B., Skriver, E., and Jørgensen, P.L. (1981) In: Intl. Cell Biol. 1980-1981 (Schweiger, H.G.) pp. 711-718, Springer-Verlag, Berlin, New York.

7. Jørgensen, P.L. (1982) Biochim. Biophys. Acta 694, 27-68.

TERTIARY STRUCTURE AND MOLECULAR SHAPE OF MEMBRANE PROTEINS

STRUCTURE-FUNCTIONAL RELATIONSHIPS IN VISUAL RHODOPSIN

Yu.A.Ovchinnikov
Shemyakin Institute of Bioorganic Chemistry, USSR Academy of Sciences,
Ul.Vavilova 32, Moscow   V-334, USSR.

INTRODUCTION

A visual pigment, rhodopsin, is a membrane protein of specialized membranes of the photoreceptor cell and consists of one polypeptide chain of MW $\sim$39 KD. A rhodopsin chromophore, 11-*cis*-retinal, is covalently bound to the ε-amino group of a lysine residue via protonated Shiff's base. Light-induced retinal isomerization into all-*trans* configuration is the most important stage in the complex of processes resulting in the visual excitation. The subsequent events change an electrical potential on the plasma membrane of the visual cell /1/.

It is generally accepted now that signal transduction from the rhodopsin molecule onto the plasma membrane is realized by the molecular transmitter system, the key elements here are calcium ions or cyclic guanosine monophosphate (GMP) /2,3/. Thus physiology problems of photoreception lie in elucidation of real processes, occurring between light quantum absorption with the rhodopsin molecule and plasma membrane polarization. Since light quantum absorption with the rhodopsin molecule switches on the total process, the knowledge of its primary structure and topography in the membrane is mostly significant.

The present report deals with the establishment of the complete amino acid sequence of rhodopsin from the bovine retina, localization of the retinal binding site and elucidation of the rhodopsin molecule topography in the photoreceptor membrane /4,5/.

MATERIALS AND METHODS

Photoreceptor membranes from the bovine retina were isolated as described in the earlier developed procedure /6/. Carboxymethylation of opsin was carried out according to Crestfield et al. /7/, the only modification being in protein solubilized in 4% sodium dodecylsulfate. The carboxymethylated protein preparation was dialyzed against water and separated from lipids and the detergent by precipitation in 80% ethanol.

Cleavage of carboxymethylated opsin with cyanogen bromide was performed in 80% formic acid by the 300-fold excess of the reagent per mole of methionine. Apomembranes (obtained upon illumination of photoreceptor membranes in the presence of hydroxylamine) were hydrolized with chymotrypsin and trypsin in 0,1 M ammonium bicarbonate buffer, pH 8,3.

Photoreceptor membranes were cleaved with papain, chymotrypsin, thermolysin, and protease from *St.aureus* at 20°C for 12-24 h in enzyme/substrate ratio I:20, in 67 mM K_2HPO_4, pH 7,0, in the presence of 5 mM cysteine and 2 mM EDTA; 67 mM sodium phosphate buffer, pH 8; 10 mM Tris-HCl buffer, pH 7,4, in the presence of 2 mM Ca^{++}; 100 mM sodium bicarbonate buffer, pH 8,0, respectively. Hydrolysis of peptides with pepsin was conducted in 10% formic acid in the presence of I mM NaCl at 20°C for 24 h.

Peptides were split at tyrosine residues with N-bromsuccinimide and at methionine residues with cyanogen bromide in 80% formic acid in the presence of the 10- and 100-fold excess of the reagent per mole of an amino acid, accordingly.

Immobilization of rhodopsin on activated thiol-glass (CPG/Thiol Pierce) and subsequent cleavage of the protein with chemical and enzymic methods were carried out as described /5/.

A peptide mixture of cyanogen bromide hydrolyzate was dissolved in a saturated guanidinium hydrochloride solution. Gradual decrease of the solvent concentration to 2 M resulted in fractions of soluble and insoluble peptides. After purification each fraction was separated on Bio-Gel P-30 in 80% formic acid with subsequent rechromatography on Bio-Gel P-30 and by HPLC on columns Nucleosil C_8 and LiChrosorb C_8 and C_{18} (acetonitrile gradient in 0,1% trifluoroacetic acid or 10 mM ammonium acetate, pH 5,7). The products of enzymic cleavage of apomembranes were separated on Bio-Gel P-30 in 80% formic acid, on cationite AG50W-x4, and by HPLC under afore mentioned conditions.

The N-terminal amino acid sequence of peptides was determined according to the Edman degradation, amino acids were identified as I-dimethylaminonaphthalene-5-sulphonyl derivatives, phenylthiohydantoins and 4-N,N -dimethylaminoazobenzene-4'-thiohydantoins. The C-terminal sequence was established by means of carboxypeptidases A and B and by the hydrozinolysis method. The N-terminal amino acids of large peptides were sequenced by the automatic degradation using a Beckman liquid-phase sequenator 890 C (program 102974) and a Rank Hilgen solid-phase sequenator AP-240.

RESULTS

Due to high solubility of delipidated rhodopsin preparations in highly concentrated formic acid and the presence of the considerable number of methionine residues in the protein molecule, cyanogen bromide cleavage became a basic fragmentation method. In the course of the structural analysis of bacteriorhodopsin it was shown that separation of cyanogen bromide hydrolyzate is considerably facilitated by step-by-step peptide extraction with different con-

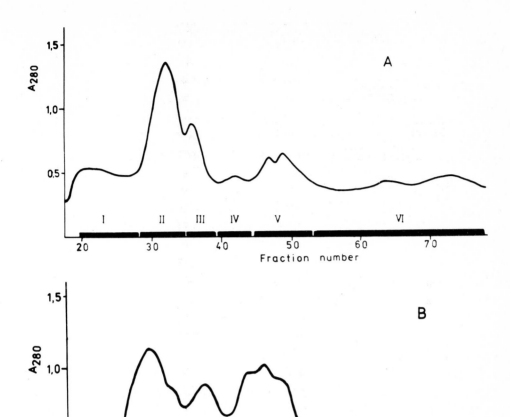

Fig. 1. Chromatography of cyanogen bromide peptides
of rhodopsin on Bio-Gel P-30. Peptides
were eluted with 80% formic acid;
column size (1,5X100 cm);
flow rate 3 ml/hr.
A - elution profile of peptides soluble
in 2 M guanidinium hydrochloride;
B - elution profile of peptides insoluble in 2 M guanidinium hydrochloride.

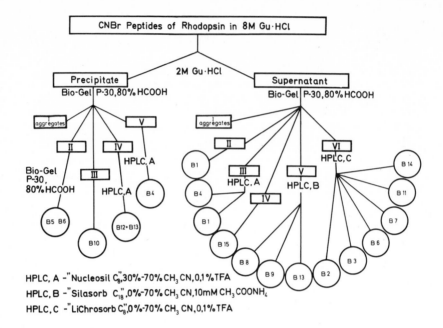

HPLC, A -"Nucleosil C$_8$", 30%-70% CH$_3$CN, 0,1% TFA
HPLC, B -"Silasorb C$_{18}$", 0%-70% CH$_3$CN, 10mM CH$_3$COONH$_4$
HPLC, C -"LiChrosorb C$_8$", 0%-70% CH$_3$CN, 0,1% TFA

Fig. 2 . Scheme for isolation of peptides of cyanogen bromide cleavage of rhodopsin molecule.

centrations of organic acids or denaturing agents /8/. The approach was also
successfully applied when separating the products of rhodopsin cleavage. The
cyanogen bromide hydrolyzate was separated into the peptide fractions soluble
and insoluble in 2 M guanidinium hydrochloride (Fig. 1).

Upon gel chromatography of the soluble one there were obtained 6 fractions
(Fig. 2.). Fraction I contained an aggregated material in negligible amounts
that was not separated further. Fractions II and IV consisted of homogeneous
peptides BI and BI5. Individual peptides B2, B3, B4, B7, B8, B9, B11, and B14
were obtained from fractions III, V, and VI by HPLC. Separation of the insoluble fraction on Bio-Gel P-30 resulted in 5 fractions. Fraction I was a peptide
aggregate. Fractions II, III, IV, and V contained mainly fragments B5+6, B10,
B12+13, and B4, correspondingly. The first two fragments were obtained in a
homogeneous state by rechromatography on Bio-Gel P-30, B12+13 and B4 - by HPLC.
When establishing amino acid sequences of fragments B5+6 and B12+13 there were
found bonds Met-Ser and Met-Thr, respectively. Cyanogen bromide cleavage of
the bonds ran with low yields, therefore only small amounts of peptides B5, B6,
B12, and B13 were detected. Thus cyanogen bromide hydrolysis of rhodopsin yielded 15 peptides, 13 peptides being isolated in preparative amounts (Fig. 3).

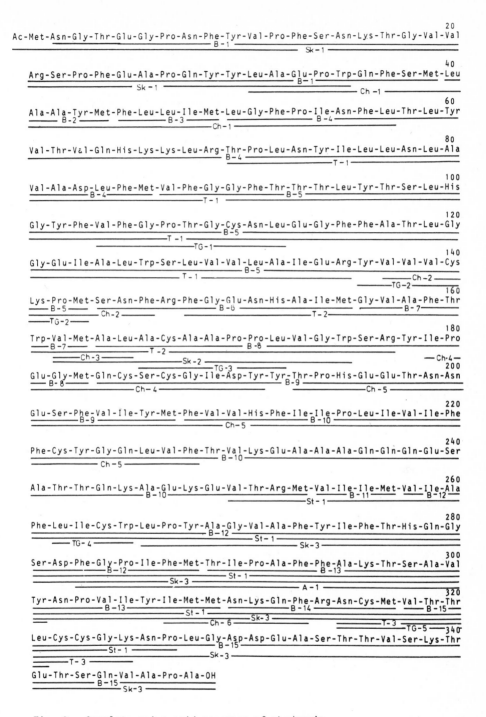

Fig. 3. Complete amino acid sequence of rhodopsin.

Peptides B2, B3, B6, B7, B11, and B14 contained from 4 to 12 amino acid residues, their structures were established routinely. In order to define the structures of large cyanogen bromide fragments they were additionally cleaved with trypsin, chymotrypsin, *St.aureus* protease, pepsin as well as with BNPS-skatole and N-bromsuccinimide at tryptophan and tyrosine residues, accordingly.

Establishment of the structures of chymotryptic peptides Ch1, Ch2, Ch3, Ch4, Ch5, and Ch6 made possible determination of amino acid sequences and disposition of cyanogen bromide fragments B1 - B2, B3 - B4, B5 - B6, B7 - B8, B8 - B9, B9 - B10, and B13 - B14, respectively, in the protein polypeptide chain. These peptides contained in total about 300 amino acid residues, that made up 85% of the rhodopsin amino acid sequence.

Thus we succeeded in obtaining 8 overlappings for cyanogen bromide fragments. From the tryptic hydrolyzate of apomembranes fragments T1, T2, and T3 were isolated. Their structural analysis allowed localization of peptides B4 - B5, B5 - B6, and B14 - B15.

To establish the rhodopsin amino acid sequence large fragments, F1 and F2, resulted from thermolytic treatment of photoreceptor membranes, were analyzed /9/. Fragments F1 and F2 were established to correspond to sequences (1-240) and (241-327) of rhodopsin. Upon the cleavage of fragment F2 with *St.aureus* protease peptide St-I was isolated, determination of its N-terminal amino acid sequence made possible overlapping 3 cyanogen bromide fragments B10 - B11 - B12. Elucidation of the structure of peptide A-1, isolated after limited acid hydrolysis of fragment F2, revealed the overlapping between cyanogen bromide fragments B12 - B13. Establishment of structures of fragments Sk2 and Sk3, isolated upon the rhodopsin cleavage at tryptophan residues by BNPS-skatole confirmed overlappings for cyanogen bromide fragments B7 - B8 and B12 - B13 - B14 - B15. The analysis of the products of the subsequent trypsin, *St.aureus* protease, and chymotrypsin cleavage of the protein immobilized on activated thiol-glass expanded the information on the rhodopsin primary structure. This approach considerably facilitated peptide separation and allowed to isolate selectively cysteine-containing peptides of rhodopsin /5/.

The completely reconstructed polypeptide chain of the rhodopsin molecule (Fig. 4) consists of 348 amino acid residues and is composed of the following amino acids: Asp (5), Asn (15), Thr (27), Ser (15), Glu (17), Gln (12), Pro (20), Gly (23), Ala (29), Cys (10), Val (31), Met (16), Ile (22), Leu (28), Tyr (18), Phe (31), His (6), Lys (11), Arg (7), Trp (5) /4,5/.

In the course of investigations into the rhodopsin structure it was essential to identify the lysine residue responsible for the retinal binding in the amino

267

acid sequence. The photoreceptor membranes were treated with NaBH$_4$ or NaBH$_3$CN and then cleaved with thermolysin /9/ into fragments (1-240) and (241-327), which were separated on a Sephadex LH-60 column equilibrated with the formic acid/ ethanol mixture (3:7). Radioactivity and retinyl fluorescence comigrated only with fragment (241-327). Then the fragment was cleaved with cyanogen bromide and the formed peptides were separated on Sephadex LH-60. Radioactivity and fluorescence accompanied only one peptide, its structural determination pointed to the ε-amino group of Lys 296 as the retinal binding site. The data on structures of shorter retinyl-containing peptides are in accordance with the obtained results /10/.

To reveal the topography of the rhodopsin polypeptide chain in the photore-ceptor membrane, accessibility of rhodopsin integrated in the photoreceptor membrane to proteolytic enzymes was analyzed as in case of bacteriorhodopsin /4/.

As mentioned above thermolysin cleaved bond 240-241 at its action onto photo-receptor membranes. The neighbouring residues were found to be accessible to the action of papain and *St.aureus* protease splitting bonds 241-242 and 239-240, respectively. One should mention that cleavage of the polypeptide chain into two large fragments was accompanied with splitting off the C-terminal fragment (322-348) and peptide (237-241).

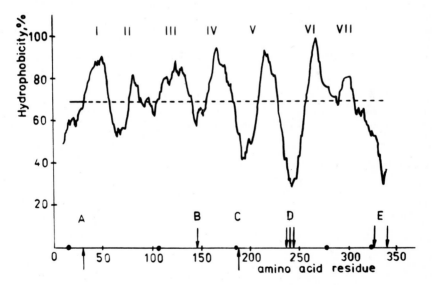

Fig. 4. Hydrophobicity profile of rhodopsin mole-cule. Arrows indicate peptide bonds acces-sible to action of various proteases onto outer and innner surfaces of photorecep-tor disks.

268

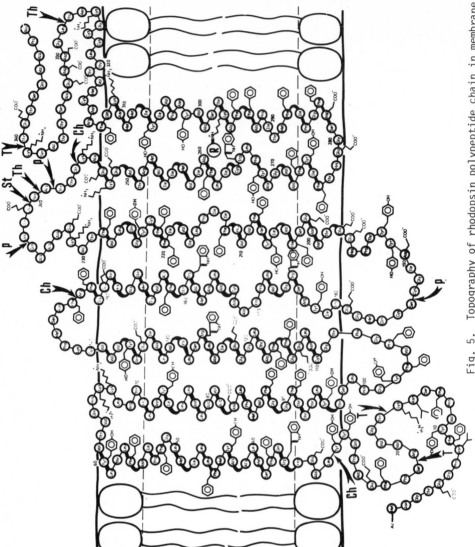

Fig. 5. Topography of rhodopsin polypeptide chain in membrane.

Upon the chymotrypsin action on photoreceptor membranes three membrane-bound fragments (1-146), (147-244), and (245-336) were formed.

Freezing and thawing of native disks resulted in membranes of inverted orientation, after the chymotrypsin action on these preparations peptides (1-10), (11-13), (14-24), and (25-30) were found in the supernatant. Upon the papain action onto inside-out membranes bond 186-187 was cleaved.

Thus according to the analysis of the protease action at least 30 residues from the N-terminus of the protein molecule and bond 186-187 were located in the intradiskal space, whereas bond 146-147 and fragment (237-244) and (322-348) were exposed on the outer membrane surface.

In order to clarify the molecule topography the disposition of hydrophobic and hydrophilic residues in the rhodopsin polypeptide chain was analysed, here hydrophobicity of the 20-membered overlapping fragments (1-20, 2-21, etc.) was calculated /11/.

Peaks in Fig. 4 correspond to hydrophobic regions of the molecule which are apparently located inside the membrane, troughs, on the contrary, conform to polypeptide chain regions in the vicinity of the membrane surface.

Comparison of the hydrophobicity profile and accessibility to the protease action permitted us to propose a model of the rhodopsin polypeptide chain arrangement in the membrane (Fig. 5). The model is characterized by seven polypeptide segments spanning the membrane width. They are composed of about 60% of all the residues of the polypeptide chain. From the segment of the average length 32 residues one could "cut out" the central more hydrophobic fragment of 20 residues, located in the middle of the membrane, as well as the regions in contact with lipid polar heads. Seven segments are linked with hydrophilic loops, three of them (between segments 3-4, 4-5, and 5-6) being accessible to proteolysis.

According to the given model the Lys 296 residue binding the chromophore is in the C-terminal (seventh) segment in the middle of the membrane (this conclusion agrees with the available experimental data on the retinal disposition relative to two membrane surfaces)/12/. The charge of the residues Asp 83, Gln 113, and Glu 122 located in the membrane are supposed to form the ion pair with the protonated aldimine as well as to interact with the polyene chain that results in the so-called "opsin shift".

Interestingly, the model for the rhodopsin topography shares common features with that for bacteriorhodopsin /4, 5/ (seven transmembrane fragments, disposition of the N- and C-termini on the opposite membrane sites, etc.). Apparently, the analogous structural organization of rhodopsin and bacteriorhodopsin could underlie their similar functioning at the initial stages.

Deciphering of the rhodopsin structure might give an impetus to the studies of light transformation by retinal-containing chromoproteins.

REFERENCES

1. Wald, G. (1968) Science, 162, 230-239.
2. Hagins, W.A. (1972) Ann.Rev.Biophys.Bioeng., 1, 131-138.
3. Bitensky, M.W., Gorman, R.E., and Miller, W.H. (1971) Proc.Natl.Acad.Sci. USA, 68, 561-562.
4. Ovchinnikov, Yu.A. (1982) FEBS Letters, 148, 179-191.
5. Abdulaev, N.G., Artamonov, I.D.,Bogachuk, A.S., Feigina, M.Yu., Kostina, M.B., Kudelin, A.B., Martynov, V.I., Miroshnikov, A.I., Zolotarev, A.S., and Ovchinnikov, Yu.A. (1982) Biochemistry International, 5, 693-703.
6. Smith, H.G., Stubbs, G.W., and Litman, B.J. (1975) Exptl.Eye.Res., 20, 211-217.
7. Crestfield, A.M., Moore, S., and Stein, W.H. (1963) J.Biol.Chem., 238, 622-627.
8. Abdulaev, N.G. and Ovchinnikov, Yu.A. (1982) in: Packer, L. (Ed.), Methods in Enzymology, Academic Press, 88, 723-729.
9. Pober, J.S. and Stryer L. (1975) J.Mol.Biol., 95, 477-481.
10. Wang, J.K., McDowell, J.H., and Hargrave, P.A. (1980) Biochemistry, 19, 5111-5117.
11. Capaldi, A.R. and Vanderkoi, G. (1972) Proc.Natl.Acad.Sci.USA, 69, 930-932.
12. Thomas, D.D. and Stryer, L. (1982) J.Mol.Biol., 154, 145-157.

STRUCTURE-FUNCTION RELATIONS IN BAND 3 PROTEIN

Z. IOAV CABANTCHIK

Department of Biological Chemistry, Institute of Life Sciences, The Hebrew
University of Jerusalem, 91904 Jerusalem (Israel)

INTRODUCTION

The exchange of chloride and bicarbonate across red cell membranes, which
subserves the removal of CO_2 from tissues to lungs, is a property for which
red blood cells are highly specialized. The function has been attributed to
an integral membrane glycoprotein that was named band 3 on the basis of its
position in SDS-polyacrylamide gel electrophoretograms (apparent Mr 95,000)
(Fig. 1)(1, 2 for reviews). The protein constitutes approximately 20% of the

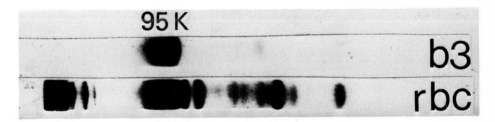

Fig. 1. SDS-PAGE of erythrocyte membranes (rbc) and of
purified and reconstituted (17) band 3 (b3) of 95,000 Mr
(95 K).

total membrane protein (by weight) or 6% of the membrane dry mass. It is
present in approximately the same amount (10^6 copies/cell) in mature red
blood cells of avians, mammals and other animals. Band 3 is absent in the
early stages of erythroid cell differentiation but, as maturation proceeds,
increasing amounts of the protein are synthesized, particularly from the
polychromic normoblast stage and on, up to the reticulocyte stage (3). The
biosynthesis of the protein on polysomes, the insertion into rough endoplasmic
reticulum (RER) membranes, and the route and kinetics of maturation to the
cell surface follow a pattern similar to those of other integral membrane
proteins (4). The insertion is apparently co-translational and the primary
translation product of band 3's mRNA is not proteolytically processed. The
configuration of the protein in RER membranes appears to be the same as that
in the plasma membrane (PM), namely, with the NH_2-terminus (N_t) facing the
cytoplasmic surface and the carbohydrate rich mannose core facing the opposite
surface (5, 6). The insertion of the proteins into RER and PM, in analogy

with that of glycophorin, is not contingent upon its glycosylation (6). From the time of appearance in the PM of the cell and until removal of the old red cell from circulation (4 months), the band 3 protein appears to retain the basic structural as well as functional features. However, a very small fraction of the polypeptides can apparently not escape *in situ* degradation to yield a 62,000 Mr (62 K) fragment. This product was postulated to serve as the senescence label which is recognized by autoimmune antibodies and signals removal of the senescent (or damaged) cell from circulation (7).

Band 3 polypeptides are the carriers of Ii antigens which have been regarded as the precursors of blood group ABH antigens (8). In umbilical cord vessels the antigen activity is of the Oi type, whereas in adult blood it is of the OI type. The antigenic change associated with development has been attributed to branching of a single exofacial oligosaccharide chain N-linked *asn* of band 3 (9). The carbohydrate (Mr 3-8 x 10^3) is of a complex type, carrying an α-Man$_2$-β-Man-β-GlcNAc-GlcNAc sequence in its core, the disaccharide Galβ1$\rightarrow$ $\rightarrow$4GlcNAcβ1$\rightarrow$3 as a variable repeating sequence with branching points at C-6 of some Gal residues (10), fucose residues both in the periphery as well as in the core portions and some sialic acid at the periphery (9).

Although the main functional role of polypeptides in the band 3 region is the transport of anions, some authors have attributed to a minor fraction of band 3 polypeptides (not necessarily the same as the former) a role in hexose transport, a matter which is currently in dispute (11). On the other hand, evidence has been presented implicating band 3 as a receptor for attachment and invasion of human red blood cells by the simian malaria *Plasmodium knowlesi* (L.H. Miller, et al., in press) and possibly the human malaria *Plasmodium vivax*.

The identification of band 3 polypeptides with anion transport (1) has been provided by a variety of studies which included selective and stoichiometric labeling of the polypeptides with specific inhibitors of the function (1, 2, 12-14), functional isolation (15), functional reconstitution into artificial membranes (15-17) and transplantation into plasma membrane of cells lacking anion exchange capacity (15).

Membrane disposition of band 3 polypeptides

The band 3 area in SDS-PAGE is highly diffuse (Fig. 1), arising from the presence of varying amounts of glycosidic moieties in the oligosaccharide linked to what is thought to be the predominant polypeptide entity in the 90-100 K region of the gel. These polypeptides are of the integral type, spanning the membrane several times. Their disposition in the membrane matrix has been studied with the aid of chemical probes in conjunction with specific polypeptide cleaving procedures. The findings provided by various groups (18-21) were integrated in the scheme depicted in Figure 2. The polypeptides have N-acetylated *met* as amino terminus (N_t) and a *val* group at the carboxy terminal (C_t). They carry two distinct domains which can be dislodged by

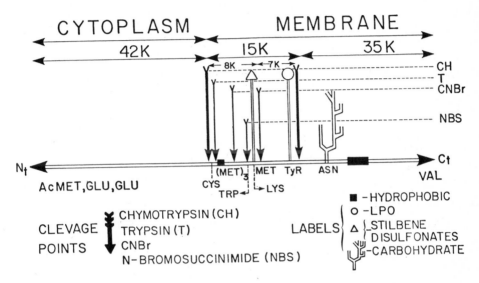

Fig. 2. Labeling and cleavage points in band 3 polypeptides adapted from refs. 1, 19, 20, and 21. LPO = lactoperoxidase mediated iodination site.

proteolytic cleavage from the inner membrane surface: the cytoplasmic water soluble domain (42 K) carrying the original N_t and the membrane-associated domain (60 K) remaining in the membrane after the above cleavage, or even after subsequent cleavages by externally applied proteases which yield the 15 K and 35 K fragments. We shall not discuss the properties of the 42 K cytoplasmic fragment as its excision from band 3 is of no demonstrable consequence to anion transport function. Although a possible role of this fragment in binding of cytosolic enzymes (22) and perhaps also hemoglobin (23) has been invoked, no functional significance has yet been attached to these findings.

The anion transport relevant 60 K, 15 K and 35 K have been studied in some detail for the past 6 years; however, the existing knowledge is still "fragmentary": the amino sequence and the spatial disposition of the fragments, particularly those of the carbohydrate containing 35 K segment, have not been fully elucidated. The basic methodology used for studying the topology of the membrane associated fragments consisted of a combination of labeling and cleaving agents of defined chemical and permeation properties and sidedness of action in cells, open membranes, isolated band 3 polypeptides or fragments derived thereof. The sites of action of these agents are depicted in Figure 2; of particular interest being the sites susceptible to the membrane impermeant specific affinity labels such as DIDS, its reduced analog H_2DIDS, or other structurally related analogs, and others (Table 1). This information,

TABLE 1

PROPERTIES OF INHIBITORY PROBES OF ANION TRANSPORT

Inhibitory probe	Amino acid conjugate	B₃ fragment labeled	Site of action	%Protection of DS*	References
DIDS or	lys (pH 7.4)	60K, 35K	transport	∿100	(1, 33)
H_2DIDS	lys-lys(pH>8.0)	60K - 35K	transport	∿100	(1, 33)
PLP-SBH	lys	60K, 35K	transport	∿ 70 (35K)	(1, 48)
FDNB	n lys	60K, 35K (?)	transport, etc.	∿100	(12)
NAP-t + hv	——	60K	modifier	∿ 70	(1, 2)
HCHO-SBH	lys	60K, 35K	transport	80 (35K)	(14)
IPS	lys	60K	transport	——	(24,25)
PI	3 lys	60K, 35K	——	100 (35K)	(25)
PG	arg	35K	——	50	(19)
CHD	arg	——	——	50	(26)
ETC	glu, asp (?)	——	——	50	(13)
FFNM	——	35K	——	90	(27)
EI	lys	——	——	50	(28)

*DS = disulfonic stilbene
DIDS = 4,4'-diisothiocyano-2,2'-stilbene-disulfonic acid; PLP = pyridoxal-5--phosphate; FDNB = 1-fluoro-2,4-dinitrobenzene; NAP-t = N-(4-azido-2-nitrophenyl)-2-aminoethyl-sulfonate; HCHO = formaldehyde; IPS = sulfophenyl-iso--thiocyanate; PI = phenylisothiocyanate; PG = phenylaglyoxal; CHD = 1,2-cyclohexanedione; ETC = 1-ethyl-3-(3-trimethylaminopropyl)-carbodiimide; FFNM = flufenamic acid; EI = eosin-isothiocyanate; SBH = Na-borohydride.

together with further localization of labels in fragments obtained from chemical cleavages (19, 21), was used to draw the disposition of the membrane associated polypeptides in the membrane matrix (Fig. 3). The 15 K fragment appears to traverse the membrane several times (at least three) and, to some extent, the same seems to apply to the 35 K segment, although that would need to be

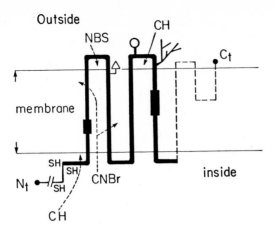

Fig. 3. Transmembrane disposition of band 3
polypeptides based on labeling and cleavage
procedures shown in Figure 2 and Table 1.
Symbols as in Figure 2.

more firmly established. From the amino acid composition it does not appear
as though any of the oligopeptides are particularly hydrophobic (19, 20).
However, labeling within the hydrophobic core with lipophilic reactants (25,
29) indicates the presence of two hydrophobic domains (sequences?), one in
the 15 K and one in the 35 K segments.

The spatial relationship of the different transmembrane segments is of
structural as well as of functional relevance. The first indication of such
a relationship has been provided by cross-linking studies with H_2DIDS, which
indicated not only a geographical proximity 15-20 Å) of the 15 K and a segment
of the 35 K fragment but also their participation in anion transport (30).
Secondly, these segments appear to interact rather tightly (21, 31) as judged
by the fact that they co-extract in non-ionic detergent and migrate as a
single charged complex (due to the carbohydrate containing 35 K segment) in
electrofocusing systems (31). Thirdly, chemical modification of one system
at the transport site affects labeling of the other segment, and *vice-versa*
(Table 1). Fourthly, proteolytic degradation of the 35 K fragment in the
intact membrane is apparently correlated with inhibition of the function (32).
These data, in conjunction with the inhibitory effect of the agents on anion
transport (Table 1), are taken to indicate that at least two transmembrane
segments of band 3 cooperate in the formation of the putative transport site
(Fig. 4).

276

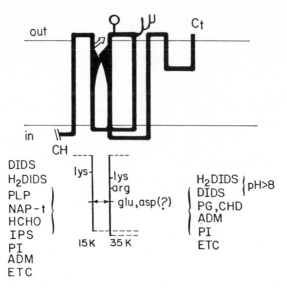

Fig. 4. Formation of anion transport sites by two
transmembrane segments of band 3 (upper figure).
Labeling of 15 K and/or 35 K segments by various
reagents and identification of target amino acids
(lower figure). The key for most abbreviations
is shown in Table 1. ADM = adamantylidine.

The anion transport site

The identification of the particular chemical groups comprising the functional sites and their tentative location in the two fragments has resided
on   labeling studies (Table 1) and structure-activity relationship studies
(1, 34, 35), and reciprocal effects of chemical modification of one segment
on labeling of the other (see above). The location of the transport site in
relation to the membrane matrix was probed from the outer surface with disulfonic stilbenes coupled to dextrans via narrow spacers (aliphatic chains or
polyethyleneglycol chains of variable length) and from the inner surface with
aminophenylmethylbenzenethiazole disulfonate (APMB) similarly coupled to dextrans (36). These as well as new studies indicate that the sites reside away
from the membrane boundaries, deeper within the matrix, but still closer to
the outer membrane surface. This interpretation is in agreement with esti-
mation of distances based on fluorescence energy transfer measurements between
fluorophores placed at the cytoplasmic domain of band 3 and disulfonic stilbenes attached to band 3 exofacially. These have placed the transport site
28-42 Å away from the inner membrane surface (37).

The transport domain comprises apparently a small area within the protein itself, flanked by two transmembrane segments and separating two aqueous compartments formed by indentations or foldings of the polypeptide into the middle of the membrane matrix (Fig. 4). Thus, translocation of substrates would not need to occur over the entire anatomical width of the membrane but rather over a small (<10 Å) distance. This is consistent with the fact that the protein, as a whole, does not rotate across the membrane (38) and that transport occurs at the remarkable speed of $2\text{-}5 \times 10^4$ ions/site/sec (37°C, pH 7.2) (2). The particular groups comprising the transport niche are two lysines (1, 14, 32, 33, 39) an arginine (1, 13), possibly a carboxyl containing amino acid (*asp* or *glu*) and an electron donating group located in a hydrophobic domain (34, 35) (Fig. 5). There is one such transport site per band 3 monomer (1, 2). Despite the fact that the protein seems to reside in the membrane as dimers with an average distance between adjacent transport sites of 28-48 Å (40), it was previously concluded that the functional form of the protein is the monomer (1, 2, 40). Although the observed half site reactivity of stilbene disulfonates in ETC, PG (13) or CHD (26) treated cells (Table 1) might have resulted from steric hindrance *per se* (40), it would seem premature at this stage to reject an intermonomer cooperativity as an essential feature of the transport mechanism (13).

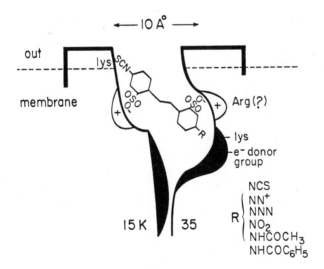

Fig. 5. Tentative configuration of anion transport sites showing the binding pocket of high affinity binding disulfonic stilbene derivatives and the chemical interactions in the pocket.

Protein-protein and protein-lipid interactions

Band 3 polypeptides were shown to interact with other membrane proteins such as spectrin (41), glycophorin, ankyrin (42), glycolytic enzymes (22, 42), and hemoglobin (23). However, no functional role in anion transport has yet been attributed to those interactions. The lipid environment, on the other hand, seems to have a significant effect on the function. This was observed both after implantation of human band 3 from the rather viscous red cell membrane environment into the more fluid plasma membrane of the mouse erythroleukemia cells (15) and after modulation of the membrane fluidity by cholesterol depletion or enrichment (43, 44). The means by which cholesterol affects anion transport is presently not understood. It could have a specific chemical effect on band 3, as recently hypothesized (45), and/or it could modulate the lateral and vertical displacement of band 3 by affecting the compressibility of the membrane matrix (46), despite the fact that the latter did not seem to affect the rotational mobility of the protein (47). Working on the idea that anion transport depends on the interaction between different transmembrane fragments of band 3 (Figs. 4 and 5), we hypothesized that modification of those interactions by membrane matrix compressibility should affect differentially the passage of large anions (e.g., pyruvate), as compared to small anions (e.g., chloride). Indeed this seems to be the case, as demonstrated experimentally in Figure 6.

Microscopic evidence for conformational changes in band 3 induced by changes in membrane fluidity was obtained in the modified binding of Cancanavalin A to the latter, taking place on an exofacial site of the 35 K fragment (Breuer et al., in press). However, before a molecular interpretation is attached to these findings, more functional and, primarily, structural studies of physical and chemical nature would have to be conducted, both in intact membranes and in reconstituted band 3 preparations.

Conclusions

The arrangement of band 3 polypeptides involved in red blood cell anion transport has been studied with a variety of labeling and cleaving agents. The single anion transport site present in each polypeptide is formed by interactions between two transmembrane fragments of the polypeptide. The interactions could be apparently modified membrane fluidity. The site is located in a niche which, although close to the external surface, is buried within the protein. It comprises a variety of positively charged amino acids, possibly also a negatively charged group and an electron donor group within

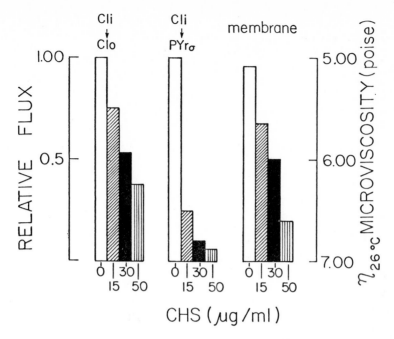

Fig. 6. Modulation of anion transport capacity by membrane
fluidity. Relative effluxes of ^{36}Cl from red cells enriched
with cholesteryl-hemisuccinate (CHS) into media containing
either NaCl - (Cl_i-Cl_o) or Na-pyruvate-(Cl_i-Pyr_o) $(0°C$, pH 7.2)
and membrane microviscosity as measured by fluorescence polarization of diphenylhexatriene (47).

a hydrophobic pocket (possibly the transport barrier). The sites are accessible
from either surface by anionic water-soluble agents, leaving the hypothetical
barrier as a rather small domain, comprising only a fraction of the membrane's
anatomical width. Transport is envisioned as a small conformational change
over the above barrier.

Acknowledgments

This work was supported, in part, by an N.I.H. grant #GM 299, and by a grant
from the United States-Israel Binational Science Foundation (BSF), Jerusalem.

References
1. Cabantchik, Z.I., Knauf, P.A. and Rothstein, A. (1978) Biochim. Biophys.
   Acta 515, 239.
2. Knauf, P.A. (1979) in: Bronner, F. and Kleinzeller, A. (Eds.) Current Topics
   in Membrane Transport, Vol. 12, Academic Press, New York, pp. 249-363.

3. Foxwell, B.M.J. and Tanner, M.J.A. (1981) Biochem. J. 295, 129.

4. Lodish, H.F., Braell, W.A., Schwarz, A.L. Strous, G.J.A.M. and Zilberstein, A. (1981) Int. Rev. Cyt. 12 (suppl.), 247.

5. Braell, W.A. and Lodish, H.F. (1981) J. Biol. Chem. 256, 11337.

6. Sabban, E., Marchesi, V.T., Adesnik, M. and Sabatini, D.D. (1981) J. Cell Biol. 91, 637.

7. Kay, M.M.B., Sorensen, K., Wong, P. and Bolton, P. (1982) Mol. Cell. Biochem. 49, 65.

8. Childs, R.A., Feizi, T., Fukuda, M. and Kakomori, S. (1978) Biochem. J. 173, 333.

9. Fukuda, M., Fukuda, M.N. and Hakomori, S. (1979) J. Biol. Chem. 254, 3700.

10. Tsuji, T., Irimura, T. and Ozawa, T. (1980) Biochem. J. 187, 677.

11. Jones, M.N. and Nickson, J.K. (1981) Biochim. Biophys. Acta 650, 1.

12. Passow, H. and Zaki, L. (1978) in: Solomon, A.K. and Karnovsky, M. (Eds.) Molecular Specialization and Symmetry in Membrane Function, Harvard University Press, Cambridge, Massachusetts, pp. 229-250.

13. Wieth, J.O., Andersen, O.S., Brahm, J., Bjerrum, P.J. and Borders, C.L. (1982) Phil. Trans. D. Soc. (Lond.) B 299, 383.

14. Jennings, M.L. (1982) J. Biol. Chem. 257, 7554.

15. Cabantchik, Z.I., Volsky, D.J., Ginsburg, H. and Loyter, A. (1980) Ann. N.Y. Acad. Sci. 34, 444.

16. Köhne, W., Haest, C.W.M. and Deuticke, B. (1981) Biochim. Biophys. Acta 664, 108.

17. Darmon, A., Zangwill, M. and Cabantchik, Z.I. (1983) Biochim. Biophys. Acta 727, 77.

18. Steck, T.L., Koziarz, J.J., Singh, M.K., Reddy, R. and Kohler, H. (1978) Biochemistry 17, 1216.

19. Drickamer, K. (1980) Ann. N.Y. Acad. Sci. 34, 419.

20. Tanner, M.J.A., Williams, D.G. and Jenkins, R.E. (1970) Ann. N.Y. Acad. Sci. 34, 455.

21. Rothstein, A., Ramjeesingh, M., Grinstein, S. and Knauf, P.A. (1980) Ann. N.Y. Acad. Sci. 34, 433.

22. Kliman, H.J. and Steck, T.L. (1980) in: U.V. Lassen, H.H. Ussing and J.O. Wieth (Eds.) Membrante Transport in Erythrocytes, Alfred Benzon Symp. 14, Copenhagen, Munksgaard, pp. 312-326.

23. Scuyare, M. and Fikiet, M. (1981) J. Biol. Chem. 256, 13152.

24. Drickamer, K. (1977) J. Biol. Chem. 252, 6909.

25. Kempf, C., Brock, C., Sigrist, H., Tanner, M.J.A. and Zahler, P. (1981) Biochim. Biophys. Acta 641, 88.

26. Zaki, L. (1980) Biochem. Biophys. Res. Commun. 99, 243.

27. Cousin, J.L. and Motais, R. (1982) Biochim. Biophys. Acta 687, 147.

28. Nigg, E., Kessler, M. and Cherry, R. (1979) Biochim. Biophys. Acta 550, 328.

29. Guidotti, G. (1980) in: U.V. Lassen, H.H. Ussing and J.O. Wieth (Eds.) Membrane Transport in Erythrocytes, Alfred Benzon Symp. 14, Copenhagen,

Munksgaard, pp. 300-311.

30. Jennings, M.L. and Passow, H. (1979) Biochim. Biophys. Acta 554, 498.

31. Ideguichi, H., Matsuyama, H. and Hamasaki, N. (1980) Eur. J. Biochem. 125, 665.

32. Passow, H., Fasold, H., Lepke, S., Pring, M. and Schuhmann, B. (1977) in: Miller, M.W. and Shamoo, A.E. (Eds.) Advances in Experimental Medical Biology, Plenum Press, New York, pp. 353-379.

33. Ramjeesingh, M., Gaarn, A. and Rothstein, A. (1981) Biochim. Biophys. Acta 641, 173.

34. Barzilay, M., Ship, S. and Cabantchik, Z.I. (1979) Membr. Biochem. 2, 227.

35. Cousin, J.L. and Motais, R. (1982) Biochim. Biophys. Acta 687, 156.

36. Barzilay, M., Jones, D. and Cabantchik, Z.I. (1979) Fed. Proc. 37, 1295.

37. Rao, A., Martin, P., Reitheimer, R.A.F. and Cantley, L.C. (1979) Biochemistry 18, 4505.

38. Cherry, R.J., Burkli, A., Busslinger, M., Schneider, G. and Parish, R.G. (1976) Nature 263, 389.

39. Cabantchik, Z.I., Balshin, M., Breuer, M. and Rothstein, A. (1975) J. Biol. Chem. 250, 5130.

40. Macara, I.G. and Cantley, L.C. (1981) Biochemistry 20, 5095.

41. Liu, S.C. and Palek, S.R. (1979) J. Supramol. Str. 10, 97.

42. Haest, C.W.M. (1982) Biochim. Biophys. Acta 694, 331.

43. Deuticke, B., Grunze, M. and Haest, C.W.M. (1980) in Lassen, U.V., Ussing, H.H. and Wieth, J.O. (Eds.) Membrane Transport in Erythrocytes, Alfred Benzon Symp. 14, Copenhagen, Munksgaard, pp. 143-160.

44. Jackson, D. and Morgan, D.B. (1982) Biochim. Biophys. Acta 693, 901.

45. Schubert, D. and Boss, K. (1982) FEBS Lett. 150, 4.

46. Nigg, E. and Cherry, R.J. (1979) Biochemistry 16, 3457.

47. Shinitzky, M. (1979) Dev. Cell Biol. 4, 173.

48. Nauri, H., Hamasaki, N. and Minakami, S. (1983) J. Biol. Chem. 258, 5985.

D-β-HYDROXYBUTYRATE DEHYDROGENASE: A MOLECULAR BIOLOGY APPROACH TO THE STUDY OF A LIPID-REQUIRING ENZYME

SIDNEY FLEISCHER, J. OLIVER MCINTYRE, PERRY CHURCHILL, EDUARD FLEER, and ANDREAS MAURER

Department of Molecular Biology, Vanderbilt University, Nashville, TN 37235, USA

INTRODUCTION

The phenomenon of lipid-requiring enzymes was discovered nearly a quarter of a century ago (1,2) and a number of such enzymes have been described. (For examples, see references 3-7). Attention is now being focused on the molecular basis of the role of lipid and the precise nature of lipid-protein interaction of such enzymes in membranes. D-β-hydroxybutyrate dehydrogenase (BDH) is one of the best documented examples of a lipid-requiring enzyme (7-10). Current knowledge of this lipid-requiring enzyme will be reviewed.

BDH is normally membrane bound, associated with the mitochondrial inner membrane where it is inserted with its functional site accessible only from the matrix face (11). It is present in most mitochondria from mammalian sources, a notable exception being the liver of ruminants (12). We could not detect this enzyme in simpler organisms such as yeast, a unicellular eukaryote (13).

When fatty acid metabolism predominates such as in starvation or diabetes, the liver produces large amounts of acetoacetate. BDH in liver converts some of the acetoacetate (AcAc), formed via fatty acid oxidation, to β-hydroxybutyrate (BOH):

$$NADH + AcAc \rightleftharpoons NAD^+ + BOH \qquad [1]$$

The β-hydroxybutyrate formed is the D-isomer. The reaction is reversible and the equilibrium is approximately 30 fold in favor of β-hydroxybutyrate (14). D-β-hydroxybutyrate is an end product which passes out from the liver into the blood circulation as does acetoacetate. These ketone bodies are utilized for energy in the peripheral tissues such as heart, kidney and skeletal muscle and

even in brain. D-β-hydroxybutyrate must be converted back to acetoacetate by
BDH in the peripheral tissues before it can be metabolized. The acetoacetate
is converted to acetoacetyl CoA in the peripheral tissues with succinyl CoA as
the CoA donor. The CoA transferase, required for this reaction, is lacking in
liver so that liver produces but does not utilize ketone bodies. In diabetic
rats, the BDH content in liver mitochondria was found to be reduced (15).

BDH can be released by phospholipase A_2 digestion of mitochondria (16)
which hydrolyzes the ester bond of the fatty acid in the sn-2 position of the
phospholipid (2).

$$\text{Phospholipid} \xrightarrow{\text{phospholipase } A_2} \text{lysophospholipid + fatty acid} \qquad [2]$$

Most of the enzyme becomes soluble when approximately 25% of the phospholipid
has been hydrolyzed (16,18). The soluble enzyme devoid of lipid, referred to
as "apoBDH", is inactive but can be reactivated by phospholipids containing
lecithin. ApoBDH has been purified (20) and when reactivated with
mitochrondrial phospholipids has essentially the same ordered sequential reaction mechanism and kinetic parameters as the membrane-bound enzyme (14).
However, apoBDH reactivated with mitochondrial lecithin instead of MPL has
tighter binding for coenzyme (14). In this regard, modulation of the phospholipid composition in cultured hepatocytes also alters the K_m for coenzyme (19).
Thus, isolation and reconstitution has not modified the enzyme according to
kinetic criteria. BDH has an absolute and specific requirement of lecithin
for function and represents an important prototype for the study of lipid-
requiring enzymes (7).

BDH has been purified to homogeneity from bovine heart (20) and rat liver
mitochondria (21-22). ApoBDH from both sources requires lecithin for
function. The requirement of lipid is in both directions of catalysis (cf.
Equation 1) (14). BDH can be activated by phospholipids in the form of
vesicles or by soluble lipid such as dioctanoyl lecithin (PC 8:0), below the
critical micellar concentration (17, 23). Hence, the bilayer organization is
not necessary for functional reactivation.

ACTIVATION BY MITOCHONDRIAL PHOSPHOLIPID (MPL) AND BY DEFINED MOLECULAR SPECIES OF PHOSPHOLIPIDS

A typical plot of activation of apoBDH with increasing concentration of mitochondrial phospholipid is shown in Figure 1. Two characteristics can be obtained from such a titration, i.e., maximal activity and efficiency of activation. The maximal activity for apoBDH with mitochondrial phospholipid of current preparations is approximately 150 μmoles/min/mg protein at 37°C, which is equivalent to a turnover number of 78 sec^{-1} based on the BDH monomer. The efficiency of activation is defined as the number of moles of lecithin per BDH monomer which gives half-maximal activation. By this definition, the smaller the number, the more efficient is the activation.

Mitochondrial phospholipid vesicles activate apoBDH with optimal acti- vity and with the best efficiency of activation (7 moles lecithin/BDH monomer). It consists largely of phosphatidylcholine (PC), phosphatidyletha- nolamine (PE) and diphosphatidylglycerol (DPG) in an approximate molar ratio

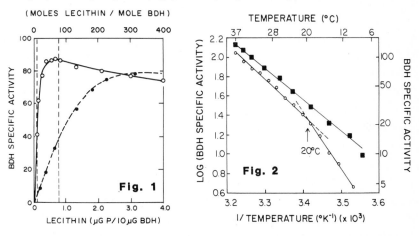

Fig. 1. Titration curves for activation of apoBDH by MPL (O) and by mitochondrial lecithin(PC)(●). The efficiency of activation, i.e. the amount of PC required for half-maximal activity of BDH, indicated by the dashed ver- tical lines is 7 and 80 mol of PC/mol of BDH subunit for MPL or PC, respec- tively (24).

Fig. 2. Arrhenius plot of BDH activated with phospholipid vesicles of dif- ferent composition. BDH was activated with either MPL (■) or the ternary mix- ture of PC (1,2-14:0)/ PE(1-16:0,2-18:1)/ PP(1-16:0,2-18:1) (O) (5/4/1 ratio by phosphorus). The mixed fatty acyl phospholipid vesicles exhibit a liquid crystalline to gel phase transition at 19 ±1 °C. (26,37).

of phosphorus 1:1:0.5 (1). Mitochondrial lecithin by itself as phospholipid vesicles in aqueous buffer activates with lower maximal activity and poorer efficiency(Fig. 1). The mixture PC and PE (1:1) gives near optimal activity but less than optimal efficiency. The ternary mixture including DPG gives both optimal activity and efficiency of activation (17,24).

What is special about mitochondrial phospholipids for the activation of apoBDH? MPL contains three main classes of phospholipids, each consisting of many molecular species due to the diversity of the fatty acids. The activation by single molecular species of mixed fatty acyl phospholipids and combinations thereof was studied. We find that: 1) the ternary mixture of PC:PE:phosphatidylpropan-1,3,-diol (PP) (1:0.8:0.2) in the liquid crystalline state mimics the optimal reactivation obtained by mitochondrial phospholipids; and 2) although some negatively charged phospholipid appears necessary for optimal efficiency of activation, diphosphatidylglycerol can be replaced by PP, another negatively charged phospholipid; and 3) unsaturation is not required for optimal activation which can be obtained by saturated mixed acyl phospholipids in the liquid crystalline state (25,26).

INFLUENCE OF LIQUID CRYSTALLINE AND GEL STATE OF THE PHOSPHOLIPID ON ENZYME ACTIVITY

Mixed fatty acyl phospholipids can be synthesized which have melting transitions over a convenient range of temperatures so that enzymic activity can readily be correlated with temperature. An example of an Arrhenius plot with MPL and with a ternary mixture of single molecular species consisting of PC:PE and PP which gives optimal activation is shown in Figure 2. MPL does not undergo a melting transition in this temperature range and does not exhibit a break in the plot. There is a break in the Arrhenius plot at 20° in the ternary mixture which coincides with the melting transition temperature. Similar studies were carried out with other molecular species of phospholipid, individually or in combinations. For each, the temperature of the break in the Arrhenius plot coincided with the melting transition temperature of the phos-pholipid (Table I). This study provides proof that the break in the Arrhenius plot is referable to the melting transition of the phospholipid (26).

TABLE I

ACTIVATION OF BDH WITH DEFINED MOLECULAR SPECIES OF PHOSPHOLIPID (26)

Lipids	Melting Transition (°Centigrade)		Activation Energy (Kcal/mole °K)	
	DPH	Arrhenius Plot	Liquid Crystal	Gel State
a) MPL	< 5	ND	16.2	-
b) Lecithins				
PC(16:0,18:1)	<10	ND	16.2	-
PC(22:0,18:1)	14.3	14.5	16.1	27.4
PC(16:0,14:0)	27.6	28	16.4	25.7
c) Ternary mixture				
PC(16:0,18:1) PE PP[16:0,18:1]	<10	ND	15.2	-
PC(16:0,14:0) PE[16:0,18:1] PP	20.7	21.8	16.6	27.4
PE(22:0,18:1) PE PP[16:0,18:1]	13.9	15.4	15.8	28.3

STRUCTURAL SPECIFICITY OF LECITHIN FOR ACTIVATION

A variety of lecithin analogues have been synthesized and were used to map the structural specificity of the lecithin molecule for activation of BDH (Fig. 3). The apoenzyme is activated by all lecithin analogues tested which contain a phosphorylcholine moiety but with modifications in the hydrophobic or glycerol portions of the molecule: 1) D and L stereoisomers of lecithin work equally well; 2) octadecyleicosylphosphorylcholine, a branched alkylphosphorylcholine, activates the enzyme to the same extent as mitochrondrial lecithin; 3) phosphono analogues of lecithin in which the glycerol to phosphorus oxygen is missing or replaced isosterically activate. Therefore, the hydrophobic region is necessary for activation but the enzyme does not exhibit specificity for this part of the molecule; 4) the phosphinate

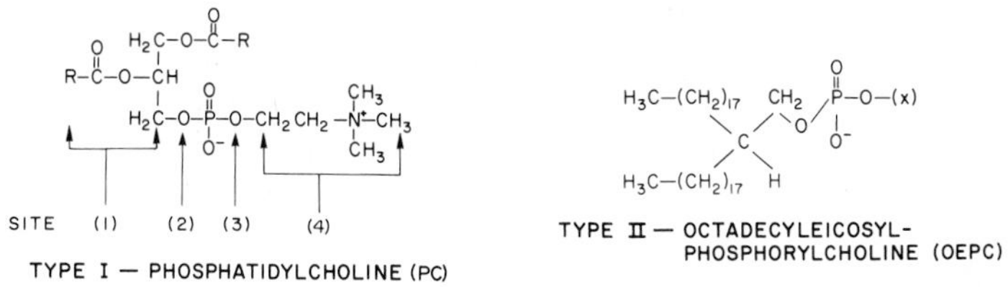

SITE   (1)   (2) (3)   (4)

TYPE I — PHOSPHATIDYLCHOLINE (PC)

TYPE II — OCTADECYLEICOSYL-
PHOSPHORYLCHOLINE (OEPC)

Fig. 3. The structure of phosphatidylcholine (type I) and octadecyleicosylphosphorylcholine (type II) indicating structural modification of the analogs (24).

analogue in which both oxygens are missing, i.e., with glycerol and with choline, does not activate (24).

The specificity of apoBDH for the polar region of the lecithin molecule
has been further studied by synthesis of a number of analogues in which the
choline group is altered. When the ethyl($-CH_2CH_2-$)group, which links the
phosphoryl and quaternary ammonium groups of the phosphorylcholine moiety, was
replaced by a propyl or butyl group, the activation of the enzyme was not
greatly altered. However, no activation was obtained with the compound which
contains an isopropyl group in this position. Thus, the distance which separates the two charged groups is not a critical aspect of the structure of the
lecithin molecule for activation of the enzyme but there are steric
constraints for this part of the molecule. The quaternary ammonium group in
lecithin is essential to obtain activation of the apoBDH since neither
phosphatidylethanolamine nor N,N-dimethylphosphatidylethanolamine activates
the enzyme. The size of the quaternary ammonium group can be increased within
defined limits; the N-ethyl analogue activates the enzyme, but no activation
was obtained by N,N,N-triethyl analogue of the choline moiety (24).

MOLECULAR SIZE AND ORIENTATION OF BDH IN THE MEMBRANE

BDH has a subunit molecular weight of 31,000 as determined by SDS
polyacrylamide gel electrophoresis (20) and 29,000 by sedimentation
equilibrium in guanidine-HCl (27). The hydrodynamic properties of soluble
BDH-PC(8:0) have been determined by active enzyme sedimentation. These studies show that the soluble complex is active as a dimer, whereas the enzyme in
the absence of lipid is inactive and highly associating even below 0.1 mg/ml
protein (27).

Recently, target analysis of radiation inactivation data has been used
to determine the oligomeric size of the active enzyme. The size approximates
a tetramer both for the purified enzyme reactivated with phospholipid as well
as for the native enzyme in the mitochondrial membrane. Irradiation leads to
decreased intensity of the BDH band due to fragmentation as observed by SDS
polyacrylamide gel electrophoresis. This enabled us to evaluate the size of
the BDH in the inactive BDH-phospholipid complex. We find that BDH in
phospholipid vesicles devoid of lecithin is also a tetramer. Hence, lecithin
which is essential for function is not the determinant of the size since the
same oligomeric size obtains in a non-activating mixture of phospholipid
vesicles (28).

When BDH is reactivated by phospholipid in soluble form or as vesicles, the reactivation requires a preincubation which is usually carried out at 30-37°C. Lower temperatures can also be used but longer preincubation times are required. What is the nature of interaction and activation of BDH with phospholipids? Gel exclusion studies showed that when soluble apoBDH interacts with the phospholipid vesicle, the enzyme becomes excluded and elutes together with the phospholipid vesicles, whereas in the absence of phospholipid it enters the internal volume of the gel. The active mixture of BDH and phospholipid in the form of vesicles cannot be significantly dissociated under the conditions of gel exclusion. The soluble BDH-PC(8:0) complex is readily dissociable on a gel exclusion column into two inactive fractions, i.e., BDH and PC(8:0). The fractions can be recombined to reform an active BDH-PC(8:0) complex which can again be separated into two inactive fractions. These studies lead to two conclusions: 1) the active form of the enzyme is the BDH-phospholipid complex; and 2) the longer the chain length of the fatty acyl ester of the phospholipid, the stronger is the association of BDH and phospholipid (20).

When increasing amounts of apoBDH are added to a constant amount of phospholipid or membrane vesicles, a titration curve for activation is obtained which reaches a plateau with respect to both binding or activation (Fig. 4A). The saturating value is proportional to the lipid content of the membrane (Fig. 4B). When the membrane is made leaky so that the

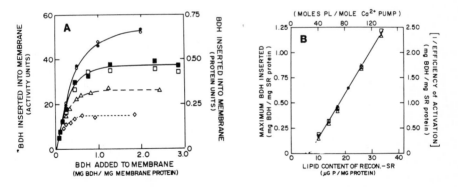

Fig. 4. Functional insertion of BDH into reconstituted sarcoplasmic reticulum membrane vesicles of different phospholipid content. Fig. 4A, Titration curves for activation of BDH activity (open symbols) and direct binding (closed symbols) are shown. The amount of BDH inserted as a function of the phospholipid content of the recipient membrane vesicles is given in Fig. 4B (29).

inner face of the vesicle is accessible, apoBDH then also adds to the inner face. These studies suggest that apoBDH inserts unidirectionally into the outer face of the vesicle (29). This is a characteristic of an amphipathic rather than a transmembrane protein. A diagrammatic representation of the orientation of BDH in the membrane is given in Fig. 5. BDH is depicted as an amphipathic molecule inserted into the phospholipid bilayer in an inlaid orientation.

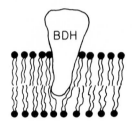

Fig. 5. Diagrammatic representation of the orientation of BDH in the membrane. BDH is depicted as an amphipathic molecule which is inlaid (11) in the phospholipid bilayer (36).

CHEMISTRY OF BDH

The amino acid composition of apoBDH from bovine heart is given in Table II (20). The determination of the primary sequence of the enzyme is in progress. A limited N-terminal sequence is given in Figure 6 for the bovine heart enzyme. The N-terminal amino acid is alanine. In the rat liver enzyme, the N-terminal amino acid appears to be blocked. Peptide maps of the enzyme from bovine heart and rat liver are superimposable except for a few spots (22).

There are six cysteines in the enzyme from heart. Of these, only two react with N-ethyl maleimide (NEM) under nondenaturing conditions. The others appear to be in disulfide linkage since they cannot be derivatized except after treatment with a reducing agent under denaturing conditions(Table III). Derivatization with one equivalent of NEM per BDH monomer leads to complete

Ala.Ala.Ser.Val.Asp[5].Pro.Val.Gly.Ser.Lys[10].

Ala.Val.Leu.Ile.Pro[15].Gly.Ser.Asp. - .Gly[20].

Phe.Gly.Phe.Ser.Leu[25].Ala.Lys.Gly.Leu.Gly[30].

- . - .Gly.Phe.Leu[35]....

Fig. 6. N-terminal sequence of bovine heart BDH.

Table II

Amino acid composition of D-β-hydroxybutyrate apodehydrogenase

Amino acid	Amino acid residues/subunit	Nearest integral no. of amino acid residues/ subunit	Amino acid mol %
Tryptophan	3.93	4	1.41
Lysine	15.9 ± 0.55	16	5.63
Histidine	5.75 ± 0.47	6	2.11
Arginine	14.8 ± 0.90	15	5.28
Aspartic acid	25.5 ± 0.82	26	9.16
Threonine	18.1 ± 1.0	18	6.34
Serine	22.8 ± 0.55	23	8.10
Glutamic acid	28.7 ± 0.84	29	10.40
Proline	11.6 ± 1.8	12	4.23
Glycine	23.4 ± 0.99	23	8.10
Alanine	24.0 ± 0.57	24	8.46
Half-cystine	5.62	6	2.11
Valine	19.3 ± 0.74	19	6.70
Methionine	9.73 ± 0.77	10	3.52
Isoleucine	11.1 ± 0.66	11	3.88
Leucine	21.8 ± 1.2	22	7.76
Tyrosine	8.99 ± 0.95	9	3.17
Phenylalanine	10.5 ± 0.16	11	3.88
Total		284	

inactivation of BDH and loss of coenzyme binding as measured by fluorescence energy transfer (30). The sulfhydryl does not seem to be involved in direct NADH binding but rather derivatization appears to sterically block NADH binding (31).

Table III

Derivatization of D-β-Hydroxybutyrate Dehydrogenase with N-Ethylmaleimide

	mole ratio (mol of MalNEt/mol of enzyme)
Inactivation Stoichiometry	
apoenzyme	1.0
enzyme-MPL complex	1.0
enzyme-lecithin complex	0.9
enzyme-phosphatidylethanolamine- diphosphatidylglycerol complex	0.9
Maximum Derivatization with Excess MalNEt	
apoenzyme	1.7
enzyme-MPL complex	1.9
apoenzyme inactivated by oxygen	0.4
apoenzyme denatured by dodecyl sulfate	5.1

Recently, we have found that one NEM can inactivate two BDH monomers suggesting half-site reactivity. This can be achieved by: 1) the presence of coenzyme or substrate ; or 2) initial derivatization of the non-essential sulfhydryl in the presence of diamide. Then, when the essential sulfhydryl is regenerated with dithiothreitol, half-site stoichiometry for inactivation with NEM is obtained (32).

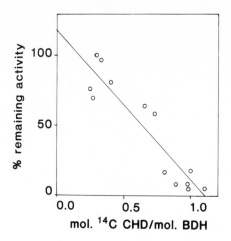

Fig. 7. Stoichiometry of inactivation of BDH by incorporation of ^{14}C-cyclo-hexanedione (CHD), an arginine specific reagent (34).

There is also an essential arginine. Arginine reagents have been shown to inactivate BDH activity (10,33,34). Cyclohexanedione, an arginine specific reagent, completely inactivates by derivatization with one equivalent per monomer (Fig. 7). Substrate analogues alone offer protection, which is nearly total when coenzyme is present. Substrate or coenzyme alone or in combination are ineffective. Since the substrate analogues are bulkier than substrate, it would appear that the essential arginine is in the vicinity of but not at the substrate binding site (34).

MOLECULAR BASIS OF THE ROLE OF LIPID

The apoBDH is incapable of tight binding of coenzyme whereas the enzyme-phospholipid complex containing lecithin binds coenzyme. A Scatchard plot of NADH binding to the active enzyme is shown in Figure 8. The Kd for NADH under these conditions is about 15 μM (35). Lecithin is specific in restoring both coenzyme binding and enzymic function. For an enzyme with an ordered sequen-tial reaction mechanism in which coenzyme must bind first, the lack of coen-zyme binding precludes catalysis (14). The role of the lecithin is related at least in part to conferring to the apoenzyme the ability to bind coenzyme. The lipid requirement by two other lipid-requiring enzymes, pyruvate oxidase from E. Coli (6) and malate-quinone reductase from mycobacterium (5), also seems to be related to conferring coenzyme binding.

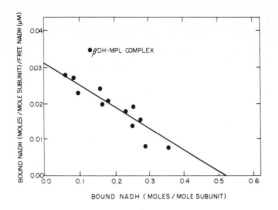

Fig. 8. Binding of NADH by BDH in the presence of MPL as measured by equilibrium dialysis (35).

MOTIONAL CHARACTERISTICS OF THE PHOSPHOLIPID AND BDH IN THE MEMBRANE

^{13}C-NMR was carried out together with the laboratory of W. Stoffel (University of Cologne) on BDH-MPL and MPL containing ^{13}C-lecithin. The lecithin was labeled in the hydrophilic (N-^{13}CH$_3$) or hydrophobic moiety (^{13}C-11 of the 18:2). Such labels monitor the motion in the hydrophilic and hydrophobic portions of the lecithin, respectively. T$_1$ relaxation times, measured using the inversion recovery technique, are summarized in Table IV. In the fast time regime, T$_1$ relaxation times are proportional to rotational mobility or rotational correlation time (τ). The motion in the polar region of the lecithin, in the presence of BDH, was found to be slowed by 37% while the motion in the hydrophobic domain was enhanced. BDH is unique in both respects as compared with apoHDL of high density serum lipoprotein and the calcium pump protein of sarcoplasmic reticulum, the other two systems which have been studied by ^{13}C-NMR in Professor Stoffel's laboratory. Of particular interest is the slowed motion in the polar moiety of lecithin which appears to reflect the specific interaction of BDH with the phosphorylcholine moiety of the lecithin. Although decreased, the motion of the polar moiety of the lecithin is not immobilized under these conditions on the time-scale of NMR ($\tau \sim 10^4$/second). Thus, the presence of BDH in the phospholipid bilayer perturbs the motion of the polar moiety of the lecithin, yet there is rapid exchange between the lipid at the boundary of BDH with bilayer phospholipid (36).

TABLE IV

Changes in T_1 Relaxation Times in [^{13}C] Lecithin after Formation of Lipid–Protein Complex

lipid–protein complex	temp (°C)	T_1 relaxation time (ms)		% ΔT_1	% $\Delta(1/T_1)$	motion
		lipid vesicles	+apo-BDH			
apo-BDH with						
PC (18:2) [N-^{13}CH$_3$]	6	182	114	−37	+60	slower
PC (18:2) [N-^{13}CH$_3$]	20	289				
PC (18:1) [11-^{13}C]	6	113	168	+49	−33	faster
PC (18:1) [11-^{13}C]	20	211				
			+apo-HDL			
apo-HDL with						
PC (18:0, 18:2) [N-^{13}CH$_3$]	37	430	485	+13	−11	faster
PC (18:0, 18:2) [14-^{13}C]	37	705	440	−38	+60	slower
			+CPP			
SR (Ca^{2+} pump protein) with						
PC (18:2) [N-^{13}CH$_3$]	37	510	480	−6	+6	ca. no change
PC (18:0, 18:2) [14-^{13}C]	37	640	380	−40	+68	slower
PC (18:2) [14-^{13}C]	37	780	620	−21	+26	slower

EPR studies using lecithin containing spin label in the 5, 7, 10, 12 or 16 positions of the esterified fatty acid in the sn-2 position reveal small changes in spectra in the presence of BDH indicating increased order and/or slowing of the motion of the hydrophobic moiety. This slowing of motion is equivalent to decreasing the temperature of the phospholipid by approximately 10 degrees centigrade. There is no indication of immobilized phospholipid on the time-scale of EPR ($\sim$10^7/sec). These studies complement the ^{13}C-NMR measurements. The motion of the phospholipid at the "boundary" of BDH in the membrane is in rapid exchange with bulk bilayer phospholipid (37).

The rotational motion of BDH in the membrane was estimated by EPR and saturation transfer EPR. The latter technique is sensitive to motion in the slower time regime (10^3 to 10^7/sec) typical of larger molecules such as proteins. As reviewed above, BDH can be selectively derivatized in either of two unique sulfhydryl groups (one is essential for function, the other appears to be non essential) and then inserted into phospholipid or membrane vesicles. The EPR spectrum of spin-labeled BDH exhibits slow motion which appears to reflect the motion of BDH in the phospholipid bilayer. By conventional EPR it has an apparent rotational correlation time (τapp) of less than 10^7/sec. Saturation transfer EPR has been used to characterize the motion of the BDH in the membrane. The motion of BDH in the liquid crystalline state has an apparent rotational correlation time of about 10^6/second. Lowering the temperature to below the melting transition temperature of the lipid mixture induces a phase transition to the gel state of the phospholipid mixture (Fig. 9). The motion of BDH is significantly diminished by the phospholipid in the gel state (τapp $\sim$10^4/sec). We conclude that: 1) phospholipid phase transitions modulate functional and motional characteristics of the enzyme; and

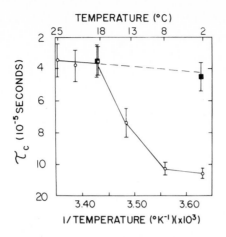

Fig. 9. Temperature dependence of apparent rotational correlation time (τ_{app}) of BDH-phospholipid prepared with each of the two phospholipid vesicle preparations shown in Fig. 2. The apoBDH was labeled with a maleimide-spin label and the τ_{app} values were determined from H"/H ratio of the high field resonance of ST-EPR spectra, and are plotted as a function of temperature (37).

2) the rotational motion of BDH is significant even when the phospholipid is in the gel state (37). The localization of the essential sulfhydryl group in BDH is being studied in terms of its localization with respect to the bilayer. ApoBDH, spin-labeled in the essential sulfhydryl, can be inserted into the phospholipid bilayer and a paramagnetic quench reagent is added. The para-magnetic ion perturbs the spectrum due to Heisenberg spin exchange which can be used to estimate the distance of the essential sulfhydryl from the aqueous domain of the bilayer surface. These studies show that the essential sulfhydryl in BDH is buried within the hydrophobic domain of the phospholipid bilayer (38).

THERMODYNAMICS OF ASSOCIATION OF ApoBDH WITH PHOSPHOLIPID (39)

The activation of apoBDH with different lecithins is shown in Figure 10. It can be observed that soluble lecithins activate apoBDH. The activity rises with increasing concentration and then when the critical micellar con-centration is reached or exceeded the activity is lost. By contrast, natural phospholipids, which form phospholipid vesicles, activate to higher specific activity under these conditions and do not inactivate the enzyme once maximal activity is reached. Phospholipids, which contain short acyl chains, behave as detergents at or above the critical micellar concentration and are gener-ally not present in living cells. The titration curves for activation of apo-

BDH by soluble lecithins were found to be consistent with a model in which the enzyme contains two identical non-interacting lecithin binding sites (Fig.11). Two lecithins per active enzyme is small enough to be considered a cofactor. From the plots in Figure 11, the dissociation constants for the binding of the soluble lecithins to apoBDH can be obtained. From these values the free energy of association of binding (ΔG^0 assoc) can be calculated. The enthalpy was determined from Arrhenius plots and, by difference, the entropy term ($T\Delta S^0$) was obtained. The plot of the thermodynamic state functions for the interaction of soluble lecithin with apoBDH as a function of acyl chain length of the lecithins (Fig. 12) leads to several conclusions (39): 1) the ΔG^0 of association increases with increasing chain length of the lecithin; 2) both enthalpy and entropy contribute to the ΔG^0 of association; 3) the free energy of association per methylene group is 2.4 KJ obtained from the slope of ΔG^0 of association vs the acyl chain length. This is a small amount of energy, approximately equal to thermal energy at room temperature. Yet, phospholipids in membranes contain two fatty acid substituents, each containing approximately 14-24 carbons, which amounts to a sizeable energy of association; 4) lecithins smaller than PC(4:0)$_2$ should not bind or activate apoBDH; this

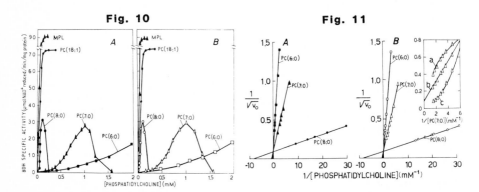

Fig. 10   **Fig. 11**

Fig. 10. Reactivation of apoBDH with different lecithins. The enzymic activity was measured by using 6.0µg of apoBDH per assay and is plotted as a function of phospholipid concentration. For each curve, the lecithins are labeled using a shorthand which refers to the chain length and unsaturation of the fatty acyl substituents. (A) Beef heart BDH; (B) rat liver BDH (39).

Fig. 11. Double reciprocal plots of the square root of initial velocities as a function of phospholipid concentration. The experimental data were obtained from the activation portion of the titration curves shown in Fig. 10. (A) Beef heart BDH; (B) rat liver BDH. (Inset) Comparison of the results obtained from the Hill equation by plotting $1/v_0$ (curve c, n=1), $1/v_0^{1/2}$ (curve b, n=2), and $1/v_0^{1/3}$ (curve a, n=3) as a function of the reciprocal PC(7:0) concentration (39).

prediction has been supported experimentally (23,24). PC(4:0)$_2$ and lecithins of longer chain length bind and activate apoBDH with increasing effectiveness (39). The association energy which increases with chain length appears to be due, at least in part, to hydrophobic interaction which stabilizes the membrane rather than merely a simple binding energy between apoBDH and phospholipids.

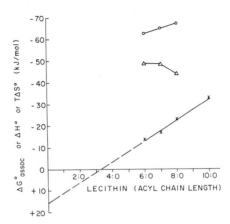

Fig. 12. Thermodynamic parameters for binding of soluble diacyl lecithins to beef heart BDH as a function of the fatty acyl chain length. ΔG assoc. (x), ΔH^0 (O), and $T\Delta S°$(△) (39).

BDH BIOSYNTHESIS AND MEMBRANE ASSEMBLY

BDH is synthesized on cytoplasmic free ribosomes as a larger size precursor than the mature form. In this regard, it is similar to malate and glutamate dehydrogenases, both matrix enzymes, as well as sulfite oxidase which is localized in the intermembrane space of mitochondria (40). That is, membrane assembly of this lipid-requiring enzyme has features in common with enzymes which have different location in the mitochondria and which do not have any apparent requirement of lipid for function. There are extensive homologies and similarities in a number of well studied soluble dehydrogenases (41). BDH is likely similar except for the nuance of the lipid requirement. The basis for the requirement is hidden in the primary sequence and chemical structure which remains to be elucidated. Progress may come, in part, from cloning approaches and obtaining crystals for structure determination which have been successful for a variety of enzymes, including membrane components (42,43).

ACKNOWLEDGEMENTS

The research described was supported in part by Grants AM 14632 and AM 21987 from the National Institutes of Health. We are pleased to acknowledge the participation of our capable and talented colleagues.

REFERENCES

1. Fleischer, S., Brierley, G., Klouwen, H., and Slautterback, D.B. (1962) J. Biol. Chem. 237, 3264-3272.

2. Fleischer, S. and Fleischer, B. (1967) in (R.N. Estabrook and M.E. Pullman, Eds.), Methods in Enzymology 10, 406-433.

3. Rothfield, L and Romeo, D. (1971) in: (Rothfield, L. ed) The Structure and the Function of Membranes, Academic Press, New York, pp 251-284.

4. Kaibuchi, K., Takai, Y., And Nishizuka, Y. (1981) J. Biol. Chem. 256, 7146-7149.

5. Imai, T. (1987) Biochim. Biophys. Acta 523, 37-46.

6. Cunningham, C.C. and Hager, L.P. (1971) J. Biol. Chem. 246, 1575-1589.

7. Fleischer, S., Bock, H-G., and Gazzotti, P. (1974) in: (G.F. Azzone, M. Klingenberg, E. Quagliariello and N. Siliprandi eds.) Membrane Proteins in Transport and Phosphorylation North-Holland Pub. Co., Amsterdam, pp 125-136.

8. Jurtchuk, P., Sekuzu, I., and Green, D.E. (1963) J. Biol. Chem. 238, 3595-3605.

9. Vidal, J.C., Guglielmucci, E.A., and Stoppani, A.D.M. (1978) Arch. Biochem. Biophys. 187 138-152.

10. El Kebbaj, M.S., Latruffe, N., and Gaudemer, Y. (1982) Biochem. Biophys. Res. Communs. 108 42-50.

11. McIntyre, J.O., Bock, H-G., and Fleischer, S. (1978) Biochim. Biophys. Acta 513, 255-267.

12. Nielsen, N.C. and Fleischer, S. (1969) Science 166, 1017-1019.

13. McIntyre, J.O., Tzagoloff, A. and Fleischer, S., unpublished studies.

14. Nielsen, N.C., Zahler, W.L., and Fleischer, S. (1973) J. Biol. Chem. 248, 2556-2562.

15. Churchill, P., McIntyre, J.O., Vidal, J.C., and Fleischer, S. (1983) Arch. Biochem. Biophys. 224, 659-670.

16. Fleischer, B., Casu, A., and Fleischer, S. (1966) Biochem. Biophys. Res. Commun. 24, 189-194.

17. Gazzotti, P., Bock, H-G., and Fleischer, S. (1975) J. Biol. Chem. 250, 5782-5790.

18. Nielsen, N.C. and Fleischer, S. (1972) J. Biol. Chem. 248, 2549-2555.

19. Clancy, R.M., McPherson, L.H., and Glaser, M. (1983) Biochemistry 22, 2358-2364.

20. Bock, H-G. and Fleischer, S. (1975) J. Biol. Chem. 250, 5774-5781.

21. Vidal, J.C., Guglielmucci, E.A., and Stoppani, A.D.M. (1977) Adv. Exp. Med. Biol. 83, 203-217.

22. Brenner, S.C., McIntyre, J.O., Latruffe, N., Fleischer, S., Tuhy, P., Elion, J., and Mann, K. (1979) Fed. Proc. 38 (Abstract 1849).

23. Grover, A.K., Slotboom, A.J., deHaas, G.H., and Hammes, G.G. (1975) J. Biol. Chem. 250, 31-38.

24. Isaacson, Y.A., Deroo, P.W., Rosenthal, A.F., Bittman, R., McIntyre, J.O., Bock, H-G., Gazzotti, P., and Fleischer, S. (1979) J. Biol. Chem. 254, 117-126.

25. Eibl, H.J., Churchill, P., McIntyre, J.O., and Fleischer, S. (1982) Biochem. Int. 4, 551-558.

26. Churchill, P., McIntyre, J.O., Eibl, H.J., and Fleischer, S. (1983) J. Biol. Chem. 258, 208-214.

27. McIntyre, J.O., Holladay, L.A., Smigel, M., Puett, D., and Fleischer, S. (1978) Biochemistry 17, 4169-4177.

28. McIntyre, J.O. Churchill, P., Maurer, A., Berensky, C., Jung, C.Y., and Fleischer, S. (1983) J. Biol. Chem. 258, 953-959.

29. McIntyre, J.O., Wang, C-T., and Fleischer, S. (1979) J. Biol. Chem. 254, 5199-5207.

30. Latruffe, N., Brenner, S., and Fleischer, S. (1980) Biochemistry, 19, 5285-5290.

31. McIntyre, J.O., Trommer, W.E., Fritsche,T.Z.,Fleer, E., Fleischer, S. unpublished studies.

32. McIntyre, J.O., Fleer, E.A.M., and Fleischer, S. (1982) Second European Bioenergetics Conference L.B.T.M-C.N.R.S. Editeur, Vitteurbanne, pp. 509-510.

33: Phelps, D.C. and Hatefi, Y. (1981) Biochemistry 20, 459-463.

34. Fleer, E.A.M. and Fleischer, S. (1983) Biochim. Biophys. Acta (in press).

35. Gazzotti, P., Bock, H-G., and Fleischer, S. (1974) Biochem. Biophys. Res. Commun. 58, 309-315.

36. Fleischer, S., McIntyre, J.O., Stoffel, W., and Tunggal, B.D. (1979) Biochemistry 18, 2420-2429.

37. McIntyre, J.O. and Fleischer, S. unpublished studies.

38. McIntyre, J.O., Dalton, L.A., and Fleischer, S. unpublished studies.

39. Cortese, J.O., Vidal, J.C., Churchill, P., McIntyre, J.O., and Fleischer, S. (1982) Biochemistry 21, 3899-3908.

40. Mihara, K., Omura, T., Harauo, T., Brenner, S., Fleischer, S., Rajagopalan K.V., and Blobel, G. (1982) J. Biol. Chem. 257, 3355-3358.

41. Rossman, M.G., Liljas, A., Branden, C-I., and Banaszak, L.J. in: (P.D. Boyer, editor) The Enzymes vol.XI, Academic Press, Inc., 1975 pp 61-102.

42. Helmer, G.L., Laimins, L.A., and Epstein, W. in: (Martonosi, A.N., editor) Membranes and Transport vol.II, Plenum Press, New York, (1982) pp 123-128.

43. Michel, H. (1983) TIBS 8, 56-59.

THE Na$^+$-GLUCOSE COTRANSPORTER OF THE SMALL INTESTINAL BRUSH BORDER MEMBRANE:
AN ASYMMETRIC GATED CHANNEL (OR PORE) RESPONSIVE TO ΔΨ.

GIORGIO SEMENZA, MARKUS KESSLER AND JAKOB WEBER
Laboratorium für Biochemie der ETH, ETH-Zentrum, CH 8092 Zürich, Switzerland

1. The asymmetry: a "channel" (or "pore")

Current ideas on the vectorial insertion of intrinsic membrane proteins during
or after their biosynthesis and obvious thermodynamic considerations justify
the expectation that intrinsic membrane proteins having on the surface hydrophilic and hydrophobic areas be inserted (during or immediately after their
biosynthesis) in the membrane asymmetrically with respect to the plane of the
membrane, and that they remain in this positioning thereafter. Transport
agencies should belong to this class of proteins, and indeed it has been con-
clusively shown even in so-called "equilibrating" transport agencies that they
are asymmetrically inserted and that they have different overall kinetic parameters at the two sides of the membrane. (For reviews on the sugar transporter
of the erythrocyte, see Widdas[1], on that of the anions in the same membrane, see
Cabantchik et al.,[2], Rothstein and Ramjeesingh[3], and on that of adenine nucleotides of the mitochondrial inner membrane, see Klingenberg et al.,[4]). Naturally,
thermodynamics sets the frame of the kinetic asymmetries possible.

As to the Na$^+$-D-glucose cotransporter of the small intestine, we have shown
previously[5-8] that this transport agency also is inserted asymmetrically with
respect to the plane of the membrane: the structures which it exposes to the two
sides of the membrane react differently with SH-reagents or proteases. It would
be odd if an asymmetric structure were to have totally symmetric kinetic parameters, and indeed we have shown that this cotransporter for Na$^+$ and glucose
does have asymmetric functional properties. This functional asymmetry was demonstrated in fact in four different aspects[9]: (i) at ΔΨ ≃ 0, trans-inhibitions
and their release ("trans-stimulations") were observed for influx only, not
for efflux (Fig. 1, Table I). (ii) Influx is more accelerated by a membrane potential difference (negative on the trans side) than efflux (Table II). (iii)
The estimated K_m for efflux appears to be larger than the K_m's determined for
influx. (iv) Efflux and influx rates may differ by a factor 10 at equivalent,
but mirrored conditions, as a consequence of these various asymmetries.

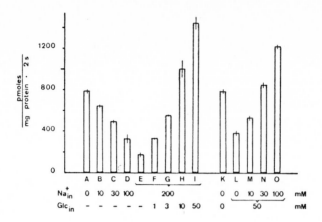

Fig. 1. Transinhibition by Na^+_{in} and by D-glucose$_{in}$ at $\Delta\Psi \simeq 0$. Small-intestinal brush border membrane vesicles were incubated at pH 7.5 in the presence of 100 mM $(Na_2 + choline_2)SO_4$, D-glucose (at the concentrations of Na^+ and/or D-glucose given under the columns), plus 100 mM choline NO_3. To start the incubation proper, the vesicle suspension was diluted into a mixture giving the following final concentrations: 100 mM choline NO_3, 100 mM Na_2SO_4, 1 mM ^3H-D-glucose (mannitol and buffer were present throughout). Incubations lasted 2 s and were done at 20 to 22°C. For further details, see original paper [9].

TABLE I

LACK OF TRANSEFFECTS ON D-GLUCOSE EFFLUX

Transsubstrate present	pmol D-glucose released in 2 s
none	1.09 ± 0.02
200 mM Na^+	1.07 ± 0.05
50 mM D-glucose	1.11 ± 0.05
200 mM Na^+ + 50 mM D-glucose	0.99 ± 0.03

The vesicles were preincubated in 50 mM Na_2SO_4 + 0.1 mM ^3H-D-glucose (+ mannitol and buffer). Efflux took place into 150 mM $(Na_2 + choline_2)SO_4$ $\pm$ D-glucose (+ mannitol and buffer). Further details, see original paper [9].

TABLE II

ASYMMETRY IN THE RESPONSE OF D-GLUCOSE INFLUX AND EFFLUX RATES TO $\Delta\Psi$, NEGATIVE AT THE TRANS-SIDE

The comparison was made between the rates of D-glucose fluxes (trans D-glucose, zero) in the presence of an initial bi-ionic gradient producing a $\Delta\Psi$ negative at the trans-side (100 mM $NaNO_3$ cis, 100 mM NaCl trans) or in the presence of equal concentrations of $NaNO_3$ (100 mM), which should produce a $\Delta\Psi \simeq 0$. Further additions were mannitol and buffer. The concentration of ^3H-D-glucose (cis only) was 0.1 mM. The incubations, at 20-22°C, lasted 3 s. For more details, see the original paper [9].

Experimental setup	Acceleration observed in the presence of the initial bi-ionic gradient expected to yield an initial $\Delta\Psi$, negative at the trans-side, as compared to the control with clamped vesicles
Influx	3- to 5-fold
Efflux	approx. 1.3-fold

TABLE III

$\Delta\Psi$-DEPENDENCE OF THE TRANS-INHIBITION OF D-GLUCOSE INFLUX BY D-GLUCOSE$_{in}$

The salt gradients were obtained by preincubating the vesicles in the appropriate salt solutions and diluting them at the start of the incubations in the salt solution given below. Further additions, both in the preincubation and in the incubation mixtures, were mannitol and buffer. The concentration of cold D-glucose in the preincubation was 50 mM (50 mM mannitol in the control). That of ^3H-D-glucose in the incubation was 1.05 mM. The incubations lasted 0.6 s. For more details, see the original paper [9].

Salt gradient at the beginning of the incubation	Trans D-glucose zero	Trans D-glucose 20 mM	% inhibition
Na_2SO_4 gradient ($\Delta\Psi$ small)	65.0 ± 1.5	40.1 ± 0.7	38%
NaSCN gradient ($\Delta\Psi \ll 0$)	338 ± 8	261 ± 5	23%

The strong trans-inhibition of influx rates by Na^+_{in} is a most prominent and reproducible feature in other Na^+-coupled transport systems also (e.g., refs. 10,11) and may well be present in all Na^+-coupled transport systems.

2. A transport agency with the kinetic properties of a "mobile carrier" a "gated channel (or pore)".

A stable structural and functional asymmetry rules out a "diffusive" or "rotative" mode of functioning and makes a "pore" or a "channel" the most probable mechanism(s). But this does not mean that transport is not accompanied, actually made possible, by the movement of a portion of the transport agency, "carrier" being defined[12] as a transport agency whose substrate binding site(s) can be exposed to both sides of the membrane, but not simultaneously. Indeed, the Na^+, D-glucose cotransporter studied here does have the kinetic properties of a "mobile carrier", the most classical one being that of counterflow accumulation (see for example, refs. 13 and 9). Also, the $\Delta\Psi$ dependence of Na^+-dependent phlorizin binding to this cotransporter[14,15] strongly indicates the motion of a portion of the cotransporter to be associated with its functioning. Finally, the fairly low turnover number (estimated to be approximately $20 \; s^{-1}$, ref. 14) is more compatible with the classical ideas on "mobile carriers" than of freely open "channels" or "pores". A "gated channel (or pore)" is thus the most likely model for this Na^+-D-glucose cotransporter.

For the purposes of the present discussion we define as "gate" that portion of the channel or pore whose movement is associated with or related to the translocation of the substrate(s).

3. The "gate" bears a negative charge of 1 or 2 (z=-1 or z=-2).

This conclusion is based on, or is compatible with, the following experimental observations:

(i) $\Delta\Psi$-dependence of trans-inhibition[9]. It can be shown[9] that in an electrogenic cation-dependent cotransport system the extent of transinhibition by the "substrate" or by Na^+ is affected by $\Delta\Psi$ (negative at the trans side). Independently of the degree of asymmetry of the cotransporter and of the Na^+:substrate stoichiometric ratio, a negative $\Delta\Psi$ leads to a _decreased_ trans-inhibition, if the mobile part of the (unloaded) cotransporter carries one or two negative charges, but to an _increased_ transinhibition if this mobile part is electrically neutral. In actual fact (Table III) the trans-inhibition by D-glucose$_{in}$ follows the prediction of the models with z=-1 or z=-2, is not compatible with model with z=0. We conclude that the "gate" of the translocator bears, in the substrate-free form at least one negative charge.

Independently of the actual mechanism, the partial relief by $\Delta\Psi$ of the trans-inhibition of influx by glucose$_{in}$, is likely to have physiological significance: it reduces or prevents the "braking" of D-glucose entry into the cell by the substrate already present at the cytosolic side of the membrane.

(ii) Lack of effect of $\Delta\Psi \ll \dot{0}$ on the rate of dissociation of phlorizin from the cotransporter. If the gate carries a negative charge of 1 and the Na^{+}:phlorizin stoichiometry of binding is one (which is made likely by the $\bar{n}$ coefficient being equal to one, for the intestinal[15] and for the renal co-transporter[16]) the ternary complex (phlorizin-Na^{+}-cotransporter^{-}) is neutral and the rate of phlorizin dissociation should not be affected by $\Delta\Psi \ll 0$. This is indeed the case for both the intestinal[15] and the renal[17] cotrans-porter.

One might object that an extremely asymmetrical (inwardly directed) cotrans-porter with z = 0 (and thus with a net charge of + 1 when bound to 1 Na^{+} plus phlorizin) would likewise not respond to $\Delta\Psi$ in its rate of dissociation of phlorizin, because a positively charged, phlorizin occupied "gate" cannot change its positioning in response to $\Delta\Psi$ (negative inside) if it is already totally oriented towards the inside surface. However, such a possibility can be ruled out, because it implies also that the cotransporter in the presence of Na^{+} (i.e., when carrying a positive charge of at least + 1) would not res-pond to a $\Delta\Psi$ (negative   inside) in any respect. But we do know that the co-transporter in the presence of Na^{+} does respond to $\Delta\Psi$ (negative inside), both as far as D-glucose transport[18] and as far as phlorizin binding[14] are con-cerned.

(iii) The apparent K_m-values for D-glucose uptake are independent of the pH in the 6.5 - 9.5 range; the apparent K_m at pH 5.5 is some fivefold larger[14].

It is quite possible, therefore, that the negative charge in the "gate" (or elsewhere in one of the sites binding either substrate) is a carboxylate group.

(iv) This conclusion would also agree with recent observations by Weber and Semenza (1983, in preparation) that D-glucose transport is inhibited irreversibly by dicyclohexenylcarbodiimide.

4. The Na^+: D-glucose flux ratio.

Kaunitz et al.[19] have recently reported that the Na^+_{out}-dependence of D-glucose uptake into small-intestinal brush border membrane vesicles is highly sigmoidal (the $\Delta\Psi$ was clamped near zero with K^+/valinomycin; Na^+-H^+ exchange was minimized with amiloride and H^+ gradients with carbonyl cyanide p-trifluoromethoxy-phenylhydrazone). The $\bar{n}$ coefficient was approximately 2. We confirmed this sigmoidicity in vesicles clamped with NO_3^- (Ref. 9) and found little or no sigmoidicity in the presence of $\Delta\Psi << 0$. This sigmoidicity by itself does not prove a 2:1 Na^+: D-glucose stoichiometry ratio, particularly if accomodated by an opposite deviation from a michaelian behaviour in the D-glucose dependence. However, flux ratios of 2 have been found by measuring directly the sugar-dependent Na^+ flux under various conditions, particularly in the presence of inhibitors depressing the D-glucose independent Na^+ flux[19,20]. Thus, at the moment of writing a Na^+ : D-glucose flux ratio of 2 must be considered as a real possibility - although an effect of the experimental conditions (e.g., of $\Delta\Psi$) on the flux ratio cannot be ruled out.

As mentioned above, the $\bar{n}$ coefficient of $\Delta\Psi$-dependent phlorizin binding to the cotransporter is equal to one[15]. If this indicates a Na^+ : phlorizin binding stoichiometry of one, it is conceivable that the Na^+-phlorizin-cotransporter complex may be non-functional for the very reason of not binding a second Na^+, if indeed only the complexes of the type Na_2^+-sugar-cotransporter are functional. This mechanism, which is presented here as a mere suggestion only, will have to be investigated experimentally in the future.

5. The translocation probabilities of partially occupied cotransporter complexes.

In view of the considerations in the preceding section, we will indicate as "fully occupied" cotransporter that form of it in which all binding sites for the substrates and cosubstrates are occupied by the respective ligands. The "fully occupied" cotransporter has, therefore, the composition D-glucose-Na_n^+-cotransporter, where n is equal to 1 or 2, depending on how large the stoichiometric flux ratio is. Likewise, we will indicate as "partially occupied" cotransporter that form of it in which one or more (but not all) substrate and cosubstrate binding sites is (are) occupied by the respective ligand(s) In the "empty" cotransporter all binding sites for substrates and cosubstrates are unoccupied.

Heinz[21] has pointed out that, to provide for an efficient energy conversion, the translocation probabilities of the partially occupied cotransporters must be low as compared with those of the fully occupied and of the empty forms. In actual fact, whatever evidence is available does indicate that the translocation probabilities of many, if not all, the partially occupied cotransporters are indeed small, as follows.

At $\Delta\Psi \simeq 0$, the trans-inhibitions by either Na^+_{in} or D-glucose$_{in}$ (Fig. 1, Table III) indicate that the in→out translocation probabilities of the respective partially occupied cotransporter forms are small. As to the out→in probabilities, that of the D-glucose-cotransporter complex must be[9] small.

At $\Delta\Psi << 0$ (negative inside) the trans-inhibitions by Na^+_{in} and by D-glucose$_{in}$[9] again indicate small in→out translocation probabilities for the respective partially occupied cotransporter forms. As to the out→in translocation probability of the D-glucose-cotransporter complex it must be small at $\Delta\Psi << 0$ also [9].

We will thus proceed under the likely assumption that all translocation probabilities of the partially occupied cotransporters are very small compared to those of the empty and of the fully occupied forms.

6. A partial Cleland kinetic analysis.

For a number of reasons discussed elsewhere[9] it is not possible to carry out a complete kinetic analysis of this type. It is possible, however, to rule out the family of (Iso) Ping Pong Bi Bi (or Ter Ter) mechanisms: in fact, both at $\Delta\Psi \simeq 0$ and at $\Delta\Psi < 0$ (negative at the trans side), increasing Na^+_{out} never leads to an _increase_ in the apparent K_m for D-glucose$_{out}$. Furthermore, the inhibition of D-glucose uptake by Na^+_{in} is of the "non-competitive" type[9]. We are thus left with the family of (Iso) Random Bi Bi (or Ter Ter), Rapid or Non-rapid Equilibrium mechanisms, which includes the "Ordered" ones, as special, limiting cases.

By analogy with the Preferred Ordered observed in phlorizin binding (i.e., Na^+_{out} first, phlorizin$_{out}$ second)[15], it seems likely that this sequence of binding is also followed in the case of poorly transported substrates (phlorizin can be regarded as an "infinitely slowly" transporter substrate).

7. A gated channel (or pore) with preferred inward orientation.

In a recent paper[15] we indicated that at $\Delta\Psi \approx 0$ the substrate binding site of the Na^+, D-glucose cotransporter is not freely accessible to the substrate from the outer, luminal, side of the membrane. The effect of $\Delta\Psi$ (negative inside the vesicles) merely consists in reorienting the mobile part of the co-transporter (i.e., the negatively charged "gate" mentioned above) towards the outside, thereby allowing - after the binding of Na^+ - the binding of phlorizin to the substrate site. Deoxycholate-disrupted membranes (naturally, at $\Delta\Psi=0$) bind phlorizin with the same K_d-value as vesicles in the presence of $\Delta\Psi<<0$.

It is remarkable that this same model - of a negatively charged "gated channel" having preferred inward orientation at $\Delta\Psi\approx0$ - can satisfactorily explain the transinhibition by substrates, their asymmetry, and the very small response of efflux rates to $\Delta\Psi<<0$ (negative at the trans side). The only addition to the model which we make here is that the translocation probabilities of the partially occupied cotransporter forms from the inside ('') to the outside (') are negligibly small in comparison with the corresponding translocation probabilities of the empty and of the fully occupied forms.

In fact, if the substrate-free cotransporter at $\Delta\Psi\approx0$ has a predominantly inward orientation (i.e., if $p_0' > p_0''$), sequestering of much cotransporter into slowly or nontranslocating partial complexes at the " side will result in a strong transinhibition of influx;(conversion of the partial complexes into the more mobile, fully occupied form would relieve the transinhibition). On the other hand, sequestering much cotransporter at the outer (') side may be difficult (due to the predominant orientation toward the inside);the trans-inhibition (from the outside) of efflux, if any, may go undetected.

Furthermore, if the empty form of the cotransporter (or, rather, its mobile, negatively charged "gate") already at $\Delta\Psi\approx0$ has a predominantly inward orientation, a $\Delta\Psi>>0$ (positive inside the vesicles) can affect its orientation only little. (The same can be said for the binary complex with glucose, if formed at all). Thus, the model of an asymmetric cotransporter with a predominantly inward orientation predicts also little or no response of the efflux rates to $\Delta\Psi$ (negative at the trans side, i.e., at the outside of the vesicles). This is actually observed experimentally (Table II).

This model of "preferred inward orientation" (at $\Delta\Psi\approx0$) is essentially based on the assumption of an asymmetry in the translocation probabilities (i.e., $p_0'>p_0''$). For thermodynamic reasons, this asymmetry must be compensated for by other asymmetries in other translocation probabilities and/or in the dissociation constants of the binary, ternary (and quaternary, if the case) complexes at both sides of the membrane. For example, in the model with a Na^+:D-glucose stoichiometry of one with negligibly small translocation probabilities of the binary complexes, the following equation holds true at $\Delta\Psi\approx0$:

$$\frac{p_0'' \; p_{GN}'}{p_0' \; p_{GN}''} = \frac{K_G^{N'} \; K_N^{'}}{K_G^{N''} \; K_N^{''}} = \frac{K_N^{G'} \; K_G^{'}}{K_N^{G''} \; K_G^{''}}$$

where the p's are the translocation probabilities of the free cotransporter (o) or of the totally occupied cotransporter (GN) and the K's are the dissociation constants (of the binary complexes; N, with Na^+, G, with glucose; of the ternary complexes: superscript indicates the substrate bound first, subscript the substrate bound second). ' indicates 'outside' or 'inward translocation'; " indicates "inside" or "outward translocation".

Similar equations can be derived for the model with a 2 : 1 Na^+, D-glucose stoichiometry.

The transeffect experiments allow one more conclusion to be drawn. The fact that both substrates, Na^+_{in} and Glc_{in} can elicit transinhibition indicates that each substrate can bind to the translocator in the absence of the other substrate. This virtually rules out any mechanism with a strictly compulsory binding sequence in the inward-facing state.

Summing up the considerations above, the small-intestinal Na^+, D-glucose cotransporter, in addition to being structurally asymmetric (with respect to the plane of the membrane) - which rules out diffusive and rotating modes of operation and makes a gated channel or pore, or a snip snap mechanism likely - is functionally asymmetric also. At $\Delta\Psi\approx0$ and in the absence of substrates its substrates' binding sites have a predominantly inward orientation (or accessibility); it has very small outward translocation probabilities of partially occupied forms; it has a negative charge of at least 1 (presumably, a carboxylate group) in the mobile part (the "gate") responding to the membrane potential difference. Kinetic mechanisms of the (Iso) Ping Pong type are ruled out. The translocation probability of the fully occupied cotransporter is probably larger in the out→in than in the in→out direction.

8. A plausible mechanistic model.

In an important paper Läuger[12] has pointed out that the kinetic and thermodynamic properties of "mobile carriers" are shown by those transport agencies which can expose the substrate binding site at both sides of the membrane, but not simultaneously (see also Patlak[22]). In the case of the Na^+-D-glucose cotransporter it is clear that it can exist in two major families of forms, the one prevailing at $\Delta\Psi\approx0$ (e.g. Form I, Fig. 2) and the other one (the "energized"

310

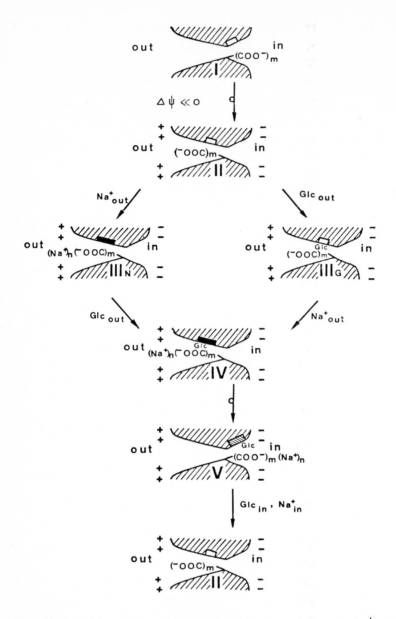

Fig. 2. A likely mode of operation of the small-intestinal Na$^+$,D-glucose co-transporter(s): an asymmetric gated channel (or pore) responsive to $\Delta\Psi$. The D-glucose (and phlorizin) binding site is indicated white (low affinity), or black (high affinity), or shaded (indefinite affinity). Subscript m: it is not known whether m is 1 or 2; subscript n: it is not known whether the Na$^+$:D-glucose flux ratio is 1 or 2. For details, see text and the original paper[9].

forms) prevailing at $\Delta\Psi <\!< 0$ (negative at the in-side) (e.g., Forms II through V). Each of the two (families of) forms operate as a "mobile carrier" in Läuger's sense, but with different energetic profiles (and thus with at least some of the individual rate constants different) in the "translocation" events. Thermodynamically, the difference between "nonenergized" and "energized" forms is apparent in the capability of the latter (but absent in the former) of producing accumulation of substrates starting from initial conditions $(Na^+)_{in} = (Na^+)_{out}$ and $(Glc)_{in} = (Glc)_{out}$.

Even at the cost of stating the obvious we want to emphasize that the displacement of the substrate binding sites depicted in Fig. 2 need not imply their actual movement. Indeed a change in their accessibility from either side of the membrane (which could be related with protein fluctuation) is a perfectly logical and likely possibility (Läuger [12]). Accordingly, "translocation" and "change in accessibility" are used for the purposes of this paper, as synonymes.

In the following we will discuss a plausible mode of operation mainly of the "energized" Na^+, D-glucose-cotransporter (i.e., in the presence of $\Delta\Psi <\!< 0$, negative inside) (see Fig. 2). In so doing, we will combine all the features which have been enlisted and discussed in the previous sections.

In addition, we will add two more features which, although plausible, do not have a direct experimental support: (i) The "gate" is (a part of) the Na^+ binding site (the alkali cations vastly prefer O over N or S as the ligand) and, (ii), the "orientation" (or "accessibility"), as well as the affinity state of the substrate binding sites must change in a concerted and identical fashion. (That is, no significant amount of the cotransporter ever has the binding site of one substrate at low and the other at high affinity; or one binding site facing one side of the membrane and the other facing the other: these conformational states would be a part of an Iso Ping Pong mechanism and/or be conductive to the transport of one substrate alone, both of which possibilities are unlikely; see previous sections.) The "gated channel (or pore)" of the Na^+, D-glucose cotransporter is asymmetric with respect to the plane of the membrane. In particular, at $\Delta\Psi \approx 0$ and at low substrate concentrations (Fig. 2) all binding sites have a spontaneously inwardly directed orientation. In Form I, then Na^+_{in} and D-glucose$_{in}$ each have access to the respective binding site and trap the cotransporter as a slowly (or non-) translocating binary complex. This is the basis of the transinhibitions by Na^+ or by D-glucose from the "in" side and for their lack from the "out"-side, as discussed in a previous section.

In the presence of a $\Delta\Psi \ll 0$ (negative inside) the "gate" (and with it, by an ill-understood mechanism, the D-glucose binding site also) "moves" towards the outside or is made otherwise accessible from the "out"-side (Form II).

Na^{+}_{out} can now bind to the gate (Form III N), which increases the affinity of the sugar binding site. The mechanism whereby the binding of Na^{+} to its site affects the affinity (or, rather, the K_m) of the sugar binding site for its own substrate is still a matter of speculation. However, the drastic change in the coulombic field near the COO^{-}-gate upon Na^{+} binding and the (even if minute) reorientation of the "gate" under $\Delta\Psi$ provide ample possibilities.

Alternatively (but less likely with poor sugar substrates, see above), D-glucose as the first may bind to its outwardly exposed binding site (Form II), leading to Form III G.Neither Form III N nor Form III G crosses fast the membrane (or, rather, expose and liberate the lone bound substrate to the in-side). We do not know why the binary complexes III N and III G have low or nil translocation probabilities. However, this is a prerequisite for efficient flux coupling [21] . Also in the case of cotransporter(s) binding 2 Na^{+} and 1 D-glucose, our model is compatible with any order of substrate binding.

From either Form III N or III G the "fully occupied" Form IV is generated, in which the gate charge is neutralized by Na^{+}, or even made positive (if the gate carries a charge of -1 and can bind 2 Na^{+}). This makes the gate snap back (Form V) in the "spontaneous" inwardly directed positioning (as in Form I), the more so if it is now positively charged. In Form V the binding sites (and thus the substrates) are exposed towards the inside, where they are liberated (we do not know in which order). Reappearance of the negative charge in the gate makes it again respond to $\Delta\Psi$ and snap towards the outside (Form II). The combined effect of Na^{+} on/off and of $\Delta\Psi$ would thus be that of modulating the function of the sugar binding site and of changing the energy profile of the channel so as to favor the exposure of the sugar binding site towards one or the other side of the membrane. An important condition for this "shoveling" mechanism to operate is that, as suggested above, the translocations (or changes in accessibility) of all substrate binding sites must take place in a concerted, identical fashion.

The model provides a unifying and realistic basis to at least three of the major characteristics of the small-intestinal Na^+, D-glucose cotransporter: the Na^+, ($\Delta\Psi$)-dependent phlorizin binding (Form III N only binds phlorizin optimally, as discussed in detail by Toggenburger et al.,[15]), the flux coupling between Na^+ and D-glucose (Form IV mainly, the fully occupied complex, liberates the substrates at the trans side), and perhaps most important of all, the accumulation of an unloaded compound (D-glucose) being driven by $\Delta\Psi$, with the proviso that Na^+ is also present. It is an attempt to explain in molecular terms the phenomenon of flux coupling in secondary active transport which was put forward for the first time by Crane, Miller and Bihler in 1961 (Refs. 23,24).

It seems very likely that the mechanism depicted in Fig. 2 be common to the kindred kidney cortex Na^+, D-glucose cotransporter(s), the mobile part of which also probably bears, in the free form, a negative charge of 1 (Aronson [17]). (For a discussion of the quantitative differences between these two transport systems, see ref. 15). Also a recent, independent observation of Hilden and Sacktor [25] on the kidney cotransporter(s) fits beautifully in the mechanistic model of Fig.2. In cortex membrane vesicles in the total absence of Na^+, $\Delta\Psi$ (negative inside) promotes phlorizin binding and a (slow) uptake of D-glucose, the K_d- and K_m-values being rather large. This is exactly what the transition between Form I and Form II (Fig. 2) predicts: a $\Delta\Psi$ (negative inside) leads to exposure of the substrates' binding sites towards the outside, leaving them, however, in the low-affinity form. In addition, Form II (when combined with D-glucose) must have a small but detectable out→in translocation probability. In addition, many of the known properties of Na^+- or H^+-coupled symport systems indicate that the "shoveling mechanism" of Fig. 2 may have a more general validity.

Needless to say, the "shoveling" model of Fig. 2 is a minimum mechanism. In addition to completion, it may well need revision in some steps, but it is remarkable that it can explain most, if not all, the known characteristics of the cotransporter. We still ignore the role, if any, of the thiol(s) which have been found to occur in the cytosolic and/or hydrophobic surface of the cotransporter [5-8]

The chemical investigation of function-structure relationships in the Na^+,D-glucose cotransporters is still at its infancy. The Na^+ binding site(s) of the

314

renal (outer cortex) cotransporter is likely to contain the phenolic group of a tyrosine residue [26], and, if it is identical with (a part of) the "gate", one or more carboxylate groups (see above).

As to the sugar binding site, the thiols mentioned above [5-7] are not likely to be a part of it, since their modification with organic mercurials or with Hg^{2+} is not delayed or prevented by sugar substrates or phlorizin. However, we have recently found [27] that a number of amino group reagents irreversibly inactivate Na^+-dependent D-glucose transport into membrane vesicles of the small-intestinal brush borders and Na^+-dependent phlorizin binding to these vesicles or to membrane fragments prepared therefrom by deoxycholate extraction. The reagents included: fluorescamin, methylacetimidate, fluorodinitrobenzene, phenyl-isothiocyanate, p-sulfophenyl-isothiocyanate; and reductive methylation. (Care was taken to make sure that the reaction studied was with amino groups rather than with other nucleophiles). Interestingly enough, the inactivation produced by these reagents was delayed by the presence of D-glucose, and the more so by that of D-glucose plus Na^+. A parallelism was found between the protecting effect of various monosaccharides and their capacity to interact with (inhibit) the cotransporter (Fig. 3). These observations make very likely the presence of an amino group within the sugar binding site; and perhaps - but of course it need not be so - that this amino group is the H-bond donor which has been suggested to form between the sugar binding site and C_6 of the monosaccharide [28] or of the sugar moiety of phlorizin [29]

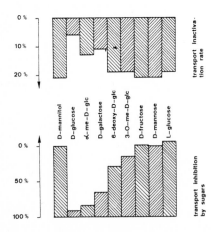

Effect of various monosaccharides (or derivatives) on the inactivation rate of D-glucose transport into brush border membrane vesicles by methylacetimidate. The vesicles were first reacted with methylacetimidate in the presence of the sugar (or sugar derivative) indicated, washed free of sugar and reagent, and tested for the transport of ^{3}H-D-glucose (top panel). The lower panel shows the reversible inhibition of ^{3}H-D-glucose transport by the same sugars in unreacted vesicles. For details, see ref. 27.

Fig. 3

ACKNOWLEDGEMENTS

The financial support of the SNSF, Berne, and of Nestlé Alimentana, Vevey, is gratefully acknowledged.

REFERENCES

1. Widdas, W.F. (1980) in: Current Topics in Membranes and Transport, Bronner, F. and Kleinzeller, A. (Eds.) Academic Press, New York, pp. 166-223.

2. Cabantchik, Z.I., Knauf, P.A., and Rothstein, A. (1978) Biochim. Biophys. Acta 515, 239

3. Rothstein, A. and Ramjeesingh, M. (1980) Ann. New York Acad.Sci. 358, 1.

4. Klingenberg, M. (1980) J. Membrane Biol. 56, 97.

5. Klip. A., Grinstein, S. and Semenza, G. (1979a) FEBS Lett. 99, 91.

6. Klip, A., Grinstein, S. and Semenza, G. (1979b) Biochim. Biophys. Acta 558, 233-245.

7. Klip, A., Grinstein, S., Biber,J. and Semenza, G. (1980a) Biochim. Biophys. Acta 589, 100-114.

8. Klip, A., Grinstein, S. and Semenza, G. (1980b) Ann. N.Y. Acad. Sci. 358, 374.

9. Kessler, M. and Semenza, G. (1983) J. Membrane Biol. 75, in press.

10. Toggenburger, G., Häusermann, M., Mütsch, B., Genoni, G., Kessler, M., Weber, F., Hornig, D., O'Neill, B. and Semenza, G. (1981) Biochim. Biophys. Acta 646, 433.

11. Förster, R. (1979) Diplomarbeit ETH Zürich

12. Läuger, P. (1980) J. Membrane Biol. 57, 163.

13. Hopfer, U., Nelson, K., Perrotto, J. and Isselbacher, K.J. (1973) J. Biol. Chem. 248, 25-32

14. Toggenburger, K., Kessler, M., Rothstein, A., Semenza, G. and Tannenbaum, C. (1978) J. Membrane Biol. 40, 269.

15. Toggenburger, G., Kessler, M. and Semenza, G. (1982) Biochim. Biophys. Acta 688, 557.

16. Turner, R.J., and Silverman, M. (1981) J. Membrane Biol. 58, 43.

17. Aronson, P.S. (1978) J. Membrane Biol. 42, 81.

18. Murer, H. and Hopfer, U. (1974) Proc. Natl. Acad. Sci. USA 71, 484.

19. Kaunitz, J.D., Gunther, R. and Wright, E. M. (1982) Proc. Natl. Acad. Sci. USA 79, 2315.

20. Kimmich, G. and Randles, J. (1980) Biochim. Biophys. Acta 596, 439.

21. Heinz, E. (1978) Mechanism and energetics of biological transport., Springer Verlag, Berlin, Heidelberg, New York.

22. Patlak, C.S. (1957) Bull. Math. Biophys. 19, 209.

23. Crane, R.K., Miller, D. and Bihler, I. (1961) In: Membrane Transport and Metabolism (Kleinzeller, A. and Kotyk, A., Eds.) pp. 439-449, Czech. Acad. Sci. Press, Prague.

24. Crane, R.K. (1977) Rev. Physiol. Biochem. Pharmac. 78, 99.

25. Hilden, S. and Sacktor,B. (1982) Am. J. Physiol. 242, F340.

26. Lin, J.T., Stroh, A. and Kinne, R. (1982) Biochim. Biophys. Acta 692, 210.

27. Weber, J. and Semenza, G. (1983) Biochim. Biophys. Acta, in press.

28. Barnett, J.E.G., Jarvis, W.T.S. and Munday, K.A. (1968) Biochem. J. 109, 61.

29. Hosang, M., Vasella, A. and Semenza, G. (1981) Biochemistry 20, 5844.

THE CALMODULIN-SENSITIVE Ca^{2+} ATPase OF PLASMA MEMBRANES. STRUCTURE-FUNCTION RELATIONSHIPS

ERNESTO CARAFOLI

Laboratory of Biochemistry, Swiss Federal Institute of Technology (ETH), ETH-Zentrum, 8092 Zurich (Switzerland)

The Ca^{2+} pumping ATPase is present in all plasma membranes studied so far. Since it has high affinity for Ca^{2+} but rather limited transport velocity, it probably plays the role of fine tuner of intracellular Ca^{2+}. The second Ca^{2+} ejecting system of plasma membranes, the Na^{+}/Ca^{2+} exchanger, cooperates with the ATPase in the homeostasis of cellular Ca^{2+}. Its properties (lower Ca^{2+} affinity, high transport velocity) qualify it as a high-capacity system, particularly in cells where physiology demands periodic increases of Ca^{2+} influx across the plasma membrane. The Na^{+}/Ca^{2+} exchanger has not yet been isolated and purified, and cannot thus be studied molecularly. The Ca^{2+}-ATPase, on the other hand, has been purified to homogeneity in 1979 [1], using a method which exploited its direct interaction with calmodulin. It is a polypeptide of M_r 138,000, which repeats the most important properties of the enzyme in situ (summarized in Table 1).

TABLE 1

THE ISOLATED PLASMA MEMBRANE Ca^{2+} ATPase

M_r	138,000
Affinity for Ca^{2+} (K_m)	< 1 μM
Ca^{2+}/ATP stoichiometry	1 (in the reconstituted system)
Inhibition	vanadate (I_{50} < 1 μM), DIDS ($I_{50} \sim$ 1 μM)
Calmodulin sensitivity	present
Effect of phospholipids	acidic phospholipids activate
Limited proteolysis	activates
Reaction, mechanism	forms acyl-phosphate

One very interesting property of the purified enzyme is its stimulation by acidic phospholipids, polyunsaturated fatty acids, and limited proteolysis [2]. The stimulation produced by these agents is alternative to that induced by calmodulin. Evidently, the active site of the enzyme is made more accessible by all these agents, as indicated schematically in Figure 1.

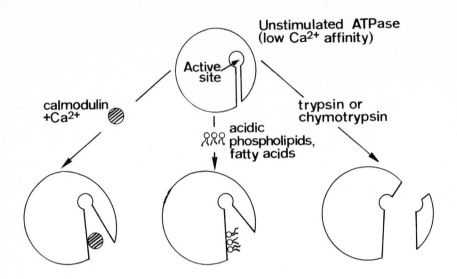

Stimulated ATPase (high Ca²⁺ affinity)

Fig. 1. Activation of the purified Ca^{2+}-ATPase by calmodulin (+ Ca^{2+}), acidic phospholipids or fatty acids, and limited proteolysis (from ref. 3).

It seems reasonable to attribute the activation by limited proteolysis to the removal of an inhibitory polypeptide that masks the active site. No conclusive information is currently available, however, on the mechanism of activation by calmodulin, acidic phospholipids, and fatty acids. A related unsolved problem is whether these three effectors act on the same domain of the ATPase molecule. Studies on the functional significance of the various domains of the enzyme molecule would thus be helpful. In the experiments to be described here, a combination of limited proteolysis and affinity label techniques has been used to provide information on this point.

Figure 2 shows the correlation between the activation of the enzyme by trypsin and the pattern of proteolytic degradation. The pattern is complex, and can be summarized as follows. At initial times, corresponding to a functional state where calmodulin still activates, and where the enzyme is only marginally activated, the molecule is degraded to a), a 90 k polypeptide, which is a transient product, b), to a 81-76 k doublet, in which the component of lower M_r becomes progressively more evident as proteolysis advances, and, c), to limit polypeptides of M_r 14,000 and 33,500. A number of minor splitting products are also visible. At advanced times of proteolysis, corresponding to loss of calmodulin

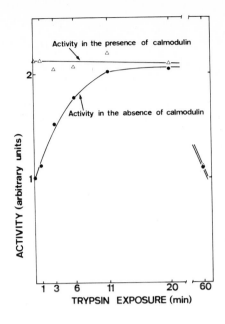

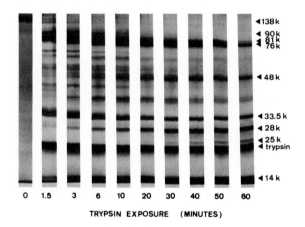

Fig. 2. Activation of the purified Ca^{2+}-ATPase by trypsin, and pattern of proteolytic degradation. The enzyme was purified according to Niggli et al. (1).
50 μl of ATPase, corresponding to about 7.5 μg of protein (3 μg in the zero time
control) were exposed to 1 μg of trypsin (1 μl) at 0° for the times indicated.
The reaction was stopped by addition of 10x concentrated soybean trypsin inhibitor. After measurement of ATPase activity with a coupled enzyme assay, 40 μl
aliquots were submitted to 10% SDS polyacrylamide gel electrophoresis. The gels
were stained with a silver impregnation method (4).

sensitivity, and to full activation of the enzyme, the pattern still shows the limit polypeptides of 14,000 and 33,500 k. The 81,000 - 76,000 k doublet decreases in intensity, in parallel with the appearance of evident bands at 28,000 and 48,000 k. The 90,000 product is no longer visible, and minor bands at M_r 60,000 and 25,000 are also visible. A provisional interpretation of the trypsin proteolysis pattern has been proposed by Zurini et al. (5), and is simplified in the scheme of Figure 3.

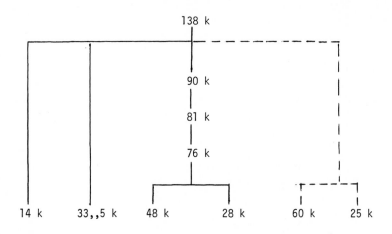

Fig. 3.   An interpretation of the trypsin proteolysis pattern.

The main proteolysis route, which leads to the formation of the most important products discussed in the preceeding paragraphs, is represented by solid lines. A second proteolysis route, probably concerning a minor conformer of the ATPase, is also proposed (dashed lines).

The experiment shown in Figure 4 identifies the fragment(s) where the high affinity site for ATP binding is located. Periodate-oxidized [14]c-ATP, a probe that has been used to label covalently specific ATP-binding site in other ATPases (6-7), binds to the intact enzyme, and, in time sequence, to the 90 k, 76-81 k, and 48 k products.

Thus the high-affinity ATP site is located in the transient 90 k product and then, as proteolysis progresses, in the 76,000 - 81,000 k fragments. Later on, as the latter doublet gives rise to the 48,000 and 28,000 k limit polypeptides, the ATP binding site is associated with 48,000 k fragment.

  Although it seems reasonable to assume that the ATP binding site, and the acyl phosphate site, are located on the same polypeptide, this need not be so. Experiments on the location of the residue responsible for the formation of the acyl-phosphate bond are currently under way.

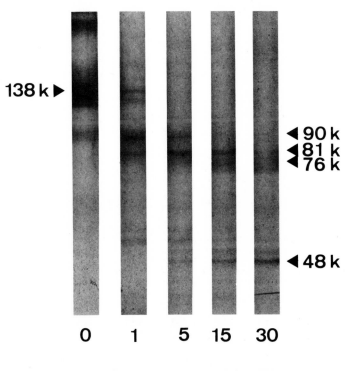

Fig. 4. Labeling of the purified Ca^{2+} ATPase and its trypsin fragments with ^{14}C-periodate-oxidized ATP (O-ATP). Uniformly labeled ^{14}C-ATP was oxidized with periodic acid (6,7). Samples of purified ATPase were incubated with ^{14}C-O-ATP (6-7 μM) for 1hr at 37°, transferred to ice, and submitted to trypsin for the times indicated (see legend for Fig. 2). Electrophoresis was on 7% gels. The gels were dehydrated in 100% dimethylsulfoxide, immersed for 3hr in 20% 2,5 di-phenyloxaxole in dimethylsulfoxide (w/w), washed 1hr with several changes of water, dried, and autiradiographed prior to staining with a silver impregnation procedure (4). Details in ref. 5.

The matter of the domain of the ATPase molecule which interacts with calmodulin has been tackled by passing a sample of trypsinized ATPase on a calmodulin affinity column. Figure 5 shows that only the 90,000 M_r fragment becomes bound to calmodulin in the column, and can be eluted in pure state with EGTA. Thus, it can be concluded that the transient 90,000 M_r product is the (main) calmodulin receptor in the ATPase molecule. Studies with azido-modified, ^{125}I-labeled calmodulin (not shown) have supported this conclusion, and have shown that the ability to interact with calmodulin becomes lost as the 90,000 M_r product is further degraded to the 81,000 M_r polypeptide.

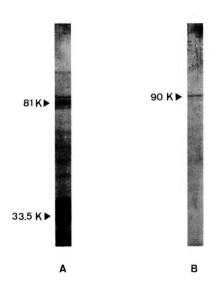

81 K▶

90 K▶

33.5 K▶

A   B

Fig. 5. Isolation of the calmodulin-binding domain of the ATPase on a calmodulin affinity chromatography column. A sample of purified ATPase, trypsinized for 20 min on ice (140 μg protein, 14 μg trypsin) was passed through a calmodulin affinity chromatography column. After several washes with Ca^{2+}-containing buffer (A), the column was eluted with 2 mM EGTA (B). SDS polyacrylamide electrophoresis (7%) and staining as in Fig. 2.

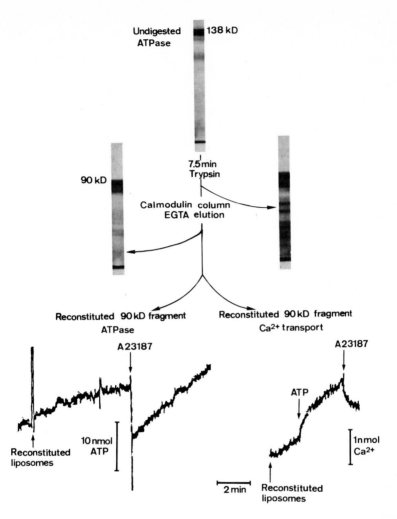

Fig. 6. Reconstitution of the 90,000 M$_r$ fragment of the ATPase in liposomes. A sample of the purified ATPase was treated with trypsin for 7.5 min (see legend for Fig. 2). Aliquots of the proteolyzed mixture were applied to a calmodulin affinity chromatography column. After Ca^{2+} washing of the column, elution with 2 mM EGTA yielded the 90,000 M$_r$, calmodulin binding fragment. (electrophoresis and staining as in Fig. 2). Aliquots of the EGTA eluate were reconstituted in asolectin liposomes (see ref. 2) by the cholate dialysis method. ATPase activity (lower left) was measured with a coupled enzyme assay, in the presence of 10 μM free Ca^{2+}. Ca^{2+} transport (lower right) was measured with a Ca^{2+}-selective electrode.

An interesting development of the work on the 90,000 M_r calmodulin-binding domain has been its reconstitution in liposomes. In the experiment of Figure 6, the fragment has been isolated from a calmodulin column, and reconstituted in asolectin liposomes by the cholate dialysis method. The reconstituted fragment has ATPase activity, which is stimulated by the Ca^{2+} ionophore A23187, indicating coupling to the translocation of Ca^{2+} into the liposome. Direct demonstration of the ability of the 90,000 M_r fragment to mediate the ATP-dependent transmembrane transport of Ca^{2+} is provided by the experiment shown in the lower panel at the right of the Figure, in which the movements of Ca^{2+} have been followed directly with a Ca^{2+}-selective electrode.

REFERENCES

1. Niggli, V., Penniston, J.T. and Carafoli, E. (1979) J. Biol. Chem. 254, 9955.

2. Niggli, V., Adunyah, E.S., Penniston, J.T. and Carafoli E. (1981) J. Biol. Chem. 256, 395.

3. Carafoli, E. and Zurini M. (1982) BBA Reviews in Bioenerg., in press.

4. Merril, C.R., Dunan, M.L. and Goldmann, D. (1981) Anal. Biochem. 110, 201-207.

5. Zurini, M., Krebs, J., Penniston, J.T. and Carafoli, E. (1983) submitted.

THE GLUCOSE TRANSPORTER OF HUMAN ERYTHROCYTES

GUSTAV E. LIENHARD[1], JOCELYN M. BALDWIN[2], STEPHEN A. BALDWIN[3] AND
FRANK R. GORGA[4]

[1]Department of Biochemistry, Dartmouth Medical School, Hanover, NH 03756
(U.S.A.); [2]Present address: Department of Experimental Pathology, School of
Medicine, University College London, London WC1E 6JJ (England); [3]Present
address: Departments of Biochemistry and Chemistry and of Chemical Pathology,
Royal Free Hospital School of Medicine, University of London, London WC1N 1BP
(England); and [4]Present address: Department of Biochemistry and Molecular
Biology, Harvard University, Cambridge, MA 02138 (U.S.A.)

INTRODUCTION

The glucose transport system of human erythrocytes is the paradigm of a
transport system of the facilitated diffusion type. In this article, we will
describe our main findings concerning the purification, structure, function,
and mechanism of this transport system. Several reviews of this topic have
recently appeared (1-3).

PURIFICATION

Assay (4). Our assay for the transporter during purification employs the
compound cytochalasin B. This compound had been shown to be a potent inhibitor
of glucose transport in the erythrocyte; the inhibition is competitive with
respect to glucose under some conditions (5). Moreover, it had been found that
the erythrocyte membrane possesses a set of high-affinity sites for cytochalasin B (dissociation constant, 10^{-7} M) from which it is displaced by the
substrate D-glucose but is not displaced by L-glucose (6,7). Taken together,
these facts indicated that the transporter binds cytochalasin B. In our assay,
we routinely determine the binding of [^{3}H]cytochalasin B to transporter-
specific sites by equilibrium dialysis in the presence of D and L glucose.

One complication in this assay procedure is that detergents, at the concentrations required for the solubilization of the membrane, are very effective
competitive inhibitors of cytochalasin B binding (4,8). Thus, it is necessary
to remove detergent and reconstitute the transporter into a bilayer before
measuring cytochalasin B binding.

Procedure (8). The membrane of the human erythrocyte is a rich source of the
glucose transporter. On the basis of the number of transporter-specific cytochalasin B binding sites and the molecular weight of the transporter (see

below), approximately 3.5% of the membrane protein is transporter.

Kasahara and Hinkle were the first to purify the glucose transporter (9), and our purification method is derived from theirs, with substantial modification to improve yield and activity.

In our procedure, human erythrocyte membranes are first stripped of peripheral proteins by treatment at pH 12. The resulting protein-depleted membranes are then partially solubilized with the detergent octyl glucoside at concentrations of protein and detergent that are critical. Unsolubilized material is removed by sedimentation, and the solubilized proteins and lipids are applied to DEAE-cellulose. The transporter, along with some phospholipid, is not retarded. Octyl glucoside in these fractions is removed by dialysis. The result is a preparation of purified transporter reconstituted into a bilayer of erythrocyte lipids, at a protein to lipid ratio of about 1 to 3 by weight. One unit of blood (400 ml) yields about 20 mg of purified transporter. Various criteria of purity, which will be evident in the subsequent discussion, indicate that the preparation consists of at least 70% of one polypeptide.

STRUCTURE

Composition and terminal analyses (8). We have determined the complete amino acid composition of our preparation. The molecule contains no cystine, 5 cysteine residues, and 7 tryptophan residues per 46,000 molecular weight (see below). Amino terminal analysis by both the dansyl and Edman techniques yielded only methionine. Carboxyl terminal analysis with carboxypeptidase A resulted in 0.65 mole of valine and 0.09 mole of alanine per 46,000 molecular weight upon limit digestion.

The transporter is a glycoprotein (see below). Sogin and Hinkle have reported the carbohydrate composition of the transporter (10). It contains about 15% carbohydrate by weight and is rich in N-acetylglucosamine and galactose.

Polypeptide molecular weight (8). When the transporter was subjected to sodium dodecyl sulfate polyacrylamide gel electrophoresis, a broad band of Coomassie blue stain in the molecular weight region from 75,000 to 45,000 was seen; and in addition, there was a lesser amount of stain extending from the 75,000 to the very high molecular weight region. When the dodecyl sulfate was replaced by a mixture of 12, 14, and 16 carbon alkyl sulfates, the band at 75,000 to 45,000 was more intense and the higher molecular weight materials were not seen. Presumably aggregates of the transporter are disrupted by the more hydrophobic 14 and 16 carbon alkyl sulfates.

The broadness of the transporter band is largely due to heterogeneity in the carbohydrate linked to a single type of polypeptide (see below, 11). Treatment of the transporter with endo-β-galactosidase removed a considerable fraction of the carbohydrate and led to substantial sharpening of the polypeptide band seen upon alkyl sulfate electrophoresis. After glycosidase treatment the mobility of the polypeptide upon 10% polyacrylamide gels corresponds to a molecular weight of 46,000. However, this value can only be considered an estimate of the true molecular weight, since the mobility of the polypeptide relative to standard water-soluble proteins varies with the percentage acrylamide in the gels.

Carbohydrate (11). The oligosaccharides of the transporter are probably entirely located on the extracellular domain of the protein. We have labeled extracellular carbohydrate by treating the intact erythrocytes with the membrane-impermeant reagent galactose oxidase, followed by reduction with tritiated borohydride. The transporter purified from these erythrocytes was found to be labeled.

In addition to the results of gel electrophoresis described above, the heterogeneity in the carbohydrate of the transporter is indicated by two other findings. First, when the transporter purified from carbohydrate-labeled erythrocytes was subjected to electrophoresis, the peak of radiolabel fell on the low mobility side of the peak of Coomassie blue stain. Second, only about half of the cytochalasin B binding sites were retained upon passage of transporter solubilized in detergent over Ricinus communis agglutinin I-agarose.

The composition of the carbohydrate and its susceptibility to endo-β-galactosidase suggests that it has largely the erythroglycan-type structure (12,13). Consequently, the transporter is probably one of the erythrocyte proteins that carries the ABH determinants. In fact, radiolabeled lectins against these determinants have been found to bind to gels of erythrocyte membrane proteins in the region where the transporter migrates (14).

Protease sensitivity (15). We have examined the susceptibility of the extracellular and cytoplasmic domains of the glucose transporter to trypsin. Treatment of intact erythrocytes had no effect upon either the transport activity, as measured by the rate of sorbose entry, nor upon the binding of cytochalasin B. On the other hand, treatment of inside-out vesicles, which are vesicles of erythrocyte membranes with the cytoplasmic surface facing outward, resulted in the loss of transport activity and of the high affinity, transporter-specific cytochalasin B binding sites. When the purified transporter was treated with trypsin, the decrease in high affinity cytochalasin B binding sites occurred in direct proportion to the disappearance of the transporter band seen upon sodium dodecyl sulfate polyacrylamide gel electrophoresis.

Langdon and his associates have suggested that in the intact erythrocyte
the transporter exists as a 100,000 dalton polypeptide and that the 46,000
dalton polypeptide is the product of a proteolytic cleavage occurring during
the purification (16). Three findings indicate that this hypothesis is incor-
rect. First, antibodies raised against the purified transporter do not react
with polypeptides of higher molecular weight in the erythrocyte membrane
(17,18). Second, photoaffinity labeling of the transporter polypeptide in
intact erythrocytes with [3H]cytochalasin B tags the 46,000 dalton polypeptide
rather than a 100,000 dalton one (19). Third, when the purification was
carried out from freshly drawn cells in the presence of protease inhibitors,
the 46,000 dalton species was still obtained (8).

Conclusion. The glucose transporter is a heterogeneously glycosylated glycoprotein with an apparent molecular weight of 46,000 for its polypeptide
portion. It possesses a carbohydrate-bearing extracellular domain and a trypsin-sensitive cytoplasmic domain, and must therefore be a transmembrane protein.

FUNCTION

Cytochalasin B binding (8). The stoichiometry for the binding of cytochalasin B to our most active preparations of transporter is 0.70 cytochalasin
B molecule per 46,000 dalton polypeptide. The transporter undergoes partial
denaturation while it is in octyl glucoside during purification; the addition
of phospholipid slows but does not completely prevent this process. We estimate that in the absence of any denaturation the stoichiometry would be 0.8.
Thus, although these values may be slightly in error due to our lack of knowledge of the true molecular weight, they suggest that each functional copy of
the transporter polypeptide binds a molecule of cytochalasin B.

Sugar binding (20). The binding of D-glucose and other monosaccharides to
the transporter cannot be measured directly by equilibrium dialysis because
these sugars have a relatively low affinity. One way in which the dissociation
constants for various sugars has been determined is through their inhibition of
the binding of [3H]cytochalasin B (see below). In addition, we have discovered
that some ligands cause about a 10% decrease in the intrinsic fluorescence of
the purified protein. D-glucose, 4,6-O-ethylidene-D-glucose (ethylidene glucose), as well as cytochalasin B, quench the fluorescence, whereas n-propyl-β-
D-glucopyranoside (propyl glucoside) does not. Dissociation constants have
been determined from the dependence of the fluorescence changes upon ligand
concentration, and the values are the same as those obtained from equilibrium

dialysis with [³H]cytochalasin B. The value of the dissociation constant for D-glucose is about 45 mM. The fact that the quenching of fluorescence is greatest at longer wavelengths suggests that an exposed tryptophan residue, possibly located at the active site, is perturbed.

Transport (21). In order to examine the transport activity of the purified transporter, we first incorporated it into homogeneous unilamellar vesicles. The transporter preparation, along with sufficient dioleoyl phosphatidylcholine to give a lipid to protein ratio in the range of 30 to 90 by weight, was solubilized with cholate. Vesicles were then formed through removal of the cholate by gel filtration. Examination of the vesicle preparations by gel filtration on agarose showed that they consisted of a relatively uniform population of about 30 nm diameter. The orientation of the transporter, as ascertained by susceptibility of cytochalasin B binding to trypsin, was at least 75% with the cytoplasmic portion external (15).

These vesicles exhibited transporter activity according to two criteria. First, the rate of entry of [¹⁴C]D-glucose was much more rapid than that of [³H]L-glucose. Second, cytochalasin B inhibited the uptake of D, but not L, glucose.

The large ratio of lipid to protein and the small size of the vesicles resulted in an average distribution of about one transporter protein per vesicle. Consequently, the number of transporters functional in transport could be calculated from the value of the intravesicular volume that rapidly equilibrated with D-glucose in the medium and the value of the internal volume for a 30 nm vesicle. The Poisson distribution was used to correct for the occurrence of more than one transporter in some vesicles. By this procedure, it was found that each transporter molecule capable of binding cytochalasin B is also functional in transport.

The transport activity of the purified transporter was determined by measuring the rate of exchange of D-glucose into vesicles equilibrated with 3 mM glucose at 0°. This rate, expressed per cytochalasin B binding site, is about 5% of that of the transporter in the intact erythrocyte under the same conditions. It should be emphasized that based on the results described above, this reduced activity must be due to 100% of the transporter molecules functioning at 5% of the in vivo rate, rather than 5% of the molecules functioning at 100% of the in vivo rate. The reason(s) for the lowered transport activity of the purified protein remain(s) to be elucidated. One explanation is that neither the composition nor the arrangement of the lipid is the same in the reconstituted system as in the erythrocyte.

330

MECHANISM

The simplest mechanism for the operation of the glucose transporter that is consistent with the major findings concerning the system is an alternating conformation model (Fig. 1) (22,23).

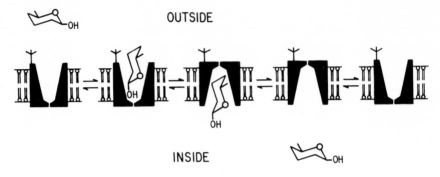

OUTSIDE

INSIDE

Fig. 1. Schematic representation of the alternating conformation model for glucose transport.

In this model, the transporter is a transmembrane protein that is largely fixed in its orientation with respect to the plane of the bilayer. Within the confines of this overall orientation, the protein alternates between two conformational states, one in which the substrate binding site faces outward and one in which it faces inward. The interconversion between these two states is envisioned as occurring by a change in the architecture of the substrate binding site of fixed structure. The complete catalytic cycle for net glucose entry is thus: binding of glucose to the conformer with the outward-facing substrate site, conformational change of the occupied transporter to yield the conformer with the inward-facing substrate site, dissociation of the bound glucose inside the cell, and conformational change of the unoccupied transporter to yield the conformer with the outward-facing substrate site.

This model for transport adequately accounts for many of the results of steady-state kinetic studies performed in the intact erythrocyte. Most notably it explains the finding that the values of V_{max} and K_m for glucose entry into cells free of glucose do not equal the values of V_{max} and K_m for glucose exit from cells into medium free of glucose and the finding that the value of V_{max} for the exchange of glucose between medium and cell is larger than the values of V_{max} for either entry or exit (24,25). Some discrepancies between certain observed and predicted kinetic parameters may exist (25), but it seems likely that these are due to the facts that D-glucose is a mixture of two substrates (40% α and 60% β anomer) and that the metabolic state of the cell affects the rate of transport (26,27).

Another feature of the model, the difference between the structure of the substrate binding site in the two conformers, is supported by studies with glucose derivatives that are nontransported, competitive inhibitors of transport (28). Derivatives of glucose with groups attached to the carbon 4 to 6 region, such as 4,6-0-ethylidene glucose (29), are much more potent inhibitors of transport when present outside the erythrocyte than when located within the cells. Conversely, glucose derivatives with groups attached to carbon 1, such as propyl glucoside, are much more effective at inhibition when within the cell than when outside it. Thus, it appears that the outward-facing substrate binding site interacts most strongly with the carbon 1 region of glucose, whereas the inward-facing site interacts most strongly with the carbon 4 to 6 region (see Fig. 1). These data suggest a conformational change in which the binding site for one end of glucose comes apart as the binding site for the other end of glucose simultaneously forms.

Our contribution in support of this mechanism has been a test of another of its features, the existence at any moment of only a single substrate binding site (30). We have found that both ethylidene glucose, which preferentially by a factor of ten forms a complex with the outward-facing substrate site, and propyl glucoside, which preferentially by a factor of ten forms a complex with the inward-facing site, inhibit the binding of [3H]cytochalasin B to the glucose transporter in competitive fashion. When both glucose derivatives were present at the same time, the observed extent of inhibition of cytochalasin B binding was equal to that calculated for the case in which only one derivative can bind at one time, rather than that calculated for the case in which the two derivatives bind simultaneously. Additional support for the concept of a single sugar binding site at any moment is our finding that the inhibitory effect of D-glucose upon cytochalasin B binding, expressed as the ratio of free to bound cytochalasin B, is proportional to the first power of the glucose concentration at concentrations up to 5 times the dissociation constant. If two glucose molecules could bind simultaneously, a dependence upon the square of the glucose concentration would have been seen.

There is evidence indicating that the binding site for cytochalasin B overlaps with the inward-facing substrate site (5). If such is the case, our finding that ethylidene glucose and D-glucose, which can bind to the outward-facing substrate site, are strictly competitive inhibitors of cytochalasin B binding reinforces the conclusion that only one site exists at any time. Moreover, neither ethylidene glucose nor glucose increased the rate of dissociation of cytochalasin B bound to the transporter (30), and thus by this criterion there is no indication of a weak interaction between these sugars and the cytochalasin B-transporter complex.

ACKNOWLEDGEMENTS

Our research has been supported by a grant (GM22996) from the National Institutes of Health.

REFERENCES

1. Baldwin, S.A. and Lienhard, G.E. (1981) Trends Biochem. Sci., 6, 208.

2. Jones, M.N. and Nickson, J.K. (1981) Biochim. Biophys. Acta, 650, 1.

3. Wheeler, T.J. and Hinkle, P.C. (1982) in: Martinosi, A.M. (Ed.), Membranes and Transport, Volume 2, Plenum Press, New York, p. 161.

4. Zoccoli, M.A., Baldwin, S.A. and Lienhard, G.E. (1978) J. Biol. Chem., 256, 6923.

5. Devés, R. and Krupka, R.M. (1978) Biochim. Biophys, Acta, 510, 339.

6. Taverna, R.D. and Langdon, R.G. (1973) Biochim. Biophys. Acta, 323, 207.

7. Lin, S. and Spudich, J.A. (1974) J. Biol. Chem., 249, 5778.

8. Baldwin, S.A., Baldwin, J.M. and Lienhard, G.E. (1982) Biochemistry, 21, 3836.

9. Kasahara, M. and Hinkle, P.C. (1977) J. Biol. Chem., 252, 7384.

10. Sogin, D.C. and Hinkle, P.C. (1978) J. Supramol. Struct., 8, 447.

11. Gorga, F.R., Baldwin, S.A. and Lienhard, G.E. (1979) Biochem. Biophys. Res. Commun., 91, 955.

12. Jarnefelt, J., Rush, J., Li, Y. and Laine, R.A. (1978) J. Biol. Chem., 253, 8006.

13. Krusius, T., Finne, J. and Rauvala, H. (1978) Eur. J. Biochem., 92, 289.

14. Finne, J. (1980) Eur. J. Biochem., 104, 181.

15. Baldwin, S.A., Lienhard, G.E. and Baldwin, J.M. (1980) Biochim. Biophys. Acta, 599, 699.

16. Shelton, Jr., R.L. and Langdon, R.G. (1983) Fed. Proc., 42, 1941.

17. Baldwin, S.A. and Lienhard, G.E. (1980) Biochem. Biophys. Res. Commun., 94, 1401.

18. Sogin, D.C. and Hinkle, P.C. (1980) Proc. Natl. Acad. Sci. USA, 77, 5725.

19. Carter-Su, C., Pessin, J.E., Mora, R., Gitomer, W. and Czech, M.P. (1982) J. Biol. Chem., 257, 5419.

20. Gorga, F.R. and Lienhard, G.E. (1982) Biochemistry, 21, 1905.

21. Baldwin, J.M., Gorga, J.C. and Lienhard, G.E. (1981) J. Biol. Chem., 256, 3685.

22. Patlak, C.S. (1957) Bull. Math. Biophys., 19, 209.

23. Vidaver, G.A. (1966) J. Theor. Biol., 10, 301.

24. Geck, P. (1971) Biochim. Biophys. Acta, 241, 462.

25. Naftalin, R.J. and Holman, G.D. (1977) in: Ellory, J.E. and Lew, V.L. (Eds.), Membrane Transport in Red Cells, Academic Press, New York, p. 257.

26. Jacquez, J.A. (1983) Biochim. Biophys. Acta, 727, 367.

27. Weiser, M.B., Razin, M.B. and Stein, W.D. (1983) Biochim. Biophys. Acta, 727, 379.

28. Barnett, J.E.G., Holman, G.D., Chalkley, R.A. and Munday, K.A. (1975) Biochem. J., 145, 417.

29. Baker, G.F., Basketter, D.A. and Widdas, W.F. (1978) J. Physiol., 278, 377.

30. Gorga, F.R. and Lienhard, G.E. (1981) Biochemistry, 20, 5108.

DETERGENT SOLUBILIZATION OF MITOCHONDRIAL SUCCINATE DEHYDROGENASE AND THE
ROLE OF BILAYER COMPONENTS IN THE REGULATION OF ENZYME ACTIVITY

M.C. BARBERO, E. RIAL, J.I.G. GURTUBAY, F.M. GOÑI AND J.M. MACARULLA
Departments of Biochemistry and Biology, Faculty of Science, P.O. Box. 644,
Bilbao (Spain)

INTRODUCTION

The present study deals with the effect of Triton X-100 on rat liver mitochondrial succinate dehydrogenase, and the influence of coenzyme Q and phospho
lipids on this enzyme activity. Succinate dehydrogenase is an integral protein
of the mitochondrial inner membrane (1); it is one of the components of the
so-called complex II. Its regulation seems to depend on many factors, both
coenzyme Q and membrane fluidity being involved (2). Triton X-100 was chosen
because of its ability for solubilizing membranes without impairment of enzyme
activities; this surfactant is often used in reconstitution studies.

MATERIALS AND METHODS

Rat liver mitochondria were isolated (3) and resuspended in 0.25 M sucrose,
0.02 M HCl-Tris, pH 7.4 buffer to a final concentration of 1 mg protein/ml.
Aliquots of this suspension were treated with the required amounts of Triton
X-100 in order to obtain final detergent concentrations ranging from 5×10^{-5}
to 5×10^{-3}M, and incubated at room temperature for 1 min before running the
assays. When required, the membrane suspensions were fractionated by centrifugation at 150000 xg for 1 h at 4°C. The supernatants from this centrifugation
were considered to contain the solubilized mitochondrial fraction. Proteins
were determined as in (3). Succinate dehydrogenase was assayed at 37°C with
phenazine methosulphate and dichlorophenolindophenol according to (4). The
enzyme (50 µl protein suspension) was added to a 3 ml cuvette containing assay
buffer (0.75 ml 0.2 M phosphate buffer, pH 7.4; 0.3 ml 0.2 M Na succinate;
30 µl 10 mM EDTA; 0.30 ml 1% bovine seroalbumin; 1.5 ml distilled water) and
incubated for 5 min; the reaction was started by adding simultaneously 0.10 ml
50 mM phenazine methosulphate and 50 µl 4.2 mM dichlorophenolindophenol. Changes
in absorbance at 600 nm were continously monitored by means of a chart recorder.
When required, enzyme assays were conducted in the presence of excess quinone;
10 µl of a duroquinone solution (12 mg/ml in ethanol) were included in the
cuvette prior to enzyme addition.

RESULTS AND DISCUSSION

In intact rat liver mitochondria, succinate dehydrogenase activity is 96.9 ± 5.6 U/mg protein (n = 7). Detergent concentrations below 1 mM cause an inhibition of enzyme activity; this inhibition is at least partially reversed by higher surfactant concentrations (Fig. 1a). We propose that the addition of Triton X-100 has a two-fold effect since it may solubilize and remove CoQ from the bilayer, thus inhibiting the enzyme, but it may also solubilize the enzyme, thus relieving any inhibitions due to lipid microviscosity.

We have shown previously that Triton X-100 solubilizes CoQ from mitochondrial membranes (3); the corresponding data have been replotted in Fig. 1b (dotted lines). At 3 mM surfactant, about 25% of the total CoQ has been solubilized, and the observed inhibition of succinate dehydrogenase is of the same order of magnitude. In addition, when detergent-treated mitochondria are assayed in the presence of excess quinone (Fig. 1a, dotted lines) no detergent-induced inhibition of the enzyme is seen. We conclude that coenzyme Q is an activator of succinate dehydrogenase, and that the inhibitory effect of 1 mM Triton X-100 is at least partially due to solubilization of this coenzyme.

Enzyme solubilization, as judged from the percent recovery of total enzyme activity in surfactants, after centrifugation of the detergent-treated mitochondria, occurs mainly at or above 1 mM Triton X-100 (Fig. 1c). Thus, enzyme solubilization takes place concomitantly with the reactivation of the detergent--inhibited enzyme (Fig. 1a). This supports the idea that Triton X-100 may activate succinate dehydrogenase by relieving restrictions imposed by the lipid bilayer.

Virtually all the succinate dehydrogenase activity is solubilized at detergent concentrations above 0.5 mM (Fig. 1c), i.e. when the bulk of mitochondrial membrane proteins has already been solubilized (Fig. 1b, solid line). This observation may be useful for the purification of the enzyme, and is consistent with the rather selective action of Triton X-100 that has been previously described by us (3), as well as by other authors.

It is interesting to note that the total activity of the 2.5 mM supernatant is about 1.5 times that of the total mitochondrial suspension: a "latent" activity has become apparent. This means that some activation process takes place during the centrifugation step. The precise nature of this activation phenomenon is not known. Succinate dehydrogenase is a dimeric enzyme composed of one catalytic subunit of molecular weight 70000, and another subunit of molecular weight 30000 (5); centrifugation may lead to a preferential solubilization of one of these subunits. On the other hand, succinate dehydrogenase is subjected to a

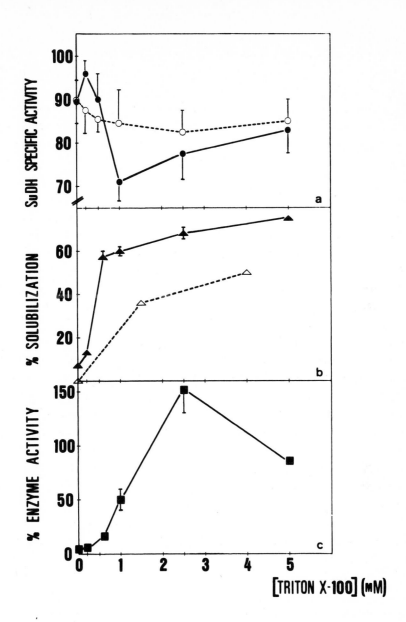

Fig. 1. (a) Effect of Triton X-100 on rat liver succinate dehydrogenase activity in the presence (○) or absence (●) of an excess duroquinone.
(b) Percent solubilized protein (▲) and ubiquinone (△) measured as described in (3).
(c) Percent enzyme activity recovered in supernatants after centrifugation of detergent-treated mitochondrial suspension. 100% = total enzyme activity (U/ml) at each detergent concentration.
   Results are means ± S.E.M. of 6 experiments.

complicated regulatory mechanism (6). It can exist under an inactive form, stabilized by binding of oxalacetate, and an active form which is stable in the presence of succinate and other compounds. A transition between the active and the inactive forms may occur during the centrifugation and thus explain the observed activation.

Another important point in this respect is the role of specific lipids in the activity of succinate dehydrogenase. The protein is linked to, and partially buried in the mitochondrial inner membrane, and a specific role has been sugges ted for cardiolipin in the assembly of the enzyme (1). However, this kind of specific requirement of phospholipids is debatable (7), and the fact that partial substitution of phospholipids by the surfactant actually enhances the enzyme activity does not speak in favour of any specific function of phospholipids.

ACKNOWLEDGMENTS

M.C.B. acknowledges financial help from the Basque Government and from "Patronato de la Universidad del País Vasco". This work was supported in part with funds from CAICYT.

REFERENCES

1. Merli, A., Capaldi, R.A., Ackrell, B.A.C. and Kearney, E.B. (1979) Biochemistry, 18, 1393.

2. Ackrell, B.A.C., Coles, C.J. and Singer, T.P. (1977) FEBS Lett. 75, 249.

3. Gurtubay, J.I.G., Goñi, F.M., Gómez-Fernández, J.C., Otamendi, J.J. and Macarulla, J.M. (1980) J. Bioenerg. Biomemb. 12, 47.

4. Baginski, M.L. and Hatefi, Y. (1969) J. Biol. Chem. 244, 5313.

5. Kennedy, W.C., Mowery, P.C., Seng, R.L. and Singer, T.P. (1976) J. Biol. Chem. 251, 2369.

6. Hatefi, Y. (1976) Enzymes Biol. Membr. 4, 1.

7. Chapman, D., Gómez-Fernández, J.C. and Goñi, F.M. (1982) Trends. Biochem. Sci. 7, 67.

ROLE OF CARDIOLIPIN ON THE PURIFICATION OF RECONSTITUTIVELY ACTIVE MITOCHONDRIAL PHOSPHATE CARRIER.

F. BISACCIA, M. TOMMASINO and F. PALMIERI
Institute of Biochemistry, Faculty of Pharmacy, University of Bari,
70126 Bari, Italy

INTRODUCTION

The basic procedure for purifying the mitochondrial Pi carrier in active form and in large amounts includes solubilization of the mitochondria in non-ionic detergents and chromatography on hydroxylapatite (HTP) (1-3). The HTP eluate contains at least 4 bands. Band 4 of 30000 Mr represents the adenine nucleotide carrier and another protein of unknown function (4). It is not yet known, on the other hand, whether only band 2 corresponds to the Pi carrier (5) or whether bands 2 and 3 are proteolytic fragments of the true Pi carrier corresponding to band 1 (4). Recently it was shown that the reconstituted Pi transport activity of the HTP eluate can be increased several fold by inclusion of cardiolipin (DPG) in the solubilization buffer (6-8). The role of cardiolipin on the purification of reconstitutively active Pi carrier has been reinvestigated by measuring the Pi transport activity in both mitochondrial extracts and HTP eluates.

The present paper shows that cardiolipin has several effects: it protects the Pi carrier against inactivation by Triton, it decreases the solubilization of mitochondrial protein and of Pi carrier, it activates the phosphate carrier when added to mitochondrial extracts, and it is needed for elution of the Pi carrier from HTP. The latter effect is of primary importance for the preparation of highly active purified Pi carrier.

METHODS

Pig heart mitochondria were solubilized as described in (8) with a buffer containing (if not otherwise stated) 2% Triton X-114 and 2 mg/ml cardiolipin. The unsolubilized material was sedimented (147000 x g, 40 min).

Abbreviations: DPG, cardiolipin; HTP, hydroxylapatite; Pi, phosphate; EYPL, egg yolk phospholipids; NEM, N-ethylmaleimide.

The extract was chromatographed on HTP columns (pasteur pipettes containing 0.6 g of dry material), eluted with the same buffer used for solubilization with or without the addition of DPG.

For reconstitution of the Pi carrier activity in liposomes, mitochondrial extracts or HTP eluates were mixed with liposomes (prepared from 80% EYPL and 20% mitochondrial phospholipids in the presence of 10 mM Pi), frozen in dry ice acetone, thawed and sonicated (2).

The Pi transport, measured as Pi/Pi exchange, was initiated by adding ^{32}Pi to the reconstituted liposomes and terminated with the addition of NEM. In the controls NEM was added before the radioactivity.

SDS-polyacrylamide gel electrophoresis (SDS-PAGE) was performed as previously described (5).

RESULTS

Effect of DPG on the solubilization of the Pi carrier.

In previous studies (6-8), the mitochondria were solubilized in the presence and in the absence of DPG, the extracts were applied to HTP columns and the activity of the reconstituted Pi carrier was measured only in the HTP eluates. In preliminary experiments of this study, the importance of the inclusion of DPG in the solubilization buffer was examined by measuring the activity in the mitochondrial extracts.

In the expt. illustrated in Fig.1 mitochondria were solubilized with increasing concentrations of Triton in the presence and absence of DPG. The amount of protein solubilized and the activity of the reconstituted Pi carrier were measured in each extract. The activity was also measured after the addition of DPG to the extracts obtained without DPG. In the absence of DPG, in agreement with the previous measurements in the eluates from HTP (6-8), the activity of the reconstituted Pi carrier exhibits a maximum at about 2% Triton and then drastically decreases at higher detergent concentrations. Inclusion of DPG in the solubilization buffer decreases the solubilization of mitochondrial protein (Fig.1). The solubilization of the Pi carrier is also decreased by the presence of DPG, as revealed by high resolution SDS-PAGE of the mitochondrial extracts (not shown). Furthermore, the presence of DPG during

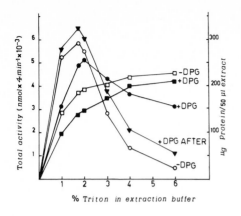

Fig.1 – Total activity of reconstituted Pi carrier in Triton extracts. Mitochondria were solubilized with the indicated Triton concentrations in the presence (+DPG) or in the absence of DPG (-DPG). Where indicated (+ DPG after), DPG was added to the extracts prepared in the absence of DPG. ●,○ and ▼ :total activity; ■ and □ : solubilized protein.

solubilization decreases the total activity of the Pi carrier at Triton concentrations lower than 2%. At higher detergent concentrations, on the other hand, the total activity is higher in the presence of DPG during solubilization than in its absence. When mitochondria are solubilized with 2% Triton (the concentration used in all the other experiments described in this paper), inclusion of DPG in the solubilization buffer has virtually no effect on the activity of the Pi carrier measured in the mitochondrial extract. These results show that during solubilization DPG has two effects: a) it decreases the solubilization of mitochondrial protein and of Pi carrier and b) it protects the Pi carrier against inactivation by Triton. The former effect prevails at lower Triton concentrations, the latter at higher concentrations. At about 2% Triton and 2 mg DPG/ml the two effects annull each other and the total activity is virtually the same. When DPG is added to Triton extracts (prepared in the absence of DPG), the solubilized Pi carrier is activated (Fig.1). This effect, probably due to replacement of some Triton bound to the carrier protein by DPG, supports the idea of a specific requirement of the Pi carrier for cardiolipin (6). The presence of DPG during solubilization has no effect on the stability of the Pi carrier in 2% Triton extract. Thus the same rate of inactivation at 25°C is observed with mitochondria solubilized in the presence and in the absence of DPG.

Effect of DPG on the purification of the Pi carrier by HTP.

Due to the lack of a significant influence of DPG on the total activity and on the stability of the Pi carrier in 2% Triton extracts, the effect of DPG was tested on the purification of the Pi carrier on HTP. It was found that much more activity is released from HTP when the mitochondrial extract is supplemented with DPG, i.e. when the elution buffer contains DPG. This result can be interpreted to show that in the absence of DPG the Pi carrier is eluted in an inactivated form or, alternatively, is retained in the column. In order to verify which of these two interpretations is correct the experiment illustrated in Fig.2 was carried out. Mitochondria solubilized with 2% Triton in the absence of DPG were applied to an HTP column and the fractions analyzed for total activity, specific activity and protein content. It can be seen that the fractions collected before the addition of DPG to the column contain a small amount of carrier. This is largely eluted after DPG addition, as shown by the

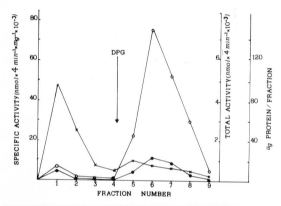

Fig.2 - Elution of Pi carrier from the HTP column before and after the addition of DPG. Where indicated 2 mg/ml DPG was added to the elution buffer. Mitochondria were solubilized at a concentration of 10 mg protein/ml in the absence of DPG. (●), total activity of reconstituted Pi carrier; (o), specific activity; (x) protein content/fraction.

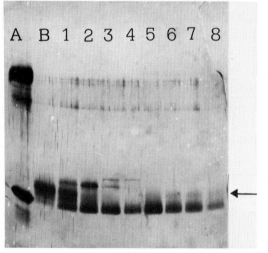

Fig.3 - SDS-PAGE of the fractions obtained from the HTP column of Fig.2 (silver staining): (A) markers, (B) mitochondria solubilized in the presence of DPG (1st fraction) used as a control, (1)-(8): fractions 1-8 of Fig.2.

increase in both the total and the specific activity. These results clearly demonstrate that DPG is required for the elution of the Pi carrier from HTP. The gel electrophoretic analysis of the various fractions from the HTP column of Fig.2 (Fig.3) allows to identify which of the usual 4-5 proteins eluted from HTP is responsible for the Pi carrier activity. Fractions 1 and 2 contain a band (marked with an arrow), which is absent in fractions 3 and 4 i.e. before the addition of DPG. This band appears in fraction 5 (after the addition of DPG) and increases in fractions 6 and 7 to decrease again in fraction 8. Since the presence of this protein (which we call band 2 since band 1 is sometimes resolved in 2 components as in the present expt.) corresponds well to the activity, it can be concluded that the Pi carrier is associated to a single protein (band 2), the others being completely unrelated to Pi transport. The assumption therefore that bands 2 and 3 present in the HTP eluate are proteolytic fragments of the protein with the highest MW thought to be the Pi carrier (4) is in contrast with the present data.

Since mitochondria contain cardiolipin, it is to be expected that when concentrated mitochondria are solubilized, the cardiolipin present in the extract is sufficient to cause elution of the Pi carrier from HTP. To check this hypothesis, different concentrations of mitochondria were solubilized in the absence of DPG. The extracts, either without or with the addition of DPG, were applied to HTP columns and the eluates were analyzed (Fig.4). When 15 mg mitochondrial protein/ml of solubilization buffer were used, the activity of Pi transport is nearly the same with or without the addition of DPG. With 20 mg protein/ml (not shown), there was no difference at all. In contrast, when less than 10 mg protein/ml were used, the Pi carrier activity is only significant if DPG is added. It is likely that at high protein concentrations the endogenous DPG is sufficient to elute the carrier from HTP, whereas at low protein concentrations the elution of the Pi carrier, and therefore the activity in the eluates, is completely dependent on the addition of DPG. This conclusion is supported by the analysis of the SDS-PAGE of this expt. (Fig.5). In the absence of DPG, band 2 is absent at 3 mg protein/ml, is present in a very small amount at 6 mg/ml and also (relative to the other proteins) at 9 mg/ml. In contrast, in the presence of DPG, band 2 is already present at

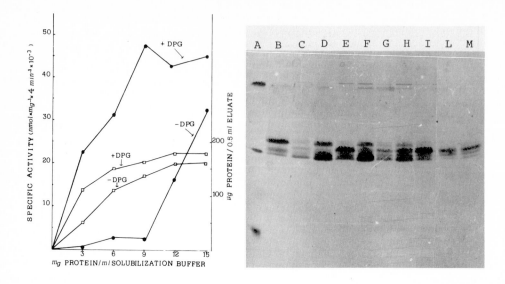

Fig.4 (left) - Dependence of the Pi carrier activity in the HTP eluate on the mitochondrial protein concentrations used during solubilization. The indicated concentrations of mitochondria were solubilized in the absence of DPG. The extracts, without or with the addition of DPG (2 mg/ml), were applied to HTP columns. (●), specific activity of reconstituted Pi carrier; (□), protein content/0.5 ml eluate.

Fig.5 (right) - SDS-PAGE of the eluates obtained from the HTP columns of Fig.4 (Coomassie blue staining): (A) markers; (B) and (C) 3 mg protein/ml of solubilization buffer; (D) and (E) 6 mg/ml; (F) and (G) 9 mg/ml; (H) and (I) 12 mg/ml; (L) and (M) 15 mg/ml. (B), (D), (F), (H), (L) without DPG; (C), (E), (G), (I), (M) with DPG added to the extracts applied to the column.

3 mg/ml. At 15 mg/ml the amount of band 2 is approx. the same in the absence and in the presence of DPG. Also in this expt. (cf. Fig.4 and Fig.5) the presence and the relative amount of band 2 corresponds to the activity of the Pi carrier. (Incidentally, it should be noted that the protein with the highest Mr, approx. 35K, a) is present in the HTP eluate before the addition of DPG to the column and disappears after DPG addition (Fig.3), b) at low protein concentrations in the extraction medium, is absent if DPG is added to the column and vice versa present if DPG is not added (Fig.5), and c) at high protein concentrations in the extraction medium, is absent both with and without the addition of DPG (Fig.5). As shown in the accompanying paper (9), the 35K protein - but not the Pi carrier - is labeled by low concentrations of dicyclohexylcarbodiimide.)

TABLE I

EFFECT OF VARIOUS PHOSPHOLIPIDS ON THE ELUTION OF THE Pi CARRIER FROM HTP

Addition to the extract applied to HTP (2 mg/ml)	Protein in HTP eluate (μg)	Total Activity (nmol x 4 min^{-1})	Specific Activity (nmol x 4 min^{-1} x mg^{-1})
-	140	2340	16700
DPG	170	9520	56000
PE	50	650	13000
DPPC	140	2800	20000
DOPC	140	3640	26000
LPC	200	7000	35000
PA	80	350	4400

Abbreviations: DPG, cardiolipin; PE, phosphatidylethanolamine; DPPC, dipalmitoyl phosphatidylcholine; DOPC, dioleyl phosphatidylcholine; LPC, lysophosphatidylcholine; PA, phosphatidic acid.

Effect of various phospholipids on the elution of the Pi carrier from HTP.

The results reported in Table I show that more carrier is eluted with DPG than with the other phospholipids tested. Thus the addition of DPG to the 2% Triton extract applied to HTP increases both the total and the specific activity about 4 times. In contrast, DPPC and DOPC increase the total activity in the eluate 1.2 and 1.5 times respectively. With PE and PA less carrier than in the control is eluted (the proteins eluted are also less than in the control). In addition to DPG, only LPC increases significantly the elution of the Pi carrier from HTP, although considerably less than DPG.

DISCUSSION

It has been suggested that the mitochondrial Pi carrier has an essential requirement for DPG (6). The results obtained with the mitochondrial extracts show that the inhibition of the Pi carrier by Triton can be prevented or partly released by DPG (Fig.1). It is likely that Triton inhibits the Pi carrier by removing the endogenous DPG from the carrier protein. It can be expected therefore that the inclusion of DPG in the solubilization buffer or its addition to the mitochondrial extracts increases the activity of the Pi carrier by preventing the removal of DPG or by substituting for some bound Triton.

Besides these effects on the activity of the Pi carrier, DPG is essential for the purification of the Pi carrier protein since it is absolutely required for its elution from HTP (Fig.2). Clearly the DPG-carrier interaction prevents the carrier protein from binding to the HTP column or detaches it from HTP if already bound. One can assume that the binding of DPG to the carrier neutralizes positive charges making the protein more hydrophobic. It can also be that the binding of DPG induces a transition of the carrier molecule from an inactive state (that binds to HTP) to an active one (unable to bind to HTP). In this case all the observed effects of DPG on the Pi carrier (protection against Triton inhibition, activation, and elution from HTP) might be the consequence of a single type of DPG-carrier interaction.

REFERENCES

1. Wohlrab, H. (1980) J. Biol. Chem. 255, 8170-8173.
2. Kolbe, H.V.J., Böttrich, J., Genchi, G., Palmieri, F. and Kadenbach, B. (1981) FEBS Lett. 124, 265-269.
3. Touraille, S., Briand, Y., Durand, R., Bonnafous, J.C. and Marie, J.C. (1981) FEBS Lett. 128, 142-144.
4. Kolbe, H.V.J., Mende, P. and Kadenbach, B. (1982) Eur. J. Biochem. 128, 97-105.
5. De Pinto, V., Tommasino, M., Palmieri, F. and Kadenbach, B. (1982) FEBS Lett. 148, 103-106.
6. Kadenbach, B., Mende, P., Kolbe, H.V.J., Stipani, I. and Palmieri, F. (1982) FEBS Lett. 139, 109-112.
7. Palmieri, F., Tommasino, M., De Pinto, V., Mende, P. and Kadenbach, B. (1982) in International Workshop on: "Membranes and Transport in Biosystems" (Bari, June 28 - July 2), pp.167-170, Laterza litostampa, Bari.
8. Mende, P., Kolbe, H.V.J., Kadenbach, B., Stipani, I. and Palmieri, F. (1982) Eur. J. Biochem. 128, 91-95.
9. De Pinto, V., Tommasino, M., Bisaccia, F. and Palmieri, F. (1983) this book.

SEPARATION OF THE 35000 Mr DCCD-REACTIVE PROTEIN FROM THE PHOSPHATE
CARRIER AND ITS PURIFICATION FROM HEART MITOCHONDRIA.

V. DE PINTO, M. TOMMASINO, F. BISACCIA and F. PALMIERI
Institute of Biochemistry, Faculty of Pharmacy, University of Bari,
70126 Bari, Italy

INTRODUCTION

It was shown that mitochondria incubated with very low concentra
tions of ^{14}C-dicyclohexylcarbodiimide (DCCD) and subjected to SDS-
-polyacrylamide gel electrophoresis exhibit three labeled bands of
Mr 9000, 16000 and 33000. The 9000 Mr polypeptide belongs to the
F_O component of the mitochondrial H^+-ATPase [1], while the 16000
polypeptide represents its aggregate. In contrast to the DCCD-
-binding protein of F_O, the 33000 Mr protein is neither extracted
by chloroform-methanol nor it is detected in the immunoprecipitated
H^+-ATPase complex [2]. More recently, it has been suggested that
this 33000 DCCD-reactive protein might be identified with the NEM
sensitive phosphate carrier on the basis of (a) the striking similarity in behaviour of both the DCCD-reactive protein and NEM-
-sensitive Pi carrier in the course of isolation and (b) the same
electrophoretic mobility of the DCCD and NEM reactive protein [3].

In this report we show that the DCCD-reactive protein of Mr more
accurately determined of 35 k can be separated from the phosphate
carrier. We also describe a procedure for its purification based
on the use of an organomercurial agarose column.

MATERIALS AND METHODS

Mitochondria (isolated according to [4] and stored at -80°C) or ^{14}C-DCCD
labeled mitochondria (2 nmol/mg protein, 0°C, 20 h) were solubilized for 30 min
in 3% Triton X-100 or X-114, 20 mM KPO_4 pH 6.5, 10 mM KCl, 1 mM EDTA at a protein concentration of 10-15 mg/ml. When indicated, 3 mg/ml cardiolipin (DPG) was
added to the solubilization buffer. The unsolubilized material was sedimented
at 147000 x g, 30 min.

Absorption chromatography: the supernatant was applied on dry hydroxylapatite
(HTP). Elution was performed with the solubilization buffer.

Organomercurial agarose column: the HTP eluate was applied to an organomercurial agarose column (Affi-gel from Bio-Rad) (6 x 0.7 cm) equilibrated with

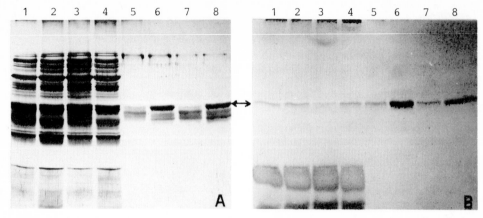

Fig.1 - Different behaviour of the DCCD-reactive protein and of the
Pi carrier on HTP in the absence and presence of DPG. 1: Triton
X-114 extract + DPG; 2: Triton X-114 extract - DPG; 3: Triton X-100
extract + DPG; 4: Triton X-100 extract - DPG; 5: HTP eluate from 1;
6: HTP eluate from 2; 7: HTP eluate from 3; 8: HTP eluate from 4.
A: silver staining; B: fluorography.

solubilization buffer containing 1 mg DPG/ml. The column was washed with 10-12
ml of the same buffer and the retained proteins were eluted by the addition of
mercaptoethanol.

The reconstitution procedure and the assay of Pi transport in proteoliposomes
were performed as in (5).

SDS-gel electrophoresis as in (6) and fluorography as in (7).

RESULTS

Fig.1 shows the SDS-gel electrophoretic pattern (A) and the fluorography (B)
of Triton extracts and HTP eluates starting from mitochondria labeled with
^{14}C-DCCD. The mitochondria were extracted with Triton X-114 or Triton X-100, in
the presence or absence of DPG. Each extract was applied to an HTP column and
the eluates were analyzed. With both Triton X-114 and Triton X-100, the addition
of DPG (lanes 5 and 7) prevents the elution from HTP of the 35K DCCD-reactive
protein indicated by the arrow, and causes the elution of the Pi carrier (see
ref.8) represented by the 2nd band of the HTP eluates in the 30-35K region. In
the absence of DPG (lanes 6 and 8), on the other hand, the 35K protein in the
HTP eluate is markedly increased. The fluorography shows that the label is
associated only to the 35K protein and gives further evidence that this protein
is eluted from HTP to a greater extent when DPG is not added to the solubilization buffer. Thus, the behaviour of the 35K protein on HTP is opposite to that

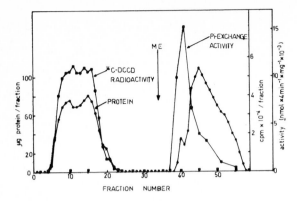

Fig.2 - Separation of the 35000 DCCD-reactive protein from the Pi carrier by an organomercurial agarose column (Affi-gel). The HTP eluate was applied to the column. The arrow shows the point of addition of 0-15 mM mercaptoethanol (ME) gradient.

of the Pi carrier, i.e. DPG prevents the elution of the former protein, whereas is needed for the elution of the second (cf ref.8). The different behaviour of these two proteins on HTP is also demonstrated by the measurements of the reconstituted Pi transport activity and the DCCD radioactivity in the HTP eluates. In the absence of DPG the HTP eluate contains more radioactivity and less Pi transport activity, and vice versa in the presence of DPG (not shown). These results demonstrate that a) the 35K protein, which corresponds in the HTP eluate to the polypeptide with the highest Mr in the 30-35K region, does not represent the Pi carrier protein (as previously suggested (9)) and b) DCCD does not bind to the Pi carrier (as suggested before (3)).

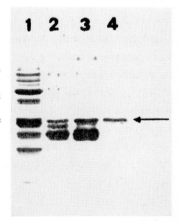

Fig.3 - Purification of the 35K DCCD-
-reactive protein. 1: Triton extract;
2: HTP eluate; 3: Affi-gel first peak
from 2; 4: HTP/celite eluate from 3.
The arrow shows the 35K DCCD-reactive
protein.

In order to separate the 35K DCCD-reactive protein from the Pi carrier, the organomercurial agarose gel, recently used to further purify the Pi carrier from the HTP eluate (6), was employed. As previously reported (6) about half of the proteins present in the HTP eluate pass through the organomercurial agarose column, whereas the remaining part is retarded and can be eluted by the addition of mercaptoethanol. As shown in Fig.2, all the DCCD radioactivity is found in the first protein peak, whereas the Pi transport activity is retarded. This result definitely shows that the 35K DCCD-reactive protein and the Pi carrier are different proteins. Further purification of the DCCD-reactive protein is achieved by applying the pass-through of the organomercurial agarose column on a celite (or hydroxylapatite/celite 1:1 w/w) column. The above reported procedure (extraction of mitochondria with Triton, chromatography on a) HTP, b) organomercurial agarose (to eliminate band 2, i.e. the Pi carrier) and c) celite) results in a pure preparation of the 35K DCCD-reactive protein (Fig.3).

REFERENCES

1. Cattel K.J., Lindop C.R., Knight I.G. and Beechey R.B. (1971) Biochem. J. 125, 169-177.

2. Houstek J., Svoboda P., Kopecky J., Kuzela S. and Drahota Z. (1981) Biochim. Biophys. Acta 634, 331-339.

3. Houstek J., Pavelka S., Kopecky J., Drahota Z. and Palmieri F. (1981) FEBS Lett. 130, 137-140.

4. Smith A.L. (1967) Methods Enzymol. 10, 81-86.

5. Mende P., Kolbe H.V.J., Kadenbach B., Stipani I. and Palmieri F. (1982) Eur. J. Biochem. 128, 91-95.

6. De Pinto V., Tommasino M., Palmieri F. and Kadenbach B. (1982) FEBS Lett. 148, 103-106.

7. Bonner W.M. and Laskey R.A. (1974) Eur. J. Biochem. 76, 83-88.

8. Bisaccia F., Tommasino M. and Palmieri F. (1983) this book.

9. Kolbe H.V.J., Mende P. and Kadenbach B. (1982) Eur. J. Biochem. 128, 97-105.

INVESTIGATIONS ON THE ROLE OF CARDIOLIPIN IN THE OPERATION OF THE NATIVE AND RECONSTITUTED PHOSPHATE-CARRIER OF BEEF HEART MITOCHONDRIA

MICHELE MUELLER , DOMINIQUE CHENEVAL AND ERNESTO CARAFOLI
Laboratory of Biochemistry, Swiss Federal Institute of Technology (ETH),
CH-8092 Zuerich , Switzerland

INTRODUCTION

The purification of the phosphate-carrier from beef heart mitochondria was achieved by Kolbe et al. (1) and further improved by de Pinto et al. (2) . Wohlrab and Flowers (3) characterized the pH dependent unidirectional phosphate uptake in a reconstituted system using a partially purified phosphate-carrier preparation . It was also demonstrated that the phosphate-carrier has an essential requirement for cardiolipin (4) , which is necessary to obtain full activity and to prevent irreversible inactivation of the carrier during its extraction from mitochondria with Triton X-114 .

It has been claimed that the anthracycline antibiotic doxorubicin (adriamycin) segregates membrane cardiolipin in a separate phase , making it inacessible to cytochrome c oxidase incorporated into cardiolipin containing liposomes (5) . It thus became interesting to investigate the effect of doxorubicin , and of its highly reactive analogue Br-daunomycin , on the reconstituted phosphate-carrier .

MATERIALS AND METHODS

Hydroxylapatite was obtained from Bio-Rad, Richmond, USA ; Celite 535 (Johns Meville) from Roth, Karlsruhe, Germany ; Soybean asolectin from Associated Concentrates, New York, USA ; Triton X-114 , Amberlite XAD-2 and Dowex AG 1 - X8 from Fluka, Buchs, Switzerland ; N-ethylmaleimide from Sigma, St. Louis, USA ; Cardiolipin from P-L Biochemicals, Milwaukee, USA ; doxorubicin and Br-daunomycin are a generous gift of Farmitalia-Carlo Erba, Milano, Italy .

The phosphate-carrier was isolated from beef heart mitochondria essentially as described by Mende et al. (6) . Asolectin liposomes were prepared as described by Kolbe et al. (1). Additional removal of Triton X-114 was accomplished with Amberlite XAD-2 beads . The unidirectional phosphate uptake was performed using the tecnique described by Wohlrab and Flowers (3). All operations were performed at $5^{o}C$. Phosphate was analyzed colorimetrically (7) . Protein was determined by the procedure of Lowry et al. (8) .

RESULTS AND DISCUSSION

Figure 1 shows the time-dependent phosphate uptake mediated by the phosphate-carrier reconstituted in asolectin liposomes . Asolectin contains up to ∿ 25% of the total lipid phosphate as cardiolipin , an amount adequate for full stimulation of the phosphate uptake . The uptake curve shows a rapid initial phase , rising the saturation within 2 min under these conditions . The uptake is limited by the dissipation of the pH gradient , since the phosphate-carrier is an electroneutral symporter (phosphate/H^+) or an antiporter (phosphate/OH^-) . Addition of 10 μg N-ethylmaleimide/mg protein to mitochondria prior to their solubilization inhibits the phosphate-carrier in the reconstituted system .

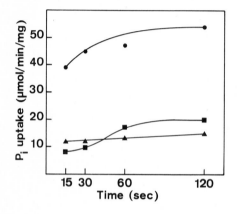

Fig. 1. Time-dependent unidirectional phosphate uptake catalyzed by the phosphate-carrier reconstituted in asolectin liposomes (●), in the presence of N-ethylmaleimide (10 μg/mg protein, added to the mitochondria before their solubilization)(▲) and in the presence of 100 μM doxorubicin (■) .

If N-ethylmaleimide is added after the reconstitution its effect (at the sameratio to protein) becomes less evident . This is due to the higher lipid to protein weight ratio in liposomes as compared to mitochondria . On the other hand the hydrophilic inhibitor mersalyl showes the same kinetic of inhibition seen in mitochondria (data not shown) even if added after the reconstitution . 100 μM doxorubicin inhibits the uptake of phosphate to the same extent as N-ethylmaleimide does . Since it has been shown that doxo-rubicin formes stables complexes with cardiolipin (9) , and since the phosphate transport activity is very sensitive to the presence of cardiolipin in the vicinity of the phosphate-carrier , it is very probable that this inhibition is due to the binding of cardiolipin by doxorubicin . The possibility nevertheless exists that doxorubicin inhibits the phosphate-carrier by direct binding to the protein itself . To test this possibility we have employed the brominated, highly reactive doxorubicin analogue, Br-daunomycin . When added to proteoliposomes, Br-daunomycin inhibits the uptake of phosphate to the same extent as doxorubicin (Figure 2) .

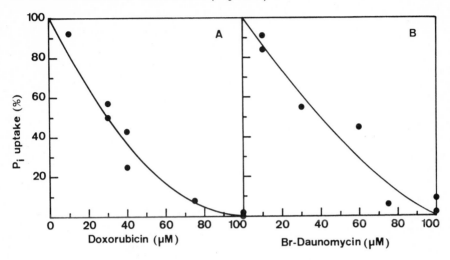

Fig. 2. Concentration-dependent inhibition of the phosphate uptake in proteoliposomes by doxorubicin (A) and Br-daunomycin (B) .

However , the phosphate-carrier isolated from mitochondria , labeled with Br-daunomycin , appears to be free from Br-daunomycin, as can be estimated colorimetrically , and is fully active after reconstitution in liposomes . The transport of phosphate in liposomes containing the phosphate-carrier isolated from Br-daunomycin treated mitochondria is fully inhibited by doxorubicin or Br-daunomycin or mersalyl (data not shown) . This supports the suggestions that doxorubicin and Br-daunomycin react specifically with cardiolipin essential for the phosphate-carrier .

Further evidence for this is presented in Figure 3 . Increasing the ionic strenght of the medium removes the inhibition of doxorubicin as expected from the fact that the interaction between cardiolipin and doxorubicin is of ionic nature (9) .

Although Br-daunomycin doesn't react with the phosphate-carrier , it reacts with other mitochondrial proteins . The labeled bands are visible on the fluorograph of Figure 4 . Their apparent M_r are $\sim$45000 (A), $\sim$35000 (B) and $\sim$30000 (C). The nature of these polypeptides is currently under investigation.

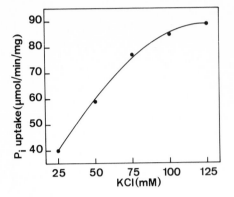

Fig. 3. Concentration-dependent reactivation of the unidirectional phosphate uptake in proteoliposomes by KCl in the presence of 100 μM doxorubicin .

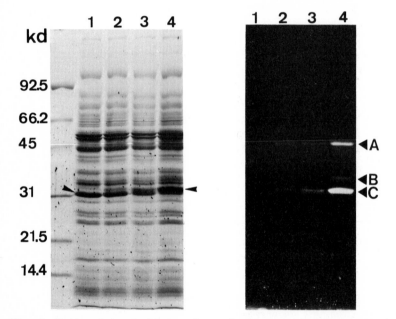

Fig. 4. Sodium dodecylsulfate polyacrylamide gel electrophoresis of mitochondria incubated with increasing amounts of Br-daunomycin (50, 100, 200, 300 μM , respectively lanes 1 to 4) for 5 min at room temperature in iso-osmotic medium at pH 7.4 . Left panel : Coomassie blue stain . Right panel : fluorography . The labeled bands have apparent M_r of ∿45000 (A) , ∿35000 (B) and ∿30000 (C) . The ADP/ATP translocator is labeled by the arrows .

A further interesting observation on Br-daunomycin-labeled mitochondria is the decreased gel mobility of the ADP/ATP translocator (arrows) with increasing Br-daunomycin concentration . The translocator itself, however , is not labeled .

ACKNOWLEDGMENTS

The auhtors wish to acknoledge many fruitfull discussions with their collegues Dr. P. Gazzotti and S. Longoni .

REFERENCES

1. Kolbe, H.J.V., Böttrich, J., Genchi G., Palmieri,F. and Kadenbach, B. (1981) FEBS Lett., 124 , 265-269

2. de Pinto, V., Tommasino, M., Palmieri, F. and Kadenbach, B. (1982) FEBS Lett., 148 , 103-106

3. Wohlrab, H. and Flowers,N. (1982) J. Biol. Chem., 257 , 28-31

4. Kadenbach, B., Mende, P., Kolbe, H.J.V., Stipani, I. and Palmieri, F. (1982) FEBS Lett., 139 , 109-112

5. Goormaghtigh, E., Brasseur, R. and Ruysschaert, J.M. (1982) Biochem. Biophys. Res. Comm., 104 , 314-320

6. Mende, P., Kolbe, H.J.V., Kadenbach, B., Stipani, I. and Palmieri, F. (1982) Eur. J. Biochem., 128 , 91-95

7. Lanzetta, P.A., Alvarez, L.J., Reinach, P.S. and Candia, O.A. (1979) Anal. Biochem., 100 , 95-97

8. Lowry, O.H., Rosenbrough, N.J., Farr, A.L. and Randall, R.J. (1951) J. Biol. Chem., 193 , 265-275

9. Goormaghtight, E., Chatelain, P., Caspers, J. and Ruysschaert, J.M. (1980) Biochim. Biophys. Acta, 597 , 1-14

EVIDENCE FOR THE EXISTENCE OF MEMBRANE RECEPTORS IN THE UPTAKE OF ASPARTATE
AMINOTRANSFERASE AND MALATE DEHYDROGENASE IN RAT LIVER MITOCHONDRIA IN VITRO

SALVATORE PASSARELLA[1], ERSILIA MARRA[1], ELISABETTA CASAMASSIMA[1], ELDA PERLINO[1],
SHAWN DOONAN[2], CECILIA SACCONE[1] AND ERNESTO QUAGLIARIELLO[1]

[1]Department of Biochemistry, University of Bari, Via Amendola 165/A, 70126
Bari (Italy); [2]Department of Biochemistry, University College, Cork (Ireland)

INTRODUCTION

It is now known that the vast majority of mitochondrial proteins are synthesized in the cell cytosol as completed translation products and subsequently imported into mitochondria (for reviews see 1,2). Although the majority of
such proteins are produced as precursors of molecular weight greater than that
of the mature forms (1,2) nevertheless mature proteins can be imported into
mitochondria in an in vitro model system. This has been shown for the
mitochondrial isoenzymes of aspartate aminotransferase (3,4) and malate dehydrogenase (5). The particular value of these systems is that manipulation of
experimental conditions and chemical modification of the proteins are straight
forward, and thus the systems provide a very valuable tool for studying the
mechanism of protein import into mitochondria (5-7). In the present paper we
describe experiments which suggest that a protein receptor exists in the
mitochondrial membrane system for uptake of aspartate aminotransferase, and
moreover that uptake of the enzyme via this receptor is inhibited by malate
dehydrogenase.

MATERIALS AND METHODS

The details of procedures for isolation of mitochondrial isoenzymes and preparation of mitochondria have been given previously (3,5).

Uptake of aspartate aminotransferase into mitochondria can be measured by
two methods, one of which involves measurement of intramitochondrial enzyme
activity and its increase after incubation of the organelles with purified iso-
enzyme (3). The other method, which is the one used here, involves radio-
labelling of the enzyme by reduction of the cofactor-apoprotein linkage with

[^{3}H]-sodium borohydride and then using liquid scintillation counting to determine the amount of the enzyme imported after incubation with the organelles and their recovery by centrifugation. The details of the method, and conditions under which it provides a true measure of protein import into mitochondria, have been described in detail (4). Essential descriptions of experimental conditions are given in the Tables.

RESULTS

Previous works, using a method in which the uptake into mitochondria of precursor proteins synthesized in cell-free translation systems was observed, provided initial evidence for receptor sites for protein import in the mitochondrial membrane system (8-10). One of our approaches to this same problem arose out of observations (5,6) that blockage of sulphydryl groups in native enzymes prevented their import into mitochondria. This suggested that the putative receptor sites might contain specific thiol-binding regions.

TABLE 1

EFFECT OF β-MERCAPTOETHANOL ON THE UPTAKE OF LABELLED MITOCHONDRIAL ASPARTATE AMINOTRANSFERASE INTO RAT LIVER MITOCHONDRIA IN VITRO

Mitochondria (1.5 mg of protein) were incubated for 1 min in 1 ml of a medium consisting of 0.25 M Sucrose, 20 mM Tris-HCl pH=7.25, 1 mM EGTA-Tris, T=20°C. The additions were as follows: A) labelled mitochondrial aspartate aminotransferase (^{3}H-mAAT): 20 μg; B) β-mercaptoethanol (ME): 17 nmoles/mg of protein, ^{3}H-mAAT: 20 μg; C) ^{3}H-mAAT (20 μg)+β-mercaptoethanol (25 μM) previously incubated for 1 min in the absence of mitochondria were added together to the mitochondrial suspension to give a final β-mercaptoethanol concentration of 0.5 μM; D) β-mercaptoethanol (31 nmoles/mg of protein), ^{3}H-mAAT (20 μg). In each case mitochondria were centrifuged 1 min after enzyme addition and the radioactivity counted as reported (4). The amount of radioactivity added in each case was 70322 cpm.

		t=0	t=1'	t=2'	t=4'	cpm	Control (%)
A	MIT	^{3}H-mAAT		–	–	12910	100
B	MIT	ME		–	^{3}H-mAAT	7430	57
C	MIT	ME + ^{3}H-mAAT		–	–	13920	107
D	MIT	ME		^{3}H-mAAT	–	9030	70

Evidence that this is so is given in Table 1 where it is shown that β-mercaptoethanol inhibits the uptake of radiolabelled aspartate aminotransferase into mitochondria when added to the mitochondrial suspension 3 min before the enzyme (compare A and B). When β-mercaptoethanol and enzyme were preincubated before addition to the mitochondria, no inhibition was produced (compare A and C); this suggests that β-mercaptoethanol inhibits uptake by interacting with the mitochondria and not with the enzyme itself. Interestingly, a higher concentration of β-mercaptoethanol incubated for a shorter time with the mitochondria produced less inhibition (compare B and D). The origin of this effect is not clear, but the results reported in Table 1 strongly support the idea that a receptor site for aspartate aminotransferase exists in the mitochondrial membrane and that interaction of enzyme and receptor involves a protein thiol. The compound β-mercaptoethanol then inhibits protein uptake by binding to this receptor.

The numerous similarities between uptake of aspartate aminotransferase and malate dehydrogenase into mitochondria (5) suggest that they may share a common receptor site. Evidence in support of this view is given in Table 2. When constant amounts of aspartate aminotranferase were added to mitochondria in the presence of increasing amounts of malate dehydrogenase there was a progressive decrease in radioactivity incorporated, that is, a progressive inhibition of uptake of aspartate aminotransferase.

Although there are other possible explanations of these results, the most likely seems to be that malate dehydrogenase decreases the rate of uptake of aspartate aminotransferase into mitochondria by interacting either with the same receptor site or with a nearby site whose occupancy is inhibitory for uptake of aspartate aminotransferase. These possibilities could, in principle, be distinguished by kinetic studies of the inhibition.

In summary, we give evidence here for the existence of a receptor site involved in uptake of aspartate aminotransferase into mitochondria. Malate dehydrogenase may interact directly with the same site, or with a distinct but interacting site.

TABLE 2

EFFECT OF MITOCHONDRIAL MALATE DEHYDROGENASE ON THE UPTAKE OF MITOCHONDRIAL
ASPARTATE AMINOTRANSFERASE IN RAT LIVER MITOCHONDRIA IN VITRO

Mitochondria (2 mg of protein) were incubated under the experimental conditions described in Table 1. After 1 min incubation, the enzymes were added in
the quantities shown and after a further 30 sec the suspension was centrifuged
and the radioactivity measured as previously described (4). The amount of
radioactivity added in each case was 163500 cpm (equivalent to 19 µg of labelled mitochondrial aspartate aminotransferase (^{3}H-mAAT)). Mitochondrial malate
dehydrogenase: mMDH.

ADDITIONS	cpm	INHIBITION (%)
^{3}H-mAAT	12100	0
^{3}H-mAAT + mMDH (0.29 µg)	4700	61
^{3}H-mAAT + mMDH (0.43 µg)	3230	73
^{3}H-mAAT + mMDH (0.57 µg)	2430	80
^{3}H-mAAT + mMDH (0.71 µg)	1466	88

REFERENCES

1. Neupert, W. and Schatz, G. (1981) Trends Biochem. Sci., 6, 1.

2. Schatz, G. and Butow, R.A. (1983) Cell, 32, 316.

3. Marra, E., Doonan, S., Saccone, C. and Quagliariello, E. (1977) Biochem.
   J., 164, 685.

4. Marra, E., Doonan, S., Saccone, C.and Quagliariello, E. (1978) Eur. J.
   Biochem., 83, 427.

5. Passarella, S., Marra, E., Doonan, S. and Quagliariello, E. (1983)
   Biochem. J., 210, 207.

6. Marra, E., Passarella, S., Doonan, S., Saccone, C. and Quagliariello, E.
   (1979) Arch. Biochem. Biophys., 95, 269.

7. Passarella, S., Marra, E., Doonan, S., Languino, L.R., Saccone, C.
   and Quagliariello, E. (1982) Biochem. J., 202, 353.

8. Zimmermann, R. and Neupert, W. (1980) Eur. J. Biochem., 109, 217.

9. Henning, B. and Neupert, W. (1981) Eur. J. Biochem., 121, 203.

10. Zimmermann, R., Henning, B. and Neupert, W. (1982) Eur. J. Biochem., 116.

© 1983 Elsevier Science Publishers B.V.
Structure and Function of Membrane Proteins,
E. Quagliariello and F. Palmieri editors.

SOLUBILIZATION OF MYELIN PROTEINS BY DETERGENTS

P. RICCIO[1], S.M. SIMONE[1], G. CIBELLI[2], A. DE SANTIS[1], A. BOBBA[1], P. LIVREA[2],
AND E. QUAGLIARIELLO[1]
[1]Istituto di Chimica Biologica, Facoltà di Scienze, Università di Bari, and
Centro di Studio sui Mitocondri e Metabolismo Energetico, C.N.R., via Amendola
165/A, 70126 Bari, Italy, and [2]Istituto di Neurologia, Facoltà di Medicina,
Università di Bari, Piazza G. Cesare, 70124 Bari, Italy.

INTRODUCTION

Myelin is a membrane surrounding the nerve axon, which is very important because implicated in some neurological diseases.

However, although the major protein components of the myelin sheath have been
isolated and characterized, their roles remain still largely unknown (1).
This might be due to the fact that myelin proteins have been usually studied
after purification with organic solvents and extreme pH values which could
alter their native conformation and thus hindering biological activity.

Since in the study of membrane protein structure and function the use of detergents seems to be most promising, to overcome above limitations we centred
our efforts on the selective extraction of myelin proteins from bovine brain
myelin by detergents.

MATERIAL AND METHODS

Myelin was prepared from bovine brain following the method of W.T.Norton (2).

Protein content was determined with a modified Lowry method in the presence
of 0.5% sodium dodecylsulfate (SDS).

The discontinuous polyacrylamide gel electrophoresis system (SDS-PAGE) for
separation of myelin proteins in SDS was adapted from the procedure of Laemmli
(3). The spacer gel and the separating gel were 3% and 15% acrylamide respectively.

The commercial detergents were used without further purification.

RESULTS AND DISCUSSION

Aliquots of thawed myelin were resuspended to 3.4-4 mg protein/ml in a medium
containing the selected detergent for solubilization. If not stated otherwise
the standard medium was buffered to pH 8.5 with Tris-HCl.

After 30 min incubation in ice the suspension was centrifuged for 40 min at
40.000 rpm in the 60Ti Beckman rotor. Then aliquots of the supernatants were

subjected to protein determination and to SDS-PAGE. Electrophoresis of whole myelin shows the presence of at least 30 bands,most of which being in the high molecular weight region (Fig. 1, row C).

The most effective detergents used were the cationic sarcosine and cetyl-trimethylammonium bromide (CTAB), the non-ionic Triton X-100 (TX-100) in the presence of 0.25-0.50 M NaCl, and the bile salt sodium deoxycholate (Na-DOC).

A detergent/protein ratio of 1 (w/w) gives usally half maximal solubilization.

We focused our attention on the selective extraction of the three major proteins of myelin: the basic protein (MBP, 19KD), the proteolipid (MPL, 25 KD) and the acidic Wolfgram protein (MWP, 52-54 KD). The results obtained are shown in Fig. 1 and in Table I.

Selective extraction of the Wolfgram protein was only possible at the lower concentration of the chaotropic agent urea (Fig. 1, row N). On the other hand a fraction of the proteolipid could be selectively solubilized only by Triton (Fig. 1, row P). The basic protein was extracted more easily and near to purity by a number of detergents: Triton X-100 at pH 3 in the absence of salts, CTAB at low concentrations, and sodium cholate (Fig. 1, rows O,Q,F respectively). The basic protein was further purified in all these detergents, the proteolipid in Triton X-100 only.

The localization of both proteolipid and basic protein in the myelin membrane can be described by studying the type of interaction of these proteins with the Triton X-100 micelles.

A particular feature of both solubilized proteins is their non-adsorption to hydroxyapatite. According to (4) this could be due to deep embedment of these proteins in Triton micelles so that a large detergent coat shields the proteins. This is a property of intrinsic membrane proteins with large hydrophobic areas and small hydrophilic domains. The non-adsorption was unexpected as the proteolipid is considered to be an intrinsic membrane protein whereas the basic protein is believed to be a peripheral membrane protein.

On the other hand the basic protein prepared in urea with the classical method of Martenson R.E. et al. (5) can bind to hydroxyapatite in a conformation which does not bind Triton.

The expected differences are detected by aid of proteolytic enzymes: if chimotrypsin is added to both proteins solubilized in Triton X-100, the basic protein is immediately destroyed, whereas the proteolipid remains practically intact.

With regards to protein properties it can be concluded that the basic protein

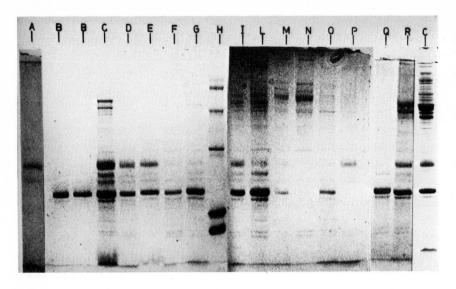

Fig. 1. SDS-PAGE of myelin protein; A: purified proteolipid; B : purified basic protein; H: standard proteins: 12.3, 17.2, 30, 45, 66.3 and 77 KD; C: myelin, starting material. Other rows as in TABLE I.

TABLE I

Myelin treated with	% solubilization	protein extracted (rows of Fig. 1)
25 mM Glycine, pH 3	3	-
1 M NaCl	5	-
0.5 M trichloroacetate, pH 3	1	-
Tx-100 (1.37 mg/mg prot), 0.25 M NaCl	56	basic protein and proteolipid (D)
Tx-100 (6.85 mg/mg prot), 0.25 M NaCl	61	basic protein and proteolipid (E)
Cholate (2.74 mg/mg prot)	7	-
Cholate (2.74 mg/mg prot), 0.25 M NaCl	20	basic protein (F)
Cholate (17.4 mg/mg prot),	20	basic protein (G)
Deoxycholate (0.66 mg/mg prot)	33	mainly basic protein and DM 20 (L)
Deoxycholate (2.63 mg/mg prot)	51	mainly basic protein and proteolipid (I)
6 M Urea, pH 10.4	25	basic and Wolfgram proteins (M)
2 M Urea, pH 10.4	15	Wolfgram protein (N)
Tx-100 (1.32 mg/mg prot), pH 3	16	basic protein (O)
Tx-100 (1.32 mg/mg prot), pH 8.5	15	proteolipid (P)
CTAB (0.7-0.85 mg/mg prot)	15-28	basic protein (Q)
Sarcosine (0.7 mg/mg prot)	77	all major protein (R)

364

is an extrinsic protein (it is completely degraded by a proteolytic enzyme),
it must be anchored to the lipidic phase by small segments (it requires detergents for solubilization, therefore it is not extracted by high salt concentrations), probably it exists as an oligomer (it requires addition of salts to
Triton X-100 for solubilization, and has, except at pH 3, a great capacity of
self association).

On the other hand the proteolipid seems to be an intrinsic protein. Preliminary data indicates that this protein exists at least as a dimer.

The question arises whether the conformation of the solubilized proteins in
detergents is close to that native.

Both the purified basic protein and the purified proteolipid contain particular bound lipids. The purified basic protein can induce experimental autoimmune
encephalomyelitis. In addition to this, the basic protein has an extremly high
capacity to bind carbohydrates depending on the type of associated detergent.
This property is only very weak in the preparation of (5) and disappears upon
addition of the denaturant SDS to the protein, that is when the basic protein
has an unordered conformation.

Further hydrodynamic studies and the use of techniques as ORD and CD will
give information on the nature of the tertiary and secondary structure of these
proteins in detergents, i.e. in an enviroment resembling that of the native
membrane.

ACKNOWLEDGEMENTS

The authors are grateful to A.M.D. Labriola, A.F. Cardinali and A. Calvario
for their skilful assistance.

REFERENCES

1. Agrawal, H.C. and Hartman, B.K. (1980) in: Bradshaw, R.A. and Schneider, D.
   M. (Ed.), Proteins of the Nervous System, Raven Press, New York, pp. 145-
   169.

2. Norton, W.T. (1974) Methods Enzymol. 31, 435.

3. Laemmli, U.K. (1970) Nature, 227, 680.

4. Riccio, P. (1983) in: Frigerio, A. (Ed.), Chromatography in Biochemistry,
   Medicine and Environamental Research, Elsevier Scientific Publishing Co.,
   Amsterdam, I, pp. 177-184.

5. Martenson, R.E., Deibler, G.E. and Kies, M.W. (1971) Res. Publ. Ass. Nerv.
   Ment. Dis., 49, 76.

MITOCHONDRIAL CYTOCHROME C OXIDASE INSERTED INTO ARTIFICIAL LIPOSOMES:
TRANSIENT KINETIC STUDIES.

M. BRUNORI, A. COLOSIMO, P. SARTI and M.T. WILSON[1]

Dept. of Biochemistry, Fac. of Medicine, II University of Rome (Tor Vergata)
and C.N.R. Center of Molecular Biology c/o Inst. of Chemistry and Biochemistry
I University of Rome (La Sapienza), Rome, Italy; [1]Dept. of Chemistry, University of Essex, Colchester, U.K.

INTRODUCTION

Artificial proteoliposomes obtained by insertion into phospholipid vesicles
of cytochrome c oxidase purified from beef heart, represent a valuable and
widely used experimental system to understand the relationships between redox
reactions and electrochemical gradients in the mitochondrion (see 1). Along
this line of research, systematic kinetic studies under pre steady-state
conditions showed to be very promising in defining the time correlation
between the oxidation of cytochrome c and the establishment of gradients
across the membrane (2). Understanding the significance of transient intermediates in the reactions of the oxidase for the regulation of the enzyme
activity "in vivo" remains a complex task.

The present report contains novel experimental information obtained along
this line of research using reconstituted proteoliposomes with high respiratory
control ratio (R.C.R.), studied by stopped-flow.

MATERIALS AND METHODS

Cytochrome c oxidase was purified following the method of Yonetani with
minor modifications (3) and inserted into phospholipid liposomes following
Casey et al. (4). The R.C.R. of proteoliposomes, defined as the relative
increase in the rate of oxygen consumption after addition of ionophores,
was measured in a Clark type oxygen electrode (Yellow Springs Instrument,
Ohio 45387, U.S.A.) under the following conditions: buffer = K^+/Hepes 10 mM
pH 7.3 containing 40 mM KCl + 50 mM Sucrose; 15 ÷ 25 μM cyt c; 15 mM ascorbate; 300 μM TMPD. Ionophores, i.e. nonactin and CCCP (5 and 10 μM respec-

tively) were added after 1 min. of oxygen consumption by proteoliposomes, the final oxidase concentration being 60 nM (total heme). Rapid mixing experiments were carried out with a Gibson-Durrum stopped flow apparatus having a dead time of 4 msec and an observation chamber of 2 cm. pathlenght. If used,ionophores were added to the proteoliposomes containing syringe. All experiments were carried out in air.

RESULTS AND DISCUSSION

The R.C.R. as obtained in the oxygen electrode is about 7 ÷ 9. The same value was obtained in rapid mixing experiments by comparing the rate of the turnover phase of cyt c oxidation measured before and after addition of ionophores. This is in agreement with the data reported by Sarti et al. (2).

As shown in fig. 1 the time course of cyt c oxidation by proteoliposomes in the absence of ionophores is characterized by two kinetic phases, the slower component being exponential when cyt c concentration is sufficiently high to permit a number of turnovers. The relative amplitude of the faster kinetic component increases as the cyt c^{2+} is decreased (see Fig. 1), while its rate remains unaffected.

Upon addition of ionophores, which are known to collapse the electrochemical gradient across the liposomes wall, the oxidation of cyt c is much faster and the time course is strictly exponential. As also seen in Fig. 1, the initial rate of the experiment after addition of ionophores is essentially identical to the initial phase of the time course in the absence of ionophores. The kinetic behaviour in the presence and absence of ionophores, as obtained at different initial concentrations of cyt c, as well as the stoichiometry of cyt c oxidation, suggest the following considerations.

The initial bimolecular electron entry into cyt a, whose rate constant is known to be very high (5), is lost in the dead time of the stopped flow apparatus. The burst (seen in Fig. 1 over the first 100 msec) should reflect the internal electron transfer between cyt a/CuA and cyt a_3/CuB which is rate limiting, according to the model proposed by us (6). The slower exponential phase which follows is the catalytic oxidation of cyt c which, in the absence of ionophores, represents the maximum turnover rate in the presence of the electrochemical gradient established by the transient electron transfer to cytochrome oxidase (i.e. the initial phase(s) commented above).

Two important conclusions seem to emerge from these preliminary experiments. First of all, the "onset" of the electrochemical gradient, slowing down the oxidase activity, has been directly observed and we tend to correlate this process to the initial reduction of the enzyme by cyt c^{2+}. Second the exponential nature of the turnover phase implies that the subsequent turnovers, albeit increasing the gradient, do not affect significantly the enzyme activity over and above the effect just described. This conclusion, if confirmed, suggests that the gradient created by the initial reduction process shifts the properties of the system to a less active-gradient controlled molecular state of the enzyme.

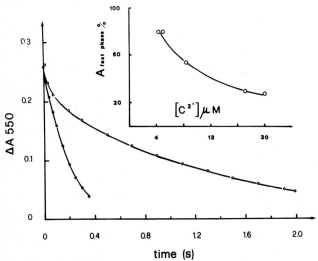

Fig. 1 . Time course of oxidation of reduced cyt c in air by proteoliposomes, respectively in the absence (°) and in the presence (•) of ionophores. The buffer is Hepes 10 mM pH 7.3 containing 40 mM KCl + 50 mM Sucrose; cyt c= 17 μM; Ascorbate = 0.5 mM; oxidase into liposomes = 2 μM (total heme); all concentrations before mixing.
Inset to fig. 1. Dependence on cyt. c^{2+} concentration of the relative amplitude of the fast phase observed in the absence of ionophores.

ACKNOWLEDGEMENT
    NATO Grant N° 1767 to M.T.W. is gratefully acknowledged.

REFERENCES

1. Wikström M., Krab K. and Saraste M. (1981) Ann. Rev. Biochem., 50, 623-655.

2. Sarti P., Colosimo A., Brunori M., Wilson M.T. and Antonini E. (1983) Biochem. J. 209, 81-89.

3. Yonetani T. (1961) J. Biol. Chem. 236, 1680-1688.

4. Casey R.P., Chappell J.B. and Azzi A. (1979) Biochem. J. 182, 149-156.

5. Brunori M., Antonini E., Wilson M.T. in "Metal Ions in Biological Systems", by H. Sigel, Marcel Dekker, New York (1981) 13, 187-228.

6. Wilson M.T., Peterson J., Antonini E., Brunori M., Colosimo A. and Wyman J. (1981), Proc. Natl. Acad. Sci. USA 78, 7115-7118.

PURIFICATION AND RECONSTITUTION OF THE MITOCHONDRIAL CITRATE CARRIER

I. STIPANI, M.C. BARBERO and F. PALMIERI
Institute of Biochemistry, Faculty of Pharmacy, University of Bari,
70126 Bari, Italy.

INTRODUCTION

Citrate is transported across the inner mitochondrial membrane
by a specific transport system called the "tricarboxylate carrier"
or "citrate carrier", which catalyzes a 1:1 exchange between
citrate or other tricarboxylates, some dicarboxylates and PEP (1).
This carrier is present in liver, but not in heart and brain (2),
is inhibited by 1,2,3-benzenetricarboxylate and less specifically
by p-iodobenzylmalonate (3). In a previous paper the transport of
citrate has been reconstituted in liposomes from Triton extracts
of liver submitochondrial particles (4).

In this report it is shown that the tricarboxylate transport
protein can be purified approx. 6 fold by chromatography on
hydroxylapatite with a recovery of 50% of the activity present in
the SMP extract. The presence of cardiolipin (DPG) during extraction
and purification is required for optimal activity of the reconstituted
tricarboxylate carrier.

METHODS

Submitochondrial particles (SMP) from rat liver, beef heart or
rat brain mitochondria were extracted as described (4) and centrifuged
at 147.000 x g, 45 min.

Absorption chromatography was carried out on dry hydroxylapatite
(HTP). The reconstitution procedure and the assay of citrate
transport were performed as described (4). Essentially, the protein
fractions were mixed with liposomes containing an appropriate
counteranion; frozen in liquid N_2, thawed and sonicated. The
citrate transport assay was initiated by adding ^{14}C-citrate to the
proteoliposomes and terminated with the addition of 1,2,3-benzene
tricarboxylate(BTA). In the control 1,2,3-BTA was added before the
radioactivity. Protein was determined by the method of Lowry in
the presence of 1% SDS. SDS-polyacrylamide gel electrophoresis was
performed as described (5).

RESULTS AND DISCUSSION

Table I shows the activity of the reconstituted citrate/citrate exchange of the extracts from liver SMP and of the eluates obtained from HTP, both in the presence and in the absence of DPG. It can be seen that absorption chromatography on HTP increases the specific activity several fold, with a recovery of the total activity of 50%. Inclusion of DPG in the solubilization medium does not influence the enhancement in purification, but increases the activity of citrate exchange both in the extracts and in the HTP eluates. A similar increase in the activity of the mitochondrial phosphate carrier in the HTP eluate has been observed and it has been suggested that DPG prevents an inactivation of the carrier occurring during its solubilization by Triton (6).

TABLE I

PURIFICATION OF THE MITOCHONDRIAL CITRATE CARRIER BY CHROMATOGRAPHY ON HYDROXYLAPATITE.

The activity of reconstituted citrate/citrate exchange is expressed as nmol/10 min x mg protein (specific activity) and nmol/10 min (total activity). The values represent the mean of at least 4 expts. SMP, submitochondrial particles; DPG, cardiolipin; HTP, hydroxylapatite.

| | | Protein (%) | Citrate exchange | | Recovery (%) |
			Specific Activity	Total Activity	
SMP EXTRACT	(-DPG)	100	40	172	-
" "	(+DPG)	100	80	292	-
HTP ELUATE	(-DPG)	10.5	196	86	50
" "	(+DPG)	8.2	486	144	49

The effect of increasing concentrations of Triton in the solubilization buffer was therefore investigated by measuring the reconstituted citrate exchange activity in both SMP extracts and HTP eluates. In both cases the total and the specific activities increase on increasing the concentration of Triton up to 3%, and decrease at higher detergent concentrations although more protein is solubilized. In the presence of 2 mg/ml DPG in the solubilization buffer, the total and the specific activities of both SMP extracts and eluates from HTP are considerably higher (2-3 fold) at 3% Triton and nearly remain constant at higher detergent concentrations up to 8% (7), showing that DPG prevents the inactivation of the citrate carrier by Triton. The inactivation is reversible since the subsequent addition of DPG to SMP extracts or eluates from HTP also causes a marked increase in the activity of the reconstituted citrate transport (7). No other phospholipid tested, including acidic (phosphatidylserine and phosphatidic acid) and unsaturated (dioleylphosphatidylcholine) ones, can substitute for DPG.

The citrate exchange activity reconstituted from the HTP eluate was characterized by studying the tissue and the substrate specificity. The activity is very small when SMP from heart or brain (instead of liver) are used (< 40 nmol/10 min x mg protein), in agreement with the observation that the citrate carrier is virtually absent in heart and brain mitochondria (2). The data reported in Table II show that virtually no citrate is transported into the proteoliposomes when they contain no anion or anions which are not substrates of

TABLE II

DEPENDENCE OF THE RECONSTITUTED CITRATE EXCHANGE ON INTERNAL ANIONS AND ON EXTERNALLY-ADDED INHIBITORS.

SMP were extracted as described in Methods in the presence of 2 mg/ml DPG. The HTP eluate was incorporated into liposomes containing the indicated anions. The listed inhibitors were added to proteoliposomes 2 min before the addition of 0.3 mM ^{14}C-citrate.

Internal anion	Inhibitor	Citrate exchange (nmol/10 min x mg protein)
-	-	11.2
PHOSPHATE	-	29.0
2-OXOGLUTARATE	-	33.5
MALATE	-	312.2
PEP	-	284.0
CITRATE	-	474.0
"	1,2,3-BTA (5.4 mM)	0
"	1,2,4-BTA (" mM)	400.5
"	1,3,5-BTA (" mM)	387.6
"	Butylmalonate (5.4 mM)	418.0
"	Phthalonate (0.4 mM)	463.4
"	α-cyanocynnamate (0.5 mM)	381.2
"	Carboxyatractylate (0.2 mM)	367.0
"	p-Iodobenzylmalonate (2.7 mM)	141.2

the citrate carrier, like Pi or oxoglutarate. Externally-added ^{14}C-citrate exchanges not only with citrate but also with malate and PEP, known substrates of the tricarboxylate carrier. Table II also reports the inhibitor sensitivity of the reconstituted purified citrate carrier. As in mitochondria, the ^{14}C-citrate/citrate exchange is inhibited by 1,2,3-BTA and p-Iodobenzylmalonate. In contrast, 1,2,4-BTA and 1,3,5-BTA have little effect. Furthermore, the citrate exchange is not inhibited by butylmalonate, phthalonate, carboxyatractylate and α--cyanocynnamate, known inhibitors of other mitochondrial carriers (1).

The absolute requirement for an appropriate counteranion, the substrate and tissue specificity and the inhibitor sensitivity of the citrate exchange reconstituted from the HTP eluate of SMP extract are distintive features of the tricarboxylate transport in mitochondria (3). Further support for the presence of the mitochondrial citrate transport protein in the HTP eluate obtained from Triton-extracted liver SMP is given by the kinetic properties of the reconstituted citrate exchange. This exchange follows first order kinetics, is highly sensitive to the temperature, exhibits a pH optimum of 7.0 and a K_m of 0.28 mM, similarly as in mitochondria (3). As far as the purity of the present preparation is concerned, the gel electrophoretic analysis of the HTP eluate from liver containing the reconstitutively active citrate carrier shows 6-8 bands, some of which are not present in the HTP eluate from heart.

REFERENCES

1. LaNoue, K.F. and Schoolwerth, A.C. (1979) Ann. Rev. Biochem. 48, 871-922.
2. Chappell, J.B. (1968) Brit. Med. Bull. 24,150-157.
3. Palmieri, F., Stipani, I., Quagliariello, E. and Klingenberg, M. (1972) Eur. J. Biochem. 26, 587-594.
4. Stipani, I., Krämer, R., Palmieri, F. and Klingenberg, M. (1980) Biochem. Biophys. Res. Commun. 97, 1206-1214.
5. De Pinto, V., Tommasino, M., Palmieri, F. and Kadenbach, B. (1982) FEBS Lett. 148, 103-106.
6. Kadenbach, B., Mende, P., Kolbe, H.V.J., Stipani, I. and Palmieri, F. (1982) FEBS Lett. 139, 109-112.
7. Stipani, I. and Palmieri, F. (1983) in preparation.

THE REDOX STATE OF THE COMPONENTS OF RESPIRATORY CHAIN IN THE PRESENCE OF N-ETHYLMALEIMIDE.

F. Zanotti, D. Marzulli e N.E. Lofrumento

Istituto di Chimica Biologica, Facoltà di Scienze e Centro di Studio sui Mitocondri e Metabolismo Energetico, Università degli Studi - Bari, Italy.

INTRODUCTION

Experimental data obtained in least few years have shown that N-ethylmaleimide (NEM) besides its inhibitory effect on the Pi transport, promotes an efflux of the cations Ca^{2+} and K^+ accumulated inside the mitochondria at expenses of the energy derived from the activity of respiratory chain (1,2). Since the effect of NEM on the distribution of cations is similar to the one obtained in the presence of respiratory inhibitors or uncouplers, a detailed study of the redox state of respiratory chain together with the distribution of protons and changes in the membrane potential, have been carried out. The results obtained will be in part presented in this report.

MATERIALS AND METHODS

Rat liver mitochondria (2 mg protein/ml) were incubated at 25°C in a standard medium at pH 7.8 containing 250 mM sucrose and 1 mM HEPES. Absorption spectra of Fig.1 were determined using the following time-course of additions in the sample cuvette: after base--line, at 3 min N-ethylmaleimide 40 nmoles/mg protein (NEM) (Fig.1-a), 0.5 mM azide (Fig.1-b); at 7 min 1 μM antimycin A (Fig.1-a) and NEM (Fig.1-b).
In the same incubation system the distribution of protons recording the pH of the medium and the electrical membrane potential with the tetraphenylphosphonium-selective electrode (3), were determined. In Expts of Fig.2 the standard medium was supplemented with 10 mM KCl and the additions were: N-ethylmaleimide 40 nmoles/ /mg protein (NEM); 0,2 nM valinomycin (VAL); 2.7 nM nigericin(NIG) and 50 nM carbonylcyanide p-trifluoromethoxyphenylhydrazone(FCCP).

RESULTS AND DISCUSSION

The absorbance of mitochondrial suspension, in the range of wave lengths 300-620 nm, is reported in Fig.1. NEM at the concentration of 40 nmoles/mg protein promotes a transition of all the components of respiratory chain in a more oxidation state. In the con ditions of Expt.(a) only the oxidation of cytochrome b and flavoproteins with maxima at 430 and 465 nm respectively, can be clearly observed. Although in the range of 300-360 nm, NEM per se exhibits an its own absorption with a maximum at 302, the oxidation of pyridine nucleotides, with the maximum at 340 nm, is also well evident. Aerobic incubation of mitochondria in the absence of added substrates, makes that the base-line rapresents already an oxi dation state of the respiratory chain. In these conditions it is expected that the a and c type of cytochromes being located at the outfall of the chain, will be in an almost complete oxidation state and their further shift, in the same direction, can be evi-

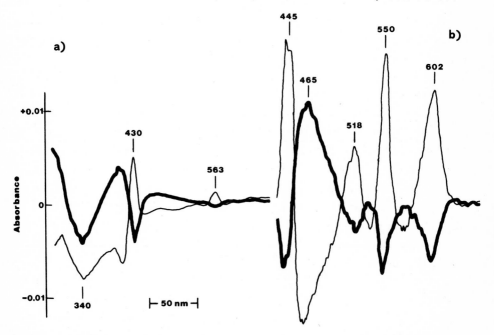

Fig. 1- Evidence for NEM induced oxidation of all the components of the respiratory chain. Heavy traces represent the spectra in the presence of NEM and ligth ones, in Expt(a) the spectrum with the addition of antimycin after NEM, in Expt(b) the spectrum with azide before the addition of NEM.

denced at very high concentration of mitochondrial protein. The
oxidation induced by NEM is not a rapid process but it takes at
least 4 min and referred to the cytochrome b it is quantitatively
the 50% of the oxidation induced by uncouplers. Light trace of
Fig.1-a shows that the oxidation of cytochrome b, flavoproteins
and pyridine nucleotides can be reversed or reduced by the addition of antimycin. Identical results have been obtained if respiratory substrates were added instead of antimycin (not shown).The-
se and other observation, indicate that mitochondria mantain their
functions and the effect of NEM cannot be ascribed to a frequently invoked unspecific alteration of the permeability of the inner
membrane. The results of Fig.1 are in agreement with the observation, obtained in a completely different experimental conditions,
that NEM promotes the oxidation of cytochrome b (4) and make clear
that its effect is not restricted to only one components of the
respiratory chain. In the Expt (b) where azide has been first added in order to have all the components in a reduced state, on addition of NEM, the transition from reduction to oxidation state
of cytochrome $a+a_3$ (602 and 445 nm), b (563 nm), c (550 and 518 nm)
and flavoproteins (465 nm), is well evident.
In Fig.2 it is reported that the uptake of K^+ induced by valinomycin, is accompanied not only by an expected decrease of transmembrane $\Delta\psi$ (Fig.2-b) but also by an efflux of protons (Fig.2-a).The
subsequent addition of nigericin which would mediate an electroneutral K^+/H^+ exchange, promotes the expected uptake of protons

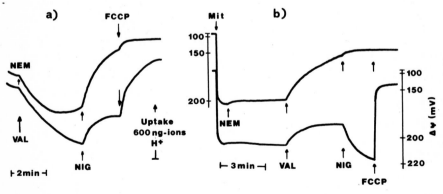

Fig.2 - Changes induced by NEM on the distribution of protons
(traces,a) and on electrical potential (traces,b) across the inner membrane. Additions as in "Materials and Methods".

and an increase of $\Delta\psi$, which may be ascribed to a disequilibrium in the rates of influx and efflux of K^+. In the presence of NEM a decrease of the efflux of protons, after the addition of valino- mycin, has been observed. Rather than to an inhibitory effect,this appears to be due to an increased rate of uptake of protons as evi dentiated by the subsequent addition of nigericin. However, as il- lustrated in Fig.2-b, the simultaneous determination of membrane potential has revealed that the redistribution of H^+ is preceded by a complete dissipation of $\Delta\psi$ and nigericin is no more able to re store it. In other Expts it has been found that the addition of respiratory substrates, completely restore the membrane potential giving further support to the view that NEM interfer with the fun- ctional activities of the membrane.

The results obtained indicate that the redistribution of protons and the more oxidation state of respiratory chain may be the con- sequence of modification of membrane potential on which NEM appears to explicate its primary effect. In addition these results appear to be consistent with the proposal, recently reported (5),that the redox state of SH groups of the inner membrane through a dithiol- -disulfide interchange, may play a relevant role in the generation and preservation of electrochemical gradient.

REFERENCES

1. Lofrumento, N.E. and Zanotti F. (1978) FEBS Lett., 87,186-190
2. Lofrumento, N.E., Pavone, A. and Zanotti, F. (1981) in: Palmie- ri, F. et al. (Eds.), Vectorial Reactions in Electron and Ion Transport in Mitochondria and Bacteria, Elsevier/North-Holland Biomedical Press, Amsterdam, pp. 311-314.
3. Kamo, N., Muratsugu, M., Hongoh, R. and Kobatake, Y. (1979) J. Membrane Biol. 49, 105-121.
4. Pozzan, M., Bernardi, P. and Di Virgilio, F. (1981) FEBS Lett., 127,263-266.
5. Robillard, G.T., and Konings, W.N. (1982) Eur. J. Biochem. 127, 597-604.

AUTHOR INDEX

Allen, G. 157
Andersen, J.P. 53, 57
Aquila, H. 145

Baldwin, J.M. 325
Baldwin, S.A. 325
Barbero, M.C. 335, 369
Berden, J.A. 177
Berthet, C. 11
Bertoli, E. 43
Bisaccia, F. 339, 347
Bjerrum, P.J. 95, 107
Bobba, A. 361
Boekema, E.J. 237
Bogner, W. 145
Bruggen van, E.F.J. 237
Brunner, J. 81, 117
Brunori, M. 365
Buse, G. 131

Cabantchik, Z.I. 271
Cantley, L. 73
Capaldi, R.A. 223
Carafoli, E. 301, 351
Casali, E. 3
Casamassima, E. 357
Cavatorta, P. 3
Cheneval, D. 351
Churchill, P. 283
Cibelli, G. 361
Cogdell, R.J. 125
Colosimo, A. 365
Crifo, C. 11

Deisenhofer, J. 191
Dencher, N.A. 23
De Pinto, V. 347
De Santis, A. 361
Doonan, S. 357
Dorset, D.L. 199

English, L. 73

Falk, G. 167
Fleer, E. 283
Fleischer, S. 283
Friedl, P. 91

Garavito, R.M. 205
Gay, N.J. 167

Georgevich, G. 223
Gerdes, U. 57
Ghersa, P. 117
Goni, F.M. 335
Gorga, F.R. 325
Grisham, C.M. 63
Gurtubay, J.I.G. 335

Hebert, H. 255
Heyn, M.P. 23
Hollemans, M. 177
Hoppe, J. 81, 91
Houstek, J. 43

Jenkins, J.A. 205
Jørgensen, P.L. 53, 245, 255

Karlish, S.J.D. 33
Kessler, M. 301
Klingenberg, M. 145

Leonard, K. 215
Lienhard, G.E. 325
Lindsay, J.G. 125
Ling, L. 73
Livrea, P. 361
Lofrumento, N.E. 373
Lottspeich, F. 191

Macarulla, J.M. 335
Malatesta, F. 223
Marra, E. 357
Marzulli, D. 373
Masotti, L. 3
Maunsbach, A.B. 211, 255
Maurer, A. 283
McIntyre, J.O. 283
Megli, F.M. 43
Meinecke, L. 131
Michel, H. 191
Miki, K. 191
Møller, J.V. 53, 57
Montecucco, C. 91
Mueller, M. 351

Ovchinnikov, Yu.A. 261

Palmieri, F. 43, 339, 347, 369
Pasquali-Ronchetti, I. 3
Passarella, S. 357